Ludmila Dymowa

Soft Computing in Economics and Finance

Intelligent Systems Reference Library, Volume 6

Editors-in-Chief

Prof. Janusz Kacprzyk
Systems Research Institute
Polish Academy of Sciences
ul. Newelska 6
01-447 Warsaw
Poland
E-mail: kacprzyk@ibspan.waw.pl

Prof. Lakhmi C. Jain
University of South Australia
Adelaide
Mawson Lakes Campus
South Australia 5095
Australia
E-mail: Lakhmi.jain@unisa.edu.au

Further volumes of this series can be found on our homepage: springer.com

Vol. 1. Christine L. Mumford and Lakhmi C. Jain (Eds.)
Computational Intelligence: Collaboration, Fusion and Emergence, 2009
ISBN 978-3-642-01798-8

Vol. 2. Yuehui Chen and Ajith Abraham
Tree-Structure Based Hybrid Computational Intelligence, 2009
ISBN 978-3-642-04738-1

Vol. 3. Anthony Finn and Steve Scheding
Developments and Challenges for Autonomous Unmanned Vehicles, 2010
ISBN 978-3-642-10703-0

Vol. 4. Lakhmi C. Jain and Chee Peng Lim (Eds.)
Handbook on Decision Making: Techniques and Applications, 2010
ISBN 978-3-642-13638-2

Vol. 5. George A. Anastassiou
Intelligent Mathematics: Computational Analysis, 2011
ISBN 978-3-642-17097-3

Vol. 6. Ludmila Dymowa
Soft Computing in Economics and Finance, 2011
ISBN 978-3-642-17718-7

Ludmila Dymowa

Soft Computing in Economics and Finance

Springer

Prof. Dr. Ludmila Dymowa
Czestochowa University of Technology
Institute of Comp. & Information Sciences
ul. Dabrowskiego 73
42-200 Czestochowa
Poland
E-mail: dymowa@icis.pcz.pl

ISBN 978-3-642-42326-0 ISBN 978-3-642-17719-4 (eBook)

DOI 10.1007/978-3-642-17719-4

Intelligent Systems Reference Library ISSN 1868-4394

Softcover re-print of the Hardcover 1st edition 2011

Typeset & Cover Design: Scientific Publishing Services Pvt. Ltd., Chennai, India.

Printed on acid-free paper

9 8 7 6 5 4 3 2 1

springer.com

This book is dedicated to Professor Cengiz Kahraman whose contribution and inspiration to implement the methods of Soft Computing in the solution of economic and financial problems should be greatly acknowledged. His papers and edited by him recent book entitled "Fuzzy Engineering Economics with Applications" is closely related to the subject of this book.

This work is also dedicated to Professor Lotfi A. Zadeh known as the "Father of Fuzzy Logic", who laid the foundations of fuzzy sets and systems, and whose outstanding scientific activity is still appreciated and admired by all. His ideas and writings have shaped and inspired the contents of this book.

Foreword

First a bit of history. Probably, the first application of soft computing in finance was the use of fuzzy sets theory in budgeting. After pioneer works by Ward (1985) and Buckley (1987), some other authors contributed to the development of fuzzy capital budgeting theory. Now the mathematical tools of fuzzy set theory and fuzzy logic are successfully used for risk analysis in: e-commerce development, portfolio selection, Black-Scholes option pricing models, corporate acquisition systems, evaluating investments in advanced manufacturing technology, interactive fuzzy interval reasoning for smart web shopping, fuzzy scheduling and in logistic. The great achievements were made in the field of linear and nonlinear fuzzy regression models. This seems to be very important as the regression analysis is one of the most powerful methods used in economic and financial applications.

An essential feature of economic and financial problems it that there are always at least two criteria to be taking into account: profit maximization and risk minimization. Therefore, the economic and financial problems are multiple criteria ones. In Prof. Dymova's book, an interesting systematization of the problems of multiple criteria decision making is proposed which allows the author to reveal unsolved problems. The solutions of them are presented as well and implemented to deal with some important real-world problems such as investment project's evaluation, tool steel material selection problem, stock screening and fuzzy logistic.

It is well known that the best results in real -world applications can be obtained using the synthesis of modern methods of soft computing. Therefore, the developed by Prof. Dymova new approach to building effective stock trading systems, based on the synthesis of fuzzy logic and the Dempster-Shafer theory, seems to be a considerable contribution to the application of soft computing method in economics and finance.

It is a common believe that almost all problems of the fuzzy evaluation of financial parameters in the capital budgeting are solved. Nevertheless, an unresolved problem is the fuzzy evaluation of the Internal Rate of Return (IRR), although this parameter plays an important role in the investments profitability assessment. In the case of real valued parameters, IRR is calculated as the solution of a nonlinear equation. So in the fuzzy environment, a fuzzy extension of this equation should be solved. Previous authors considered this equation and stated that it cannot be applied to the fuzzy case because its left hand side is fuzzy, 0 in the right hand side is a crisp value and an equality is impossible. In Prof. Dymova's book, this problem is solved using a new method which makes it possible to solve linear and nonlinear interval and fuzzy equations and systems of them. The developed new method allows the author to obtain an effective solution of the Leontjev', input-output problem in the interval setting.

Soft Computing in Economics and Finance makes a major contribution to a better understanding of how fuzzy logic can be applied to solve important problems in economic and financial applications. The author deserves our thanks and congratulations for presenting an interesting book.

Czestochowa, June, 2010 Leszek Rutkowski

Contents

Chapter 1
Introduction

The initial idea behind writing this book was to present new applications of soft computing in the solution of economic and financial problems developed during the last ten years as the result of research project supervised by the author of this book in the Institute of Theoretical and Applied Informatics, Czestochowa University of Technology, Poland. Some results have been published in the papers contributed by the author, as well as Prof. Pavel Sevastjanov, Dr. Pawel Figat, Pawel Bartosiewicz, M.Ph (he is the author's Ph.D. student), and the author's former Ph.D. students Dr. Krzysztof Kaczmarek and Dr. Marek Dolata.

However, it became clear that the scope of this book should go far beyond the initial idea of its contents. It is obvious that the book ought to provide much more information, not only concentrate on the subject of the research into economic and financial applications of soft computing, but also exhibit a wider view within the general framework of soft computing methods.

In addition to the first idea, concerning the above mentioned research, the intention was to incorporate some results from the author's other papers and book on applications of soft computing methods in industry and economy. The main reason for this was that some of these papers and the book were published in Russian and Polish, and are not accessible to many interested readers.

The review of recent achievements in the field of implementation of the soft computing methods in the solution of economic and financial problems is presented in Chapter 2, where the problems which were revealed in the process of these methods implementation are performed as well.

In Chapter 3, an overview of modern methods based on fuzzy sets, including type 2 and level 2 fuzzy sets, intuitionistic fuzzy sets, interval analysis, and the Dempster-Shafer theory of evidence (*DST*) is presented. The interrelations between these methods are shown and some problems that impede their applications are emphasized as well. The author has no intention to make in this chapter a comprehensive overview of all modern methods as now they are well presented in numerous books and handbooks. Therefore, in this chapter the modern methods for uncertainty modeling are presented only on the extent needed for understanding the applications performed in the following chapters.

L. Dymowa: Soft Computing in Economics and Finance, ISRL 6, pp. 1–5.
springerlink.com

It is shown that one of the most undesirable negative features of interval arithmetic is the fast increasing of width of intervals obtained as the results of interval calculations (excess width effect). Another important problem of interval analysis is the so-called natural interval extension. If we have to make interval extension of real valued function, all argument of this function should be replaced with corresponding intervals and all operations should be replaced with corresponding operations on intervals. Such approach to interval extension seems to be justified enough and intuitively clear. Nevertheless, the so-called dependency problem is a major obstacle to the application of extension principle in interval arithmetic. Although interval methods can determine the range of elementary arithmetic operations and functions very accurately, this is not always true with more complicated functions. If an interval occurs several times in a calculation, and each occurrence is taken independently then this can lead to an unwanted expansion of the resulting intervals.

An important problem of interval extension is also that the accuracy of resulting interval strongly depends on the algebraic form of function chosen for extension. It is worthy no note that as the fuzzy arithmetic operations are usually based on the α -cut representation of fuzzy numbers, the above mentioned problems of interval analysis are the problems of fuzzy arithmetic as well. It is noted in this chapter that one of the most important applications of Atanassov's intuitionistic fuzzy sets (*A-IFS*) are the multiple criteria decision making problems (*MCDM*).

It is shown that there exist two important problems in *MCDM* in the intuitionistic fuzzy setting: aggregation of local criteria without intermediate defuzzification in the case when criteria and their weights are intuitionistic fuzzy values (*IFVs*); comparison of *IF* valued scores of alternatives basing on the degree to which one *IFV* is grater/smaller than the other. In this chapter, it is shown that there exist a strong link between *DST* and *A-IFS* which makes it possible to reformulate the basic definitions of *A-IFS* in terms of *DST*. We show that using the *DST* semantics, it is possible to enhance the performance of *A-IFS* when dealing with *MCDM* problems. Particularly, this approach allows us to use directly the Dempster's rule of combination to aggregate local criteria presented by *IFVs* and develop a method for *MCDM* without intermediate defuzzification when local criteria and their weights are *IFVs*. As the result we get final alternative's evaluations in the form of belief intervals. Hence, an appropriate method for such intervals comparison is needed. Therefore, a method for interval and fuzzy numbers comparison based on *DST*, which provides the result of comparison in the form of belief interval is presented.

It is noted that now besides the classical Dempster's rule of combination a number of other methods for the combination of evidence are proposed in the literature. All of them have own merits and drawbacks and the problem of choosing the best method is now open.

In Chapter 4, the problems typical for *MCDM* are analyzed and new solutions of them are proposed as well. The problem of appropriate common scale for representation of objective and subjective criteria is solved using the simple subsethood measure based on the α-cut representation of fuzzy values. To develop an appropriate method for aggregation of aggregating modes, the synthesis of tools of type-2 and level-2 fuzzy sets is used. As the result the final assessments of compared alternatives are presented in the form of fuzzy valued membership function defined on the support composed of considered alternatives. To compare the obtained fuzzy assessments, the probabilistic approach to fuzzy values comparison is used. In is shown that the investment evaluation problem is frequently a hierarchical one and a new method for solving such problems, different from commonly used fuzzy analytic hierarchy process (*AHP*) method, is proposed. The developed methods are used for the solution of the stock ranking problem based on *MCDM* and optimization in the fuzzy setting, and for the multiple criteria fuzzy evaluation and optimization in budgeting.

Chapter 5 deals with the so-called distribution problem, which belong to the wide class of the logistic problems. It is known that distribution transportation and transportation problems have similar mathematical structures and are usually treated as particular cases of the general linear programming problem. There are many effective algorithms for the solution of transportation and distribution problems proposed in the scientific literature and in the textbooks. So we can say that these problems in the case of real valued parameters are, generally, solved. Nevertheless, in practice, we often meet different kinds of uncertainty when the parameters of these optimization problems are presented by intervals or fuzzy values. The known approaches to the solution of fuzzy transportation and distribution problems are usually based on some restrictions imposed on the form of membership functions. These restrictions make it possible, using analytical procedures, to transform the initial fuzzy problem to the set of usual linear programming problems with real valued parameters. Nevertheless, in practice, membership functions representing the parameters of the problem may have substantially complicated forms and analytical procedures can not be used. Therefore, in this chapter a new approach to the solution of fuzzy distribution problem is developed. In the framework of this approach, all parameters and variables may be fuzzy values without any additional restrictions. It is important that real-world distribution problems are usually multiple criteria ones. In this chapter, the results obtained as the solution of fuzzy single criterion distribution problem are used as the base for the formalization and solution of multiple criteria fuzzy distribution problem.

Chapter 6 is devoted to the synthesis of fuzzy logic and *DST* in stock trading decision support systems. Modern computerized stock trading systems (the so-called mechanical trading systems) are based on the simulation of the decision making process and generate advices for the traders to buy or sell stocks or other financial tools by taking into account the price history, technical analysis indicators, accepted rules of trading and so on. There are many approaches to building stock trading systems proposed in the literature. The applications of the methods of soft computing in this field of researches are analyzed in Chapter 2. It is noted that the source of many failures when building really profitable stock trading systems is the ignoring of human factor. It was recognized in [7], after obtaining a negative result that “The trading system loses money and gets a negative Sharpe Ratio. We believe that if expert’s experience is available, it will generate more promising results”.

We can say that the last statement was the pivotal idea on which the methods presented in this chapter are based. We believe that the wisdom accumulated by generations of traders in the form of well-known trading rules of technical analysis are an adequate base on which it is possible to build optimal fuzzy expert systems for stock trading. Our starting point was the paper [1], where the authors presented an expert system based on a fuzzy logic representation of technical analysis trading rules which are usually used by investors for decision making. Since technical analysis provides indicators used by experts to predict stock price movements, the method proposed in [1] maps these indicators into new inputs that can be used in a fuzzy logic system. This chapter generalizes our experience in building stock trading systems. Some results we have obtained are partially presented in [2, 6].

Here we present and compare three different expert systems for stock trading based on the synthesis of fuzzy logic and technical analysis. The first one is a special adaptation of classical Mamdani approach. Another method is based on the so-called "logic-motivated fuzzy logic operators" [3]. The third system that will be presented is based on the synthesis of fuzzy logic and DST.

In Chapter 7, the application of interval and fuzzy analysis in economical modeling is presented. In this chapter, a new concept of the solution of interval and fuzzy equations based on the generalized procedure of interval extension called "interval extended zero" method is proposed. The central for the proposed approach is the treatment of "interval zero" as an interval symmetrical with respect to 0. It is shown that such proposition is not of heuristic nature, but is a direct consequence of the interval subtraction operation. Some methodological problems concerned with this definition of interval zero are discussed. It is shown that the resulting solution of interval linear equations based on the developed method may be naturally treated as a fuzzy number. An important advantage of a new method is that it substantially decreases the excess width effect. On the other hand, we show that it can be used as the reliable practical tool for solving the linear interval and fuzzy equations and the systems of them. The fundamentals of the proposed approach were presented in [5, 4]. In this chapter, we present the generalization of these obtained and some new results. The applications of a new approach are performed by the solution of well known Leontief's input-output problem in the interval setting and the solution of the problem of fuzzy Internal Rate of Return in budgeting.

Many researchers contributed to these results in the fields related to the contents of this book. The long list of reference includes their names associated with the publications cited in the book. Professor L.A. Zadeh, who is known as the "Father of Fuzzy Logic", and the pioneers who initiated research into soft computing applications, are mentioned with regard to their publications. The outstanding contributions of Prof. J.J Buckley and Prof. C. Kahraman to application of soft computing methods in the solution of economic and financial problems should also be emphasized. However, a number of researchers who made significant contributions to the solution of these problems are not cited in this book, since it is impossible to refer to all of them. The interested reader can find the related papers and books in other bibliography lists.

The author would like to especially acknowledge the contribution of Prof. C. Kahraman to fuzzy modeling in finance and economics. His papers and the edited by him recent book entitled "Fuzzy Engineering Economics with Applications" is closely related to the subject of this book. Therefore, the author dedicates her book to Prof. Cengiz Kahraman. There are more persons whose important contributions to the contents of this book should be acknowledged. There are many people whom the author would like to thank for their help, encouragement, and understanding. Presented below is a special acknowledgment.

Acknowledgments. As the author, I would like to express my sincere gratitude to Prof. Janusz Kacprzyk, the Editor of the book series "Studies in Fuzziness and Soft Computing", for his encouragement to publish the book. The author thanks Prof. Leszek Rutkowski for his Foreword. The author also gratefully acknowledges Dr. Marek Dolata (her former Ph. D student) Pawel Bartosiewicz, M.Ph (her Ph. D student), for the material being a part of our joint research and included in the book. The author thanks Prof. D. Rutkowska for her valuable consultations. The words of special thankfulness are directed to my husband Prof. P. Sevastjanov for his help in the process of preparing the final form of this book.

References

1. Dourra, H., Siy, P.: Investment using technical analysis and fuzzy logic. Fuzzy Sets and Systems 127, 221–240 (2002)
2. Dymova, L., Sevastianov, P., Bartosiewicz, P.: A new approach to the rule-base evidential reasoning: stock trading expert system application. Expert Systems with Applications 37, 5564–5576 (2010)
3. Santiprabhob, P., Nguyen, H.T., Pedrycz, W., Kreinovich, V.: Logic-Motivated Choice of Fuzzy Logic Operators, pp. 646–649. FUZZ-IEEE (2001)
4. Sevastjanov, P., Dymova, L.: Fuzzy solution of interval linear equations. In: Wyrzykowski, R., Dongarra, J., Karczewski, K., Wasniewski, J. (eds.) PPAM 2007. LNCS, vol. 4967, pp. 1392–1399. Springer, Heidelberg (2008)
5. Sevastjanov, P., Dymova, L.: A new method for solving interval and fuzzy equations: linear case. Information Sciences 17, 925–937 (2009)
6. Sevastianov, P., Dymova, L.: Synthesis of fuzzy logic and Dempster-Shafer Theory for the simulation of the decision-making process in stock trading systems. Mathematics and Computers in Simulation 80, 506–521 (2009)
7. Shen, L., Loh, H.T.: Applying rough sets to market timing decisions. Decision Support Systems 37, 583–597 (2004)

Chapter 2
Applications of Modern Mathematics in Economics and Finance

Nowadays the dominating paradigms of economic theories are based on the classical mathematics and presented in terms of probabilistic and statistical methods. These methods may be treated as the traditional ones. As the applications of them in finance and economics are well presented in numerous papers, books and textbooks, the detailed description of these applications is out of scope of this book. It should be emphasized that in applications, the probabilistic and statistical methods are often and successfully used in the synthesis with modern methods of soft computing. Now it is understood that in applications we often deal with different types of uncertainty (not only of probabilistic nature). Therefore, this chapter presents a brief overview of the applications of modern methods of soft computing in economics and finance and the problems which were revealed in the process of these methods implementation.

2.1 Fuzzy Set Theory and Applied Interval Analysis in Economical and Financial Applications

This section presents an analysis of applications of the theory of fuzzy sets and its generalizations such as Atannassov's intutionistic fuzzy sets in the different fields of economics and finance. The economical and financial applications of interval analysis are presented as well. The main problems concerned with the use of fuzzy set theory and interval analysis methods in economical and financial applications are discussed.

The most salient feature of the late 20th century was the scale of changes affecting social, economic and corporate life. Our environment is changing at a rate which would once have been unthinkable, and the speed of events has now become astonishing. The problems posed by these new situations are increasingly complex and changeable and traditional models based on determinism and chance are no longer able to cope with this reality. Therefore, in the last four decades the group of new methods (the methods of soft computing) which make it possible to operate

L. Dymowa: Soft Computing in Economics and Finance, ISRL 6, pp. 7–39.
springerlink.com

with non-probabilistic uncertainty was developed. Among them the tools of fuzzy set theory [199] are most frequently used in economic and financial applications.

Zadeh's main motivation for the invention of fuzzy sets was the imprecision in human decision-making. Fuzzy sets emerged from the need to bridge the gap between mathematical models and their empirical interpretations. The capability of fuzzy sets to express gradual transitions from membership to non-membership and vice versa has a wide utility.

One of the major thrusts of economic science is to describe the behavior of individual units such as consumers, households, firms, government agencies and their interactions. But a large number of concepts, which we use in everyday life, are vague. Fuzziness can be found in many areas.

In the literature, almost all the economic theories are explained in the classical mathematics frame. In this perspective, fuzzy mathematics seems to be more suited in explaining the concepts of economics than the classic one.

Therefore, the international association for fuzzy-set management and economy (*SIGEF*) has been set up to encourage research and study relating to all aspects of the economy in general, and corporate management in particular. The international journal of this association "Fuzzy economic review" plays an important role by bringing together the most important projects devised by specialist and offering them a fundamentally practical opening into the business world, in pursuit of university-business cooperation which is both necessary and beneficial to both sides. Different applications of fuzzy sets theory and fuzzy logic in economics and finance are presented in the books [21, 26, 59, 82, 83, 117] and in the thousands of scientific papers.

It is worthy to note that the best results can be obtained in the cases when dealing with direct fuzzy extension of considered economical of financial concept presented by some mathematical expressions.

Therefore, the tools of fuzzy set theory were most successfully used in budgeting. There are a lot of financial parameters proposed in literature [15, 23, 35, 106] for budgeting. The main ones being: Net Present Value (*NPV*), Internal Rate of Return (*IRR*), Payback period (*PB*), Profitability Index (*PI*). It is shown in [20] that the most important parameters are *NPV* and *IRR*. A good review of other useful financial parameters may be found in [9].

Net Present Value is usually calculated as follows:

$$NPV = \sum_{t=1}^{T} \frac{P_t}{(1+d)^t} - KV, \tag{2.1}$$

where d is a discount rate, KV is a starting capital investment, P_t is a total income (cash flow) in a year t, T is a duration of an investment project in years. Usually the discount rate is equal to the average bank interest rate in a country of an investment or other value corresponding to a profit rate of alternative capital investments.

The value of *IRR* is a solution of the following non-linear equation with respect to d:

$$\sum_{t=1}^{T} \frac{P_t}{(1+d)^t} - KV = 0. \tag{2.2}$$

If P_t, KV (or at least one of them) are fuzzy numbers then with the use of fuzzy extension of Eq. (2.2), i.e., by replacement of its parameters and variables with fuzzy ones and all arithmetic operations with relevant fuzzy operations, Eq. (2.2) can be transformed to a fuzzy equation. The problem is to find a fuzzy solution of such fuzzy equation, i.e., to obtain a fuzzy *IRR*.

The economic nature of *IRR* can be explained as follows. If an actual bank discount rate or return of any other alternative investment under consideration is less than *IRR* of considered project, then investment in this project is more preferable. An estimation of *IRR* is frequently used as a first step of the financial analysis. Only projects with *IRR* above some accepted threshold value (usually 15–20%) can be chosen for further consideration.

Nowadays traditional approaches to the evaluation of *NPV*, *IRR* and other financial parameters is subjected to quite deserved criticism, since the future incomes P_t and rates d are rather uncertain parameters. Uncertainties which one meets in capital budgeting cannot be adequately described in terms of probability. Really, in budgeting we usually deal with a business-plan that takes a long time — as a rule some years — for its realization. In such cases, the description of uncertainty via probability representation of P_t, KV and d usually is impossible due to a lack of information about probabilities of future events. Thus, what we really have in such cases are some subjective expert's judgments. In real-world situations, investors or experts involved are able to estimate only intervals of possible values P_t and d and the expected (more probable) values inside these intervals.

That is why during the last two decades the growing interest to applications of the interval arithmetic [122] and fuzzy set theory methods [199] in budgeting has being observed. After pioneer works by Ward [188] and Buckley [24], some other authors contributed to the development of fuzzy capital budgeting theory [30, 41, 44, 45, 51, 77, 78, 79, 80, 81, 94, 135, 151].

It is safe to say now that almost all problems of the fuzzy *NPV* estimation are solved. An unresolved problem is the fuzzy estimation of the *IRR*. Ward [188] considered Eq. (2.2) and stated that such an expression cannot be applied to the fuzzy case because the left hand side of Eq. (2.2) is fuzzy, whereas 0 in the right hand side is a crisp value and an equality is impossible. Hence, the Eq. (2.2) is senseless from the fuzzy viewpoint. Kuchta [94] proposed a method for fuzzy *IRR* estimation where α-cut representation of fuzzy numbers [86] was used. The method is based on an assumption (see [94, p. 380]) that a set of equations for *IRR* determination on each α-cut may be presented (in our notation) as follows:

$$(KVV^{\alpha})_1 + \sum_{t=1}^{T} \frac{(P_t^{\alpha})_1}{(1+IRR_1^{\alpha})^t} = 0, \quad (KVV^{\alpha})_2 + \sum_{t=1}^{T} \frac{(P_t^{\alpha})_2}{(1+IRR_2^{\alpha})^t} = 0, \tag{2.3}$$

where $KVV = -KV$, indexes "1", "2" stand for the left and right bounds of corresponding intervals respectively, $P_t^{\alpha} = [(P_t^{\alpha})_1, (P_t^{\alpha})_2]$ are crisp interval representations of fuzzy cash flows at time t on α -cuts. Of course, from the equations (2.3) all crisp intervals $d^{\alpha} = [d_1^{\alpha}, d_2^{\alpha}]$, expressing the fuzzy valued IRR may be obtained. On the other hand, Eqs.(2.3) are not a direct consequence of conventional fuzzy and interval arithmetic rules. Eqs.(2.3) were obtained in [94] using fuzzy extension of (2.2) assuming that P_t, KV (or at least one of them) are fuzzy numbers and representing them by the sets of α-levels. Since Eqs.(2.3) should be verified on each α -cut, it is quite enough to consider only crisp interval extension of (2.2), which is the particular case of more general equation

$$F(d) - B = 0, \tag{2.4}$$

where B is an interval ($B = KV$ in our case) and $F(d)$ is an interval valued function of interval argument d (in our case $F(d) = \sum_{t=1}^{T} \frac{P_t}{(1+d)^t}$).

Using regular interval arithmetic [75], this equation can be transformed to $[F_1(d) - B_2, F_2(d) - B_1] = 0$, and finally we get two equations $F_1(d) - B_2 = 0$, $F_2(d) - B_1 = 0$. Of course, if we deal with a linear interval function $F(d) = A \cdot d$ (A is an interval) , then $F_1(d) = A_1 \cdot d_1$ and $F_2(d) = A_2 \cdot d_2$, but if $F(d) = \frac{A}{d}$ we have $F_1(d) = \frac{A_1}{d_2}$, $F_2(d) = \frac{A_2}{d_1}$ since F_1 is the left bound (minimal value in interval) and F_2 is the right bound (maximal value in interval) of interval value $F(d)$. Hence, the use of the regular interval arithmetic rules leads to the following equations:

$$(KVV^{\alpha})_1 + \sum_{t=1}^{T} \frac{(P_t^{\alpha})_1}{(1+IRR_2^{\alpha})^t} = 0, \quad (KVV^{\alpha})_2 + \sum_{t=1}^{T} \frac{(P_t^{\alpha})_2}{(1+IRR_1^{\alpha})^t} = 0. \tag{2.5}$$

There is no way to get a correct not inverted interval solution of (2.5). Only inverted intervals IRR, i.e., such that $IRR_1^{\alpha} > IRR_2^{\alpha}$ can be obtained from (2.5). Since it is hard or even impossible to interpret reasonable such results, they can not be used in practice. It is shown in [51] that only approximate real valued IRR (represented by usual non interval numbers) may be obtained from (2.5).

In Chapter 7, we show that the main problem is that the conventional interval extension (and the fuzzy as well) of usual equations, which leads to the interval or fuzzy equations such as (2.5) is not a correct procedure. A new approach to the solution of fuzzy IRR problem is proposed in [152], but rather as a heuristic method. In Chapter 7, this method is presented as a part of more general problem of the solution of nonlinear fuzzy equations.

Now the mathematical tools of fuzzy set theory and fuzzy logic are successfully used for the risk analysis in e-commerce development [128], in portfolio selection [19, 40, 71, 180], in Black-Scholes option pricing models [98, 192], in corporate acquisition systems [118], in evaluating investments in advanced manufacturing technology [1], in interactive fuzzy interval reasoning for smart web shopping [105], in fuzzy scheduling [113], in logistic [32, 55].

The great achievements were made in the field of linear and nonlinear fuzzy regression models [27, 69, 76, 170]. This seems to be very important as the regression analysis is one of the most powerful methods used in economic and financial applications.

Nowadays it is rather impossible to make a comprehensive review and analysis of applications of fuzzy set theory and fuzzy logic in economics and finance as the methods of soft computing are used almost in all fields of economics. Nevertheless, in some branches of modern economics and finance these methods still are waiting for their application. For example, in insurance only first steps were made and adoption of tools of fuzzy sets theory is now on the stage of problem formulation [153, 154].

There are some problems in implementation of fuzzy set theory tools in logistic [32, 55]. Basing on the literature analysis, we can say that now there are no such general approaches that make it possible to obtain the solution of fuzzy linear programming problem, fuzzy transportation or fuzzy distribution problems without additional restrictions on the form of fuzzy parameters and variables. Obviously, such restrictions substantially limit the ability of known methods to solve practically important problems when additional restrictions disturb the initial structure of the problem.

Therefore, in Chapter 5, we propose a new numerical approach to the solution of fuzzy distribution problem based on the direct fuzzy extension of the simplex method.

Another important challenge for fuzzy set theory is input-output analysis developed by V. Leontief [99]. Input-output analysis is an extremely effective tool used in more than 70 countries over the world for manufacturing processes optimization, economy condition improvement and intersectors costs allocation analysis. The fuzzy extension of Leontief's model in a spirit of L. Zadeh was proposed by J. Buckley in [25], but now only a few papers are devoted to this problem. The fuzzy extension of Leontief's model in practice needs a solution of a system of thousands linear fuzzy equations. Therefore, we are faced with the problems of computational and methodological nature which complicate the successful solution of such systems.

In Chapter 7, the solution of interval Leontiev's input-output problem with the use on a new approach to interval extension is proposed.

Intuitionistic fuzzy sets (*A-IFS*) proposed by Atanassov [6] is one of the possible generalizations of fuzzy set theory and appears to be relevant and useful in some applications. The concept of *A-IFS* is based on the simultaneous consideration of membership μ and non-membership ν of an element of a set to the set itself such that $0 \leq \mu + \nu \leq 1$.

The similar approach, the so-called vague sets, proposed by Gau and Buehrer in [58] is proved to be equivalent to the *A-IFS* (see [28]). Since vague sets were proposed later than *A-IFS*, here we shell always speak of *A-IFS*. There are many papers devoted to the theoretical problems of *A-IFS* in the scientific literature (see [130] for an overview).

Nevertheless, only a few papers presenting applications of *A-IFS* in economics and finance can be cited. In all cases *A-IFS* are not used solely, but in synthesis with other methods including the methods of soft computing.

In [31], the authors propose a new approach for fuzzy inference in intuitionistic fuzzy systems, which combines the outputs of two traditional fuzzy systems to obtain the final conclusion of the intuitionistic fuzzy system. A new method is illustrated using the example of monitoring a nonlinear dynamic plant. Guo and Zhang [64] proposed a new approach to project risk evaluation based on intuitionistic fuzzy sets and the so-called *TOPSIS* method. In [182], the authors investigate the multiple attribute decision making problems to deal with the supplier selection in supply chain management with intuitionistic fuzzy information, in which the information about attribute weights is completely known, and the attribute values take the form of intuitionistic fuzzy numbers.The use of *TOPSIS* method with intuitionistic fuzzy information is proposed. Then, based on the traditional *TOPSIS* method, calculation steps for solving intuitionistic fuzzy multiple attribute decision-making problems with known weight information are given. Finally, an illustrative example about supplier selection is given to verify the developed approach and to demonstrate its practicality and effectiveness. In [138], an intuitionistic fuzzy goal programming approach for a quality control problem is proposed. The method for multi-objective intuitionistic fuzzy linear programming and its application to the solution of transportation problem is proposed in [73]. In [74], a multi-objective transportation problem under intuitionistic fuzziness is considered. In [138], the intuitionistic fuzzy goal programming and its application for solving the multi-objective transportation problem are presented.

It is worth noting that, unfortunately, in all papers cited above, the applications seem to by rather numerical examples than real-world applications.

Let as consider the applications of interval analysis in economics and finance.

Interval mathematics was developed by M. Warmus in his paper [189] where the basic rules of interval arithmetic were formulated. But the birth of modern interval arithmetic was marked by the appearance of the book "Interval Analysis" by Ramon E. Moore in 1966 [122]. He had the idea in Spring 1958, and a year later he had published a report [121] on how interval arithmetic could be implemented on a computer. Now the modern interval analysis is well presented in the book [123] and some important applications of interval analysis can be found in the book [75].

The journal "Reliable Computing" (originally Interval Computations) has been published since the 1990s, dedicated to the reliability of computer-aided computations. We can say that most of papers devoted to internal mathematical problems of interval analysis are now published in this journal. Moreover, the literature analysis makes it possible to conclude that "interval community" is now focused mainly on theoretical researches. Probably, a relatively low number of applications of interval analysis can be explained by the great popularity of fuzzy set theory. Really, an interval may be naturally treated as a particular case of a fuzzy value.

Although Lodwick in his book [115] makes an effort to eliminate the gap between interval and fuzzy analysis, we can say that now interval analysis and fuzzy set theory are developing almost independently, whereas theoretical results obtained by "interval community" may be successfully exploited in the framework of fuzzy set theory.

A relatively low number of papers was devoted to the solution of linear programming problems in the interval setting with application to the transportation and other problems of logistic. Steuer [165], Tong [173], Chanas and Kuchta [33] proposed the solutions of the linear programming problem with the interval target function. The generalization of these solution was presented by Kuchta in [95]. Similarly, in [48] the authors proposed the procedure for the solution of transportation problem with interval parameters of the objective function and constraints. The generalized linear fractional programming under interval uncertainty is developed in [68]. The proposed method reduces the problem to solving from two to four real-valued generalized linear fractional programs. The method is illustrated by a simple von Neumann economic growth model. It should be noted that in all above papers, the developed methods are based on analytical procedures that makes it possible to transform the initial interval problem to the set of usual linear programming problems with real valued parameters. In such a way, the authors avoid the direct use of methods of applied interval analysis based on the rules of interval arithmetic. Therefore, such approaches can be only formally treated as applications of interval analysis.

Liu [114] developed a method to derive the iinterval profit of the inventory model when the demand quantity and unit cost are intervals. A pair of two-level mathematical programming model to derive the upper bound and lower bound of the profit is formulated. It is shown that the interval profit contains more information for determining inventory policy. In [43, 49], a method for evaluating interval forecasts with applications to the financial risk management was developed. In [4], this method is used for the comparison of a number of alternative autoregressive conditional duration models using a sample of data for three major companies traded on the Australian Stock Exchange. Li et al. [104] proposed a dual-interval vertex analysis method through incorporating the vertex method within an interval-parameter programming framework. A management problem in terms of regional air pollution control is studied to illustrate applicability of the proposed approach. The results indicate that useful solutions for planning the air quality management practice have been generated. They can help decision makers to identify desired pollution-abatement strategies with minimized costs and maximized environmental efficiencies.

Summarizing we can say that using the α-cut presentation of fuzzy numbers, the problems of fuzzy arithmetic can be reduced to the problems of interval analysis. Hence, the methods of interval analysis can be successfully used in applications of fuzzy set theory. Therefore, in Chapter 3 we present the basics and some important problems of interval analysis, in the following chapters (especially in Chapter 7) we present some new methods of interval analysis and show how they can be used in the economic and financial applications in the interval and fuzzy environments.

2.2 Economic and Financial Applications of Rough Sets Theory

Rough sets (*RS*) theory was introduced more than 20 ago by Pawlak [133] and has emerged as a powerful technique for the automatic classification of objects [177].

The popularity of *RS* theory stems primarily from its operational processes, which adhere closely to the notions of knowledge discovery and data mining [110].

Most *RS* applications are designed to deal with classification problems of one form or another. In constructing such applications, *RS* theory is generally integrated with other theories such as fuzzy set theory, grey systems theory and so on.

Based on the notion of the existence of indiscernibility relation between objects, *RS* deal with the approximation of sets or concepts by means of binary relations. Compared to other methods used in financial area, such as discriminant analysis, univariate statistical method and linear probability model, this new method has the following advantages [50, 62]: it is based on the original data only and does not need any external information; it is a tool suitable for analyzing not only quantitative attributes, but also qualitative ones; it discovers important facts hidden in data and expresses them in the natural language of decision rules; the set of derived decision rules gives a generalized description of the knowledge contained in the database, eliminating any redundancy typical of the original data; the obtained decision rules are based on facts, because each decision rule is supported by a set of real examples; the results of *RS* are easy to understand, while the results from other methods (credit scoring, utility function and outranking relation) usually require an interpretation of the technical parameters, with which the user may not be familiar.

The applications of *RS* in economic and financial prediction can be divided into three main areas [155]: business failure prediction [2, 50, 62, 67, 161, 162, 163, 164], database marketing [92, 93, 120, 125, 136, 137] and financial investment including stock market analysis algorithms and portfolio selection [10, 11, 12, 60, 61, 70, 108, 146, 155, 160, 167, 185, 201].

The detailed review of applications of *RS* in financial domains can be found in [172]. The methods of *RS* theory were used in the solution of scheduling problem [109], in the model for extracting payment rules of vehicle license tax [196] and for the solutions of other problems of economic and financial analysis.

It is worthy to note that the best and useful in practical applications results have been obtained when the tools of *RS* theory were used in the synthesis with other methods of soft computing.

2.3 Artificial Neural Network-Based Applications in Economics and Finance

In this section, the economical and financial applications of artificial neural networks are presented and discussed. Nowadays the tools of artificial neural networks (*ANN*) are successfully used in almost all branches of economic and financial analysis. Therefore, it seems to be impossible to make a comprehensive review of all applications of *ANN* in these fields. Therefore, here we present only the most important and successful applications especially concerned with the research issues we deal with in the following chapters of this book.

Basing on the bibliography of neural network business applications research [191] we can see that in 1994-1998 years production/operations had the largest number of *ANN* applications, followed by finance, marketing/distribution, information systems, accounting/auditing, and human resources. In production/ operations, the most popular research areas were the part family/machine grouping, job shop scheduling, cellular manufacturing system design, and equipment/machine fault diagnosis/detection. Bankruptcy prediction of banks/firms was the most common application in the area of finance. In particular, the neural network model trained by the back-propagation learning algorithm is the most popular tool used for the solution of financial decision-making problems, whose prediction accuracy outperforms that of other models, such as logistic regression (*LR*), linear discriminant analysis (*LDA*), multiple discriminant analysis (*MDA*), *k*-nearest neighbor (*k-NN*), decision trees, *etc*. This indicates that choosing an appropriate learning model/classifier is a major factor affecting the classification or prediction result.

The analysis of the modern scientific literature make it possible to say that these basic tendencies are still actual.

The state of the art in application of *ANN* for bankruptcy forecasting is presented in [134], where the author noted that the first studies on the use of *ANNs* were carried out in the early 1940s, but the financial applications are much more recent. As for experiments within the scope of the forecast of bankruptcy, they did not come out until 1990. The *ANNs* can be compared with the traditional statistical tools of forecast: "The neural networks make it possible to approximate both linear and nonlinear functions and achieve outstanding performances within the applications concerning classification, the prediction of future values as well as the modeling and the control of processes" [134]. They have three main features that allowing them to adapt the issue of bankruptcy forecast. On the one hand, the failure of a company is a somewhat complex field which is still not the subject of a complete theory. It is important that the large availability of the companies accountancy on the data-processing data bases meets the need for the neural networks. Additionally, bankruptcy forecast is not a problem that can be separated linearly, however, an artificial neural network does not require a linear assumption of separation of the classes. Therefore, their capacities can only be expressed efficiently within the scope of both a significant base of examples and a non-linear complex problem since it has not yet been the subject of mathematical modeling. The bankruptcy of companies shows these three characteristics. In [29], *ANNs* are used for auditing and risk assessment. The paper focuses on the use of *ANNs* as an enabler of the new business risk auditing framework and provides insight into future research opportunities. In general, *ANNs*, which are classifiers by nature, offer the capacity to consider multiple types of evidence simultaneously and can assist auditors in assessing risks and making judgments.

The important applications of *ANNs* are credit scoring, prediction of bankruptcy of banks and firms, country investment risk estimation and financial crisis prediction.

The neural network ensembles were used for bankruptcy prediction and credit scoring in [174]. A bank failure early warning system based on a novel localized pattern learning and semantically associative fuzzy neural network is presented in [127]. In [13], three different *ANN* models serve as a cross validation of the efficacy of using and consequent performance of neural network models that categorize country investment risk. Two main *ANN* models named *RBF* and *MLP* were applied in [171] to model the risky economical projects and performance results were compared. Analysis of the neural network outputs proved that more predictive capability can be achieved by *MLP*. The proposed network can be applied to any risky economical project analysis if the project inputs are normalized with a Gaussian distribution function. The application of *ANN* models for predicting US recessions is presented in [139]. This paper examines the relevance of various financial and economic indicators in predicting US recessions via neural network models. The author shares the view that business cycles are asymmetric and cannot be adequately accommodated by linear constant-parameter single-index models. Therefore the author uses a novel neural network to recursively model the relationship between the leading indicators and the probability of a future recession.

The usefulness of *ANNs* for early prediction of economic crisis is analyzed in [90]. The authors noted that during the 1990s, the economic crises in many parts of the world have sparked a need in building early warning system (*EWS*) which produces signal for possible crisis, and accordingly various *EWSs* have been established. In this paper, the authors focus on an interesting issue: "How to train *EWS*?" To study this, various aspects of the training data (i.e., the past crisis related data) were discussed and then several data mining classifiers including artificial neural networks were probed as a training tool for *EWS*. In [197], a multiscale neural network learning paradigm for financial crisis forecasting is proposed. For illustration purpose, the proposed multiscale neural network learning paradigm is applied to exchange rate data of two Asian countries to evaluate the state of financial crisis. Experimental results reveal that the proposed multiscale neural network learning paradigm can significantly improve the generalization performance relative to conventional neural networks.

The methods based on *ANNs* are successfully used in production prediction. In [129], the authors describe a decision support system that incorporates both curve fitting and *ANN* approaches, and present a range of possible solutions for oil production prediction. The synthesis of seasonal time series *ARIMA* method and neural networks with genetic algorithms for predicting the production value of the mechanical industry in Taiwan is presented in [107]. The paper [7] presents an approach based on an integrated genetic algorithm and artificial neural network to estimate and predict the electricity demand using stochastic procedures. The economic indicators used in this paper are price, value added, number of customers and consumption in the previous periods. This model can be used to estimate the energy demand in the future by optimizing the values of parameters. Modified neo-fuzzy neuron-based approach for economic and environmental optimal power dispatch is developed in [36]. It is shown that the proposed model is capable of achieving accurate results and the training is observed to be faster than other popular neural networks. In [142], the synthesis of *ANNs* and multiple criteria analysis is used for sustainable irrigation planning. In the paper [57], the authors provide a microeconomic application of *ANNs* by input-output mapping for 82 US major investor-owned electric utilities using fossil-fuel fired steam electric power generation for the year 1996. They construct a multilayer feed-forward neural network with back-propagation to represent the relationship between a set of inputs and an electricity production as an output.

One of the classical tools for characterizing an economic system is input-output analysis, invented by Leontief. The original input-output analysis is a static model [100], and it has been extended to a dynamic input-output model [101].

In [184], an alternative method of input- output analysis is proposed. This method is based on the layered neural network model. It shows that neural networks method can be useful for input-output analysis for a dynamic economic system.

The problem of scheduling is a traditional application of *ANNs*. In [156], the authors first propose an attribute selection algorithm based on the weights of neural networks to measure the importance of system attributes in a neural network-based adaptive scheduling (*NNAS*) system. Next, the *NNAS* system is combined with the attribute selection algorithm to build scheduling knowledge bases. In [157], a real-time scheduling algorithm is proposed. This algorithm first makes a fuzzy classification for the operations of jobs in real-time and then schedules using the heuristic. The heuristic is obtained by training a neural network off line with the use of genetic algorithm. Based on these ideas a real-time scheduler is built with neuro-fuzzy network.

The job-shop scheduling problem is one of the most difficult problems in scheduling. It aims to allocate a number of machines over time to perform a set of jobs with certain constraint conditions in order to optimize a certain criterion. To solve this problem, in [194] the neural network is constructed based on the constraint conditions of a job-shop scheduling problem. Its structure and neuron connections can change adaptively according to the real-time constraint satisfaction situations that arise during the solving process. In [3], the authors give a comprehensive overview on *ANNs* approaches for solution of production scheduling problems, discuss both theoretical developments and practical experiences, and identify research trends. More than 50 major production and operations management journals published in years 1988-2005 have been reviewed. Existing approaches are classified into four groups, and additionally a historical progression in this field was emphasized. An application of *ANNs* for the solution of flowshop problem is presented in [54]. This paper considers the n-job, m-machine permutation flowshop with the objective of minimizing the mean flowtime. Initial sequences that are structured to enhance the performance of local search techniques are constructed from job rankings delivered by a trained neural network. The network's training is done by using data collected from optimal sequences obtained from solved examples of flowshop problems.

ANNs are successfully used for the solution of different financial problems.

The paper [91] introduces a resource allocation neural network model to optimize investment weight of portfolio. This model can dynamically adjust the investment weight as a basis of 100 percent of summing all of asset weights in the portfolio. A model based on the combining neural networks and decision trees for earnings management prediction is presented in [175]. This paper is a preliminary study focusing on the development of neural network and decision trees models to predict the level of earnings management. It is believed that the predictive models can help users of financial statements who make decisions depending on the earnings amounts.

Besides, building a prediction model to indicate the level of earnings management in advance is a new application for neural networks.

One of the most important and fruitful applications of *ANNs* is the forecasting in different branches of finance.

The distribution forecasting of high frequency financial time series is presented in [132]. The availability of high frequency data sets in finance has allowed the use of very data intensive techniques using large data sets in forecasting. In this study, an algorithm requiring fast *k-NN* type search has been implemented using a binary neural network based upon correlation matrix memories. This work has also constructed probability distribution forecasts. In assistance to standard statistical error measures, the implementation of simulations has allowed actual measures of profit to be calculated.

One of the major difficulties in investment strategy is to integrate a supply chain with finance for controlling the marketing timing. In [42], a hybrid forecast marketing timing model based on the probabilistic neural network, rough set and decision tree is developed. The authors use not only the different indexes of fundamental and technical analysis, but also the rough set theory and artificial neural networks inference system to construct three investment market timing classification models. This includes the probabilistic neural network classification model, rough set classification model and hybrid classification model combining probabilistic neural networks, rough sets and decision tree. An approach for the financial performance prediction based on the *ANNs* and the data of fundamental and technical analysis is proposed in [97]. This research project investigates the ability of neural networks, specifically, the back propagation algorithm, to integrate fundamental and technical analysis for the financial performance prediction. The predictor attributes include 16 financial statement variables and 11 macroeconomic variables. A hybrid approach to the Japanese candlestick method for financial forecasting is developed in [84]. This paper discusses an experimental study of the Japanese candlestick method used in hybrid stock market forecasting models. Two models are presented in this paper. The first of them is a committee machine with simple generalized regression neural networks experts. This model also has a simple gating network. The second model has a similar committee machine along with a hybrid type gating network that contains fuzzy logic. A method for forecasting stock market returns with the use of data mining and neural networks is presented in [56]. This paper introduces an information gain technique used in machine learning for data mining to evaluate the predictive relationships of numerous financial and economic variables. Neural network models for level estimation and classification are then examined for their ability to provide an effective forecast of future values. A cross-validation technique is also employed to improve the generalization ability of several models. The forecasting model of global stock index based on the stochastic time effective neural network is proposed in [111]. This paper introduces a new stochastic time effective function to model a stochastic time effective neural network model. The effectiveness of the model has been analyzed by performing a numerical experiment on the data of *SAI*, *SBI*, *HSI*, *DJI*, *IXIC* and *S&P*500, and the validity of the volatility parameters of the Brownian motion is tested. Further, the paper shows some predictive results on the global stock indices using the stochastic time effective neural network model. In [143], the forecasting the volatility of stock price index is analyzed. This paper proposes a hybrid model with neural network and time series analysis for forecasting the volatility of stock price index in two view points: deviation and direction. This model demonstrates the utility of the neural network combined with time series analysis for the forecasting financial goods. A regression neural network for error correction in foreign exchange forecasting and trading is used in [39]. The authors propose an adaptive forecasting approach which combines the strengths of neural networks and multivariate econometric models. This hybrid approach contains two forecasting stages. In the first stage, a time series model generates estimates of the exchange rates. In the second stage, general regression neural network is used to correct the errors of the estimates. In [89], the stock market prediction using *ANNs* with optimal feature transformation is considered. This paper compares a feature transformation method using a genetic algorithm (*GA*) with two conventional methods for building *ANNs*. In this study, the *GA* is incorporated to improve the learning and generalizability of *ANNs* for stock market prediction. Daily predictions are conducted and prediction accuracy is measured. In this study, three feature transformation methods for *ANNs* are compared. A new method for stock market prediction of S&P 500 based on *ANNs* is presented in [198]. The authors propose an improved bacterial chemotaxis optimization, which is then integrated into the back propagation artificial neural network to develop an efficient forecasting model for prediction of various stock indices. Experiments show its better performance than other methods in learning ability and generalization. The use of multiple classifiers for stock market prediction is proposed in [140]. In this paper, the authors investigated the predictability of the Dow Jones Industrial Average index to show that not all periods are equally random. They used the Hurst exponent to select a period with great predictability. Parameters for generating training patterns were determined heuristically by auto-mutual information and false nearest neighbor methods. Some inductive machine-learning classifiers-*ANN*, decision tree, and *k*-nearest neighbor were then trained with these generated patterns. Through appropriate collaboration of these models, the authors achieved prediction accuracy up to 65 percent.

The paper [124] is devoted to the important methodological problem "Is the predictability of emerging and developed stock markets really exploitable?". The authors have analyzed the daily and weekly forecastability of stock returns of a large number of markets during several years. They employed different information sets as well as model specifications; they also considered linear and nonlinear forecasts to assess the validity of the results. They use a variety of linear and nonlinear artificial neural networks models and perform a computationally demanding forecasting experiment to assess the predictability of returns. The results suggest that nonlinear models do not provide superior predictions than the linear ones and that emerging and developed stock markets are generally nonpredictable when total transaction cost are considered. It is very important that according to obtained results, *ANNs* do not provide superior performance than the linear models.

An important application of *ANNs* is the development of intelligent stock trading decision support systems. Modern computerized stock trading systems (or the so-called mechanical trading systems) are based on the simulation of the decision making process and generate advices for the trader to buy or sell some stocks or other financial tool he/she deals with taking into account the price history, technical analysis indicators, accepted rules of trading and so on.

An intelligent stock trading decision support system based on fuzzy neural network and artificial neural network is developed in [96]. The authors proposed a genetic algorithm based fuzzy neural network to formulate the knowledge base of fuzzy inference rules which can measure the qualitative effect on the stock market. Next, the effect is integrated with the technical indexes through the *ANN*. An example based on the Taiwan stock market is utilized to assess the proposed intelligent system. Evaluation results indicate that the neural network considering both the quantitative and qualitative factors excels the neural network considering only the quantitative factors both in the clarity of buying-selling points and buying-selling performance.

The trading system based on *ANNs*, forecasting and trading strategy is proposed in [47]. It is noted that previous researches in learning methods has focused on predictability based on comparative evaluation and these techniques may be employed to forecast financial markets as a prelude to intelligent trading systems. This paper explores the effect of a number of possible scenarios in this context. The alternative combinations of parameters include the selection of a learning method, whether a neural net or case based reasoning; the choice of markets, whether in one country or two; and the deployment of a passive or active trading strategy. When coupled with a forecasting system, however, a trading strategy offers the possibility for returns in excess of a passive buy-and-hold approach. In this study, the authors investigated the implications for portfolio management using an implicit learning technique (neural nets) and an explicit approach. In [141], a novel recurrent neural network-based prediction system for option trading and hedging is developed. The authors propose a novel non-parametric method using an ad-hoc recurrent neural network for estimating the future prices of war commodities such as gold and crude oil as well as currencies, which are increasingly gaining importance in the financial markets. The price predictions from the network shown to be accurate and computationally efficient, are used in a hedging system to avoid unnecessary risks. Experiments with actual gold and currency trading data show that the developed system using the proposed network and strategy can construct portfolios yielding a great return on investment.

The decision-making model for stock markets based on *ANNs* is proposed in [126]. The paper introduces an intelligent decision-making model which is based on the application of *ANNs* and swarm intelligence technologies. The proposed model is used to generate one-step forward investment decisions for stock markets. The *ANNs* are used to make the analysis of daily stock returns and to calculate one day forward decision for purchase of the stocks. Subsequently the particle swarm optimization algorithm is applied in order to select the "the best" *ANN* for the future investment decisions and to adapt the weights of other networks toward the weights of the best network.

A novel fuzzy neural network intelligent stock trading system is developed in [169]. It combines the superior predictive capability of *ANNs* and the use of widely accepted moving average and relative strength indicator trading rules. The system is demonstrated empirically using real live stock data to achieve significantly higher multiplicative returns than a conventional technical analysis rule-based trading system.

The fuzzy adaptive decision-making for rational traders in speculative stock markets is developed in [14]. This paper introduces a hybrid neurofuzzy system for decision-making and trading under uncertainty. The efficiency of a technical trading strategy based on the neurofuzzy model is investigated in order to predict the direction of the market for 10 of the most prominent stock indices of USA, Europe and Southeast Asia. It is demonstrated via an extensive empirical analysis that the neurofuzzy model allows technical analysts to earn significantly higher returns by providing valid information for a potential turning point on the next trading day.

The system developed in [34] is a first attempt in the literature to predict the sell/buy decision points instead of stock price itself. In this study, an integrated system based on the combining dynamic time windows, *CBR*, and neural network for stock trading prediction is developed and there are three stages in this system: (1) screening out the potential stocks and the important influential factors; (2) using a back propagation network to predict the buy/sell points (wave peak and wave trough) of stock price; (3) adopting case based dynamic windows to further improve the forecasting results. The empirical results show that the proposed model can reduce the false alarm of buying or selling decisions.

In [37], intelligent technical analysis based equivolume charting for stock trading using neural networks is presented. Two technical indicators, namely the volume adjusted moving average (*VAMA*) and the easy of movement (*EMV*) indicator, are developed from equivolume charting. This paper explores the profitability of stock trading by using a neural network model developed to assist the trading decisions based on *VAMA* and *EMV*. The generalized regression neural network is chosen and utilized on past S&P 500 index data. The results show that the stock trading using the neural network with the *VAMA* and *EMV* outperforms the results of stock trading generated from the *VAMA* and *EMV* without neural network assistance, the simple moving averages (*MA*) in isolation, and the buy-and-hold trading strategy.

A hybrid stock trading system for intelligent technical analysis-based equivolume charting is proposed in [38]. The authors present the use of an intelligent hybrid stock trading system that integrates neural networks, fuzzy logic, and genetic algorithms techniques to increase the efficiency of stock trading when using a volume adjusted moving average (*VAMA*), a technical indicator developed from equivolume charting. For this research, a neuro-fuzzy-based genetic algorithm utilizing a *VAMA* membership function is introduced. The results show that the intelligent hybrid system takes advantage of the synergy among these different techniques to intelligently generate more optimal trading decisions based on *VAMA*, allowing investors to make better stock trading decisions.

An empirical methodology for developing stock market trading systems using artificial neural networks is presented in [179]. It is noted that a great deal of work has been published over the past decade on the application of neural networks to stock market trading. Individual researchers have developed their own techniques for designing and testing these neural networks, and this presents a difficulty when trying to learn lessons and compare results. This paper aims to present a methodology for designing robust mechanical trading systems using soft computing technologies such as artificial neural networks. The paper describes the key steps involved in creating a neural network for the use in stock market trading, and places particular emphasis on designing these steps to suit the real-world constraints the neural network will eventually operate in. Such a common methodology brings with it a transparency and clarity that should ensure that previously published results are both reliable and reusable.

Summarizing we can say that during the last two decades powerful mathematical methods have been used to find a way to predict stock prices accurately, but they have produced less than successful results in practice [66]. In [96], it is shown that numerous studies addressing stock price prediction have generally employed the time series analysis techniques [87] and multiple regression models. Recently, artificial intelligence techniques like *ANNs* and genetic algorithms have been applied in this area. However, the above-mentioned concern still exists [8, 116, 119].

In [88], the use of *ANNs* had some limitations in learning the patterns because stock price data have tremendous noise and complex dimensionality. In [96], it is pointed out that numerous factors such as macro-economical and political events may have a major influence on stock prices. As it was noted in [185] "in recent times interest has turned to the use of neural networks for this task, but had less than successful results". That is why, a growing interest of researchers to the application of *RS* theory for trading rules extraction is observed [155, 185]. The synthesis of fuzzy logic and Dempster-Shafer theory of evidence (*DST*) also seems to be challenging enough in the building stock trading systems. Chapter 6 of this book is devoted to this issue.

2.4 Applications of Multiple Criteria Decision Making in Economics and Finance

It is safe to say that almost all real-world economic and financial problems are multiple criteria ones since usually at least two controversial criteria should be taken into account: the profit maximization and risk minimization. In reality, these problems can be solved taking into account a great number of conflicting local criteria of different importance, often ill defined and organized into the complex hierarchical structures. Therefore, a large number of applications of multiple criteria decision making (*MCDM*) in economics and finance can be found in the literature.
In this section, the more important applications are presented with the classification of them.

The authors of [166] presented the widest review of this problem based on more than 250 literature indices. They have showed that *MCDM* is now actively used in such areas as portfolio analysis, capital budgeting, general financial planning, working capital and commercial bank management, auditing, accounting, insurance and pension fund management, interest rate and risk analysis, prediction and classification, government and nonprofit organizations, strategic planning, mergers, and acquisitions.

The first mathematically formalized decision theory was developed by John von Neumann and Oskar Morgenstern. In 1944, they published their book "Theory of Games and Economic Behavior". In this book, they moved on from Bernoulli's formulation of an utility function over wealth, and defined an expected utility function over lotteries or gambles.

The classical utility theory deals with the probabilistic type of uncertainty and was used in many economical and financial applications. It makes it possible to aggregate local criteria presented by utility functions to rank the competing alternatives.

There are different approaches to formulation of *MCDM* proposed in the literature. The first attempt of classification of these approaches was made in [144]. The author separates tree main classes of *MCDM* problems, but more scientifically grounded seems to be the classification presented ten years later in [131], where the following classes of *MCDM* were distinguished: multiobjective mathematical programming, multiattribute utility theory, outranking relations approach, preference disaggregation approach.

Multiattribute utility theory is the generalization of classical utility theory and is based on the aggregated utility functions. The problems concerned with formulation of aggregated multiple criteria utility functions under probabilistic uncertainty are described in [200].

An approach based on the preference relations was implemented in the *ELECTRE* method. This method is flexible enough to analyze alternatives when it is hard to discriminate them. The methods based on the preference relations are presented in [145, 148, 181].

The methods of preference disaggregation are based on the ordinal regression analysis. The broad bibliography about these methods is presented in [72, 131].

The methods described above were used for the solution of different economic and financial problems. The comprehensive review of such applications may be found in [202].

The main limitations of the described above methods are the use of classical utility theory with strongly monotone utility functions representing local criteria, the weighted sum used for aggregation of local criteria and the probabilistic approach to modeling uncertainty.

Nevertheless, the nature of economic and financial *MCDM* problems usually needs more flexible methods for modeling different kinds of uncertainties.

For example, let us consider an investment evaluation problem. It is well known that evaluation of important investment projects usually can not be successfully carried out using only financial parameters since the possible ecological, social and even political factors of project's implementation should be evaluated as well. The role of these factors rises along with the project's importance. Obviously, such factors, as a rule, can not be predicted with a high accuracy, moreover their estimations are usually based on the expert's opinions expressed in a verbal form. So the proper mathematical tools are needed to incorporate such ill defined estimations into the general evaluation of investment project. Since the applicability of traditional probabilistic methods is often limited by absence of objective probabilistic information about future events, during last two decades the growing interest to the application of the methods of soft computing such as interval analysis, fuzzy sets theory methods, rough sets and the tools of *DST* to the solution of *MCDM* problems in economics and finance has been observing. As a rule, in applications the different combinations of these methods are used.

Nowadays the most popular in the applications *MCDM* method is the analytic hierarchy process (*AHP*) developed by T. Saaty [147].

Vaidya and Kumar [178] presented a broad overview of applications of *AHP*. A total of 150 application papers are referred to in this paper, 27 of them are critically analyzed. It is shown that the main applications of *AHP* and fuzzy *AHP* are concerned with the following problems: selection, evaluation, benefit-cost analysis, allocations, planning and development, priority and ranking, decision making, forecasting, medicine and related fields, quality function deployment.

In [52], a fuzzy *AHP* approach is used for computer-aided machine-tool selection. Both economic evaluation criterion and strategic criteria such as flexibility, quality improvement, which are not quantitative in nature, are considered for evaluation. Much has been written about the deficiencies of traditional models for justifying advanced manufacturing systems. It is emphasised that the use of fuzzy set theory allows us for incorporating unquantifiable, incomplete and partially known information into the decision model. In this paper, a fuzzy *AHP* is used for the evaluation and justification of an advanced manufacturing system. Finally, an example of machine tool selection is used to illustrate and validate the proposed approach. In [65], an *AHP* based approach was used for the solution of personnel selection problem. Due to the increasing competition in the process of globalization and fast technological improvements, world markets demand companies to have quality and professional human resources. This can only be achieved by employing potentially adequate personnel. In this paper, the authors proposed a personnel selection system based on fuzzy analytic hierarchy process (*FAHP*). The *FAHP* is applied to evaluate the best adequate personnel dealing with the rating of both qualitative and quantitative criteria. The result obtained using *FAHP* is compared with results produced by Yager's weighted goals method. In addition, a practical computer-based decision support system is introduced to provide more information and help manager to make better decisions under fuzzy circumstances.

The hierarchical structure of the problem of evaluating alternative production cycles is considered in [190] and *AHP* method is used for its building. Nevertheless, the drawback of this work is that the simple normalization of financial parameters (dividing them by their maximal values) is applied instead of natural local criteria.

An interesting example of practical application of the multiple criteria hierarchical analysis is presented in [103]. The generalized *AHP* method has been used for estimation of 103 mutually dependent investment projects proposed for the Tumen river region (China) industrial development.

In the last decade, the grooving interest to the application of *DST* is observed. It should be noted that *DST* usually is not used solely, but with combination with some other methods. As now only a few papers devoted to the applications of *DST* in economics and finance are published, we analyze them with pertinent details.

Beynon et al. [16] developed a new method based on the synthesis of *DST* as a promising improvement on "traditional" approaches to decision analysis and *AHP*. The method is illustrated by simple example. In this example, the decision involves buying a new car, from a choice of 3 known types of car (decision alternatives), say *A*, *B* and *C*. In the *DST* terminology *A*, *B*, *C* is then the frame of discernment. The criteria to help us judge each of these cars are: price, fuel, comfort and style. The overall objective (focus) is to decide which is the best car to buy.

The paper [17] outlines a new software system that utilizes the newly developed method (DS/AHP) which combines aspects of the *AHP* with D*DST* for the purpose of multi-criteria decision making. The method allows a decision maker considerably greater level of control (compared with conventional *AHP* methods) on the judgments made in identifying levels of favouritism toward groups of decision alternatives. More specifically, the DS/AHP analysis allows decision makers for additional analysis, including levels of uncertainty and conflict in the decisions made, for example. In this paper, an expert system is introduced which enables the application of DS/AHP to *MCDM*. The expert system illustrates further the usability of DS/AHP, also including new aspects of analysis and representation offered through using this method. In [18], the author exposits a novel technique for the ranking and classification of objects to a particular state. Each object is described by measurements from a number of variables which may offer different levels of support for the individual objects to be associated with the two states: a given hypothesis and not the hypothesis. The *DST* is a central component of this technique. This makes it possible to estimate the measure of concomitant ignorance, which may encompass the precision of the individual measurements as well as the possible ambiguity of their influence on the subsequent classification of objects. The level of ignorance influences the utilization of the technique as a tool for the ranking or classification of objects. A simplex plot method for representing data provides a clear visual representation (interpretation) to the degree of interaction of the support from the variables to the ranking or classification of the objects. To illustrate the proposed technique, the application considered in the paper is the elucidation of the risk of corporate failure of a number of companies. Subsequently, each variable (financial and non-financial) may offer support for the ranking and classification of companies between the extreme states of being a failed or non-failed company. A comparison on the ranking and classification of companies is made with a traditional multivariate discriminant analysis.

Wang et al. [186] have used the so-called evidential reasoning (*ER*) approach (based on the *DST*) for environmental impact assessment (*EIA*) problems. These problems are often characterized by a large number of identified environmental factors that are qualitative in nature and can only be assessed on the basis of human judgments, which inevitably involve various types of uncertainties such as ignorance and fuzziness. All assessment information, quantitative or qualitative, complete or incomplete, and precise or imprecise, is modelled using a unified framework of a belief structure. The *ER* approach is used to aggregate multiple environmental factors, resulting in an aggregated distributed assessment for each alternative policy. A numerical example and its modified version are studied to illustrate the detailed implementation process of the *ER* approach and demonstrate its potential applications in *EIA*.

In [159], the authors have developed a rough set and *DST* based formalism to objectively represent uncertainty inherent in the process of service discovery, characterization, and classification. Rough set theory is ideally suited for dealing with limited resolution, vague and incomplete information, while *DST* provides a consistent approach to model an expert's belief and ignorance in the classification decision process. Integrating these two formal approaches in spatial domain provides a way to model an expert's belief and ignorance in service classification. In an application scenario of the model, the authors have used a cognitive map of retail site assessment, which reflects the partially subjective assessment process. Thus, the authors provide a naturalistic means of incorporating both qualitative aspects of intuitive knowledge as well as hard empirical information for service management within a formal uncertainty framework.

In [193], the evidential reasoning approach (*ER*) is used in combination with multiple attribute decision analysis (*MADA*). In this paper, the *ER* approach is further developed to deal with *MADA* problems with both probabilistic and fuzzy uncertainties. In this newly developed *ER* approach, precise data, ignorance and fuzziness are all modelled under the unified framework of a distributed fuzzy belief structure, leading to a fuzzy belief decision matrix. In contrast to the existing *ER* algorithm that is of a recursive nature, a new fuzzy *ER* algorithm provides an analytical means for combining all attributes without iteration, thus providing scope and flexibility for sensitivity analysis and optimization. A numerical example (ranking the cars) is provided to illustrate the detailed implementation process of the new *ER* approach and its validity and wide applicability.

XXX

Nowadays, a new ranking method called technique for order preference by similarity to ideal solution (*TOPSIS*) becomes very popular in economic and financial application. It is a typical *MCDM* method and seems to be flexible enough for its exploiting in applications in the synthesis with other methods. Since *TOPSIS* is a relatively new method, here we analyze its applications with some details.

In [46], a fuzzy *TOPSIS* approach for selecting plant location is proposed, where the ratings of various alternative locations under various criteria and the weights of various criteria are assessed in linguistic terms represented by fuzzy numbers. In the proposed method, the ratings and weights assigned by decision makers are averaged and normalized into a comparable scale. The membership function of each normalized weighted rating can be presented using arithmetic of fuzzy numbers. Using the suggested method, the decision maker's fuzzy assessments with different rating viewpoints and the trade-off among different criteria are considered in the aggregation procedure to assure more convincing decision making. A numerical example demonstrates the feasibility of the proposed method.

Yong [195] developed a new fuzzy *TOPSIS* method. Compared with existing fuzzy *TOPSIS* methods, the proposed method can deal with group decision-making problems in a more efficient manner. A numerical example of plant location selection is used to illustrate the efficiency of the proposed method.

Shyur [158] models the commercial-off-the-self (*COTS*) evaluation problem as an *MCDM* problem and proposes a five-phase *COTS* selection model combining the technique of *ANP* (analytic network process) and modified *TOPSIS* method. Given the high interest in motivation to the use of commercially available software, the evaluation and selection of (*COTS*) products is an important activity in the software development projects. Selecting an appropriate *COTS* product is often a non-trivial task in which multiple criteria need to be careful considered. The case study demonstrates the effectiveness and feasibility of the proposed evaluation procedure.

A new approach to multi-objective inventory planning using multi-objective particle swarm optimization (*MOPSO*) and *TOPSIS* is presented in [176]. This paper first employs the (*MOPSO*) algorithm to generate the non-dominated solutions of a reorder point and order size system. The *TOPSIS* method is then used to sort the non-dominated solutions by the preference of decision makers. That is, a two-stage multi-criteria decision framework which consists of *MOPSO* and *TOPSIS* is presented to find out a compromise solution for decision makers. By varying the weights of various criteria, including minimization of the annual expected total relevant cost, minimization of the annual expected frequency of stock-out occasions, and minimization of the annual expected number of stock-outs, managers can determine the order size and safety stock simultaneously which fits their preference under different situations.

In [85], the interpretive structural modeling and fuzzy *TOPSIS* are used for the selection of reverse logistics provider. It is worthy to note that this paper is devoted to the solution of the real-world problem. For industries, the management of return flow usually requires a specialized infrastructure with special information systems for tracking and dedicated equipment for the processing of returns. Therefore, industries are turning to third-party reverse logistics providers (3*PRLPs*). In this paper, a multi-criteria group decision making (*MCGDM*) model in fuzzy environment is developed to guide the process of selection of the best 3*PRLP*. The interactions among the criteria are also analyzed before arriving at a decision for the selection of the best 3*PRLP* among 15 alternatives. The analysis is done through interpretive structural modeling (*ISM*) and fuzzy *TOPSIS*. Finally the effectiveness of the model is illustrated using the case study of battery manufacturing industry in India.

A two step fuzzy *AHP* and *TOPSIS* methodology is developed in [63] for the evaluation of hazardous waste transportation firms. Hazardous wastes are likely the source of danger to human health and/or environment. The safe transportation of them is then a very important problem. Consequently, the selection of the right and most appropriate transportation firm is an important problem for hazardous waste generators. In this paper, a two step methodology is structured to evaluate hazardous waste transportation firms using the methods of fuzzy *AHP* and *TOPSIS*. A numerical example is presented to clarify the methodology.

An interesting real-world application of *TOPSIS* in combination with fuzzy *AHP* is presented in [149]. The aim of this study is to propose a fuzzy multi-criteria decision model to evaluate the performances of banks. The largest five commercial banks of Turkish banking sector are examined and these banks are evaluated in terms of several financial and non-financial indicators. Fuzzy analytic hierarchy process (*FAHP*) and *TOPSIS* methods are integrated in the proposed model. After the weights of local criteria are determined based on the opinions of experts using the *FAHP* method, these weights are input to the *TOPSIS* method to rank the banks. The results show that not only financial performance, but also non-financial performance should be taken into account in a competitive environment.

In [183], the fuzzy hierarchical *TOPSIS* is used for the solution of supplier selection problem. This study proposes a new fuzzy hierarchical *TOPSIS* method, which not only is well suited for evaluating fuzziness and uncertainty problems, but also can provide more objective and accurate criterion weights, while simultaneously avoiding the problem of Chen's fuzzy *TOPSIS*. For application and verification, this study presents a numerical example and build a practical supplier selection problem to verify the proposed method and compare it with other methods.

Ashtiani et al. [5] developed an extension of fuzzy *TOPSIS* method based on interval-valued fuzzy sets. In this paper, the interval-valued fuzzy *TOPSIS* method is presented aiming at solving *MCDM* problems in which the weights of criteria are unequal, using interval-valued fuzzy sets concepts. The method is illustrated by the example of choosing a manager for *R&D* department of telecommunication company.

The solution of the real-world problem of choice of location for direct foreign investment in new hospitals in China using *ANP* and *TOPSIS* is presented in [112]. This study models location choices for foreign direct investments in new hospitals in China as the *MCDM* problem and develops a multidirectional relationship decision model which combines the techniques of analytic network process (*ANP*) and *TOPSIS*. This study discusses applying *ANP* to the relative weighting of multiple assessment criteria. The *TOPSIS* approach is employed to rank 15 counties of China's Yangtze River Delta region (without the Zhoushan) in terms of their overall performance under the decision model. To illustrate how the proposed approach is applied to the problem of selecting locations for new hospitals in China the empirical study of the real case is performed.

There are only a few papers devoted to the application of intuitionistic fuzzy sets *A-IFS* to the solution of *MCDM* economic and finance problems in the literature, whereas as it is stated in [168], *A-IFS* provide a richer apparatus to grasp imprecision than the conventional fuzzy sets and they seem to be a promising tool for extended decision making models.

In [102], the methods of intuitionistic fuzzy sets are used for the solution of an air-condition system selection problem which is a multiattribute decision making problem. Unfortunately, the application seems to be only a numerical example.

Boran et al. [22] combined the method of multi-criteria intuitionistic fuzzy group decision making with *TOPSIS* method to select an appropriate supplier in group decision making environment. Intuitionistic fuzzy weighted averaging operator is utilized to aggregate individual opinions of decision makers for rating the importance of criteria and alternatives. Finally, a numerical example for supplier selection is given to illustrate application of intuitionistic fuzzy *TOPSIS* method.

In [187], the authors proposed an approach to multiattribute decision making with interval-valued intuitionistic fuzzy assessments and incomplete weights. By employing a series of optimization models, the proposed approach derives a linear program for determining attribute weights. An illustrative investment decision problem is employed to demonstrate how to apply the proposed procedure and comparative studies are conducted to show its overall consistency with existing approaches.

2.5 Summary and Discussion

In this chapter, we have shown that almost all modern soft computing methods can be successfully applied to the solution of economic and financial problems in different fields. In applications (especially in *MCDM*), the modern methods usually are not used solely, but in different combinations. Nevertheless, in many cases the applications seem to be rather numerical examples, not the solutions of real-world problems. Especially, the applications of intuitionistic fuzzy sets and the Dempster-Shafer theory are performed in the literature by such numerical and sometimes artificial examples.

There is a problem of interconnection of applied interval analysis and fuzzy set theory. Basing on the literature analysis, we can say that now interval analysis and fuzzy set theory are developing almost independently, whereas theoretical results obtained by "interval community" may be successfully exploited in the framework of fuzzy set theory. Moreover, the literature analysis makes it possible to conclude that "interval community" is now focused mainly on the theoretical researches. Probably, the relatively low number of applications of interval analysis can be explained by the great popularity of fuzzy set theory. Really, an interval may be naturally treated as a particular case of a fuzzy number.

Nevertheless, we can say that using α-cut presentation of fuzzy values, the problems of fuzzy arithmetic can be reduced to the problems of interval analysis. Hence, the methods of interval analysis can be successfully used in applications of fuzzy set theory. Therefore, in Chapter 3 we present the basic definitions and some important problems of interval analysis, and in the following chapters (especially in Chapter 7) we present some new methods of interval analysis and show how they can be used in the economic and financial applications in the interval and fuzzy environments.

Recently, a new approach called belief-rule-based systems has been developed. A belief-rule-base inference methodology combines evidential reasoning based on the tools of *DST* and traditional IF-THEN rules which can be presented using fuzzy logic. This approach requires the use of some system parameters including rule weights, attribute weights, and belief degrees. These parameters need to be determined with care for reliable system simulation and prediction. Some off-line optimization models have been proposed in the literature. In our opinion, this new approach does not match yet to the traditional classification of soft computing methods. Therefore, we desist from its presentation in this chapter. We shall describe this approach with new results and application to building stock trading decision support systems in Chapter 6. Nevertheless, we can say that this approach may be successfully used in financial applications [53, 150].

References

1. Abdel-Kader Magdy, G., David, D.: Evaluating investments in advanced manufacturing technology: a fuzzy set theory approach. British Accounting Review 33, 455–489 (2001)
2. Ahn, B.S., Cho, S., Kim, C.: The integrated methodology of rough set theory and artificial neural network for business failure prediction. Expert Systems with Applications 18, 65–74 (2000)
3. Akyol, D.E., Bayhan, G.M.: A review on evolution of production scheduling with neural networks. Computers and Industrial Engineering 53, 95–122 (2007)
4. Allen, D., Lazarov, Z., McAleer, M., Peiris, S.: Comparison of alternative ACD models via density and interval forecasts: Evidence from the Australian stock market. Mathematics and Computers in Simulation 79, 2535–2555 (2009)
5. Ashtiani, B., Haghighirad, F., Makui, A., Montazer, G.: Extension of fuzzy TOPSIS method based on interval-valued fuzzy sets. Applied Soft Computing 9, 457–461 (2009)
6. Atanassov, K.T.: Intuitionistic fuzzy sets. Fuzzy Sets and Systems 20, 87–96 (1986)
7. Azadeh, A., Ghaderi, S.F., Tarverdian, S., Saberi, M.: Integration of artificial neural networks and genetic algorithm to predict electrical energy consumption. Applied Mathematics and Computation 186, 1731–1741 (2007)
8. Baba, N., Kozaki, M.: An intelligent forecasting system of stock price using neural networks. In: Proceedings of IJCNN 1992, pp. 317–377 (1992)
9. Babusiaux, D., Pierru, A.: Capital budgeting, project valuation and financing mix: Methodological proposals. Europian Journal of Operational Research 135, 326–337 (2001)
10. Baltzersen, J.K.: An attempt to predict stock market data: a rough sets approach. Diploma thesis, Knowledge Systems Group, Department of Computer Systems and Telematics, The Norwegian Institute of Technology, University of Trondheim (1996)
11. Bazan, J.G., Skowron, A., Synak, P.: Market data analysis: a rough set approach. ICS Research Reports 6/94, Warsaw University of Technology (1994)
12. Bazan, J.G., Szczuka, M.S.: RSES and rSESlib - A collection of tools for rough set computations. In: Ziarko, W.P., Yao, Y. (eds.) RSCTC 2000. LNCS (LNAI), vol. 2005, p. 106. Springer, Heidelberg (2001)
13. Becerra-Fernandez, I., Zanakis, S.H., Walczak, S.: Knowledge discovery techniques for predicting country investment risk. Computers and Industrial Engineering 43, 787–800 (2002)
14. Bekiros, S.D.: Fuzzy adaptive decision-making for boundedly rational traders in speculative stock markets. European Journal of Operational Research 202, 285–293 (2010)
15. Belletante, B., Arnaud, H.: Choisir ses Investissements. Chotar et Assosies Editeurs, Paris (1989)
16. Beynon, M., Curry, B., Morgan, P.: The Dempster-Shafer theory of evidence: an alternative approach to multicriteria decision modeling. Omega 28, 37–50 (2000)
17. Beynon, M., Cosker, D., Marshall, D.: An expert system for multi-criteria decision making using Dempster Shafer theory. Expert Systems with Applications 20, 357–367 (2001)
18. Beynon, M.: A novel technique of object ranking and classification under ignorance: An application to the corporate failure risk problem. European Journal of Operational Research 167, 493–517 (2005)
19. Bilbao-Terol, A., Perez-Gladish, B., Antomil-lbias, J.: Selecting the optimum portfolio using fuzzy compromise programming and Sharpe's single-index model. Applied Mathematics and Computation 182, 644–664 (2006)

20. Bogle, H.F., Jehenck, G.K.: Investment Analysis: US Oil and Gas Producers Score High in University Survey. In: Proceedings of Hydrocarbon Economics and Evaluation Symposium, Dallas, pp. 234–241 (1985)
21. Bojadziev, M., Bojadziev, G.: Fuzzy Logic for Business, Finance, and Management. World Scientific Publishing Company, Singapore (1997)
22. Boran, F.E., Gen, S., Kurt, M., Akay, D.: A multi-criteria intuitionistic fuzzy group decision making for supplier selection with *TOPSIS* method. Expert Systems with Applications 36, 11363–11368 (2009)
23. Brigham, E.F.: Fundamentals of Financial Management. The Dryden Press, New York (1992)
24. Buckley, J.J.: The fuzzy mathematics of finance. Fuzzy Sets and Systems 21, 257–273 (1987)
25. Buckley, J.: Solving fuzzy equations in economics and finance. Fuzzy Sets and Systems 48, 289–296 (1992)
26. Buckley, J.J., Buckley James, J., Eslami, E.: Fuzzy Mathematics In Economics And Engineering. Springer, Heidelberg (2002)
27. Buckley, J., Feuring, T.: Linear and non-linear fuzzy regression: Evolutionary algorithm solutions. Fuzzy Sets and Systems 112, 381–394 (2000)
28. Bustince, H., Burillo, P.: Vague sets are intuitionistic fuzzy sets. Fuzzy Sets and Systems 79, 403–405 (1996)
29. Calderon, T.G., Cheh, J.J.: A roadmap for future neural networks research in auditing and risk assessment. International Journal of Accounting Information Systems 3, 203–236 (2002)
30. Calzi, M., Li: Towards a general setting for the fuzzy mathematics of finance. Fuzzy Sets and Systems 35, 265–280 (1990)
31. Castillo, O., Alanis, A., Garcia, M.A., Arias, H.: An intuitionistic fuzzy system for time series analysis in plant monitoring and diagnosis. Applied Soft Computing 7, 1227–1233 (2007)
32. Chanas, S., Kuchta, D.: A concept of the optimal solution of the transportation problem with fuzzy cost coefficients. Fuzzy Sets and Systems 82, 299–305 (1996)
33. Chanas, S., Kuchta, D.: Multiobjective programming in optimization of interval objective functions - A generalized approach. European Journal of Operational Research 94, 594–598 (1996)
34. Chang, P.-C., Liu, C.-H., Lin, J.-L., Fan, C.-Y., Ng Celeste, S.P.: A neural network with a case based dynamic window for stock trading prediction. Expert Systems with Applications 36, 6889–6898 (2009)
35. Chansa-ngavej, C., Mount-Campbell, C.A.: Decision criteria in capital budgeting under uncertainties: implications for future research. Int. J. Prod. Economics 23, 25–35 (1991)
36. Chaturvedi, K.T., Pandit, M., Srivastava, L.: Modified neo-fuzzy neuron-based approach for economic and environmental optimal power dispatch. Applied Soft Computing 8, 1428–1438 (2008)
37. Chavarnakul, T., Enke, D.: Intelligent technical analysis based equivolume charting for stock trading using neural networks. Expert Systems with Applications 34, 1004–1017 (2008)
38. Chavarnakul, T., Enke, D.: A hybrid stock trading system for intelligent technical analysis-based equivolume charting. Neurocomputing 72, 3517–3528 (2009)
39. Chen, A.-S., Leung, M.T.: Regression neural network for error correction in foreign exchange forecasting and trading. Computers and Operations Research 31, 1049–1068 (2004)

40. Chen, L.-H., Huang, L.: Portfolio optimization of equity mutual funds with fuzzy return rates and risks. Expert Systems with Applications 36, 3720–3727 (2009)
41. Chen, S.: An empirical examination of capital budgeting techniques: impact of investment types and firm characteristics. Eng. Economist 40, 145–170 (1995)
42. Cheng, J.-H., Chen, H.-P., Lin, Y.-M.: A hybrid forecast marketing timing model based on probabilistic neural network, rough set and C4.5. Expert Systems with Applications 37, 1814–1820 (2010)
43. Christoffersen, P.: Evaluating interval forecasts. International Economic Review 39, 841–862 (1998)
44. Chiu, C.-Y., Park, C.S.: Fuzzy cash flow analysis using present worth criterion. Eng. Economist 39, 113–138 (1994)
45. Choobineh, F., Behrens, A.: Use of intervals and possibility distributions in economic analysis. J. Oper. Res. Soc. 43, 907–918 (1992)
46. Chu, T.-C.: Selecting Plant Location via a Fuzzy *TOPSIS* Approach. International Journal of Advanced Manufacturing Technology 20, 859–864 (2002)
47. Chun, S.-H., Steven, H.K.: Data mining for financial prediction and trading: application to single and multiple markets. Expert Systems with Applications 26, 131–139 (2004)
48. Das, S.K., Goswami, A., Alam, S.S.: Multiobjective transportation problem with interval cost, source and destination parameters. European Journal of Operational Research 117, 100–112 (1999)
49. Diebold, F., Gunther, A., Tay, S.: Evaluating density forecasts with applications to financial risk management. International Economic Review 39, 863–883 (1998)
50. Dimitras, A.I., Slowinski, R., Susmaga, R., Zopounidis, C.: Business failure prediction using rough sets. European Journal of Operational Research 114, 263–280 (1999)
51. Dimova, L., Sevastianov, D., Sevastianov, P.: Application of fuzzy sets theory, methods for the evaluation of investment efficiency parameters. Fuzzy Economic Review 5, 34–48 (2000)
52. Durarn, O., Aguilo, J.: Computer-aided machine-tool selection based on a Fuzzy-AHP approach. Expert Systems with Applications 34, 1787–1794 (2008)
53. Dymova, L., Sevastianov, P., Bartosiewicz, P.: A new approach to the rule-base evidential reasoning: stock trading decision support system application. Expert Systems with Applications 37, 5564–5576
54. El-Bouri, A., Balakrishnan, S., Popplewell, N.: A neural network to enhance local search in the permutation flowshop. Computers and Industrial Engineering 49, 182–196 (2005)
55. El-Wahed, W.F.A., Lee, S.M.: Interactive fuzzy goal programming for multi-objective transportation problems. Omega 34, 158–166 (2006)
56. Enke, D., Thawornwong, S.: The use of data mining and neural networks for forecasting stock market returns. Expert Systems with Applications 29, 927–940 (2005)
57. Erbas, B.C., Stefanou, S.E.: An application of neural networks in microeconomics: Input-output mapping in a power generation subsector of the US electricity industry. Expert Systems with Applications 36, 2317–2326 (2009)
58. Gau, W.L., Buehrer, D.J.: Vague sets. IEEE Trans. Systems Man Cybernet 23, 610–614 (1993)
59. Maria, G.-L.A.: Fuzzy Logic in Financial Analysis (Studies in Fuzziness and Soft Computing). Springer, Heidelberg (2005)
60. Golan, R.: Stock market analysis utilizing rough set theory. PhD thesis, Department of Computer Science, University of Regina, Canada (1995)
61. Golan, R., Edwards, D.: Temporal rules discovery using datalogic/R+ with stock market data. In: Proceedings of the International Workshop on Rough Sets and Knowledge Discovery (RSKD 1993), pp. 74–81. Springer, New York (1993)

62. Greco, S., Matarazzo, B., Slowinski, R.: A new rough set approach to evaluation of bankruptcy risk. In: Zopounidis, C. (ed.) Operational Tools in the Management of Financial Risks, pp. 121–136. Kluwer Academic Publishing, Boston (1998)
63. Gumus, A.T.: Evaluation of hazardous waste transportation firms by using a two step fuzzy-AHP and TOPSIS methodology. Expert Systems with Applications 36, 4067–4074 (2009)
64. Guo, Z., Zhang, Q.: A New Approach to Project Risk Evaluation Based on Intuitionistic Fuzzy Sets. In: Proc. of Sixth International Conference on Fuzzy Systems and Knowledge Discovery, vol. 6, pp. 58–61 (2009)
65. Güngör, Z., Serhadlioğlu, G., Kesen, S.E.: A fuzzy AHP approach to personnel selection problem. Applied Soft Computing 9, 641–646 (2009)
66. Haefke, C., Helmenstein, C.: Predicting stock market averages to enhance profitable trading strategies. In: Proceedings of the Third International Conference on Neural Networks in the Capital Markets, London, pp. 378–389 (2000)
67. Hashemi, R., Le Blanc, L.A., Rucks, C.T., Rajaratnam, A.: A hybrid intelligent system for predicting bank holding structures. European Journal of Operational Research 109, 390–402 (1998)
68. Hladk, M.: Generalized linear fractional programming under interval uncertainty. European Journal of Operational Research 205, 42–46 (2010)
69. Hun, H.D., Changha, H.: Extended fuzzy regression models using regularization method. Information Sciences 164, 31–46 (2004)
70. Huang, K.Y., Jane, C.-J.: A hybrid model for stock market forecasting and portfolio selection based on ARX, grey system and RS theories. Expert Systems with Applications 36, 5387–5392 (2009)
71. Huang, X.: Risk curve and fuzzy portfolio selection. Computers and Mathematics with Applications 55, 1102–1112 (2008)
72. Jacquet-Lagreeze, E., Siskos, J.: Mthode de dcision multicritre. Hommes et Techniques, Paris (1983)
73. Jana, B., Roy, T.K.: Multi-objective intuitionistic fuzzy linear programming and its application in transportation model. Notes on Intuitionistic Fuzzy Sets 13, 34–51 (2007)
74. Jana, B., Roy, T.K.: A multi-objective transportation problem under intuitionistic fuzziness. Journal of Technology 34, 17–25 (2007)
75. Jaulin, L., Kieffir, M., Didrit, O., Walter, E.: Applied Interval Analysis. Springer, London (2001)
76. Kacprzyk, J., Fedrizzi, M.: Fuzzy Regression Analysis. Physica-Verlag, Heidelberg (1992)
77. Kahraman, C., Ulukan, Z.: Continous compounding in capital budgeting using fuzzy concept. In: Proceedings of 6th IEEE International Conference on Fuzzy Systems (FUZZ-IEEE 1997), Bellaterra, pp. 1451–1455 (1997)
78. Kahraman, C., Ulukan, Z.: Fuzzy cash flows under inflation. In: Proceedings of 7th IEEE International Fuzzy Systems Association World Congress (IFSA 1997), Univeristy of Economics, Prague CZech Republic Bellaterra, pp. 104–108 (1997)
79. Kahraman, C., Tolga, E., Ulukan, Z.: Justification oof manufacturing technologies using fuzzy benefit/cost ratio analysis. Int. J. Product. Economy 66, 45–52 (2000)
80. Kahraman, C.: Fuzzy versus probabilistic benefit/cost ratio analysis for public work projects. Int. J. Appl. Math. Comp. Sci. 11, 705–718 (2001)
81. Kahraman, C., Ruan, D., Tolga, E.: Capital budgeting techniques using discounted fuzzy versus probabilistic cash flows. Information Sciences 142, 57–76 (2002)
82. Cengiz, K. (ed.): Fuzzy Applications in Industrial Engineering. Springer, New York (2006)

83. Cengiz, K. (ed.): Fuzzy Engineering Economics with Applications. Springer, Heidelberg (2008)
84. Kamo, T., Dagli, C.: Hybrid approach to the Japanese candlestick method for financial forecasting. Expert Systems with Applications 36, 5023–5030 (2009)
85. Kannan, G., Pokharel, S., Kumar, P.S.: A hybrid approach using ISM and fuzzy TOPSIS for the selection of reverse logistics provider. Resources, Conservation and Recycling 54, 28–36 (2009)
86. Kaufmann, A., Gupta, M.: Introduction to fuzzy-arithmetic theory and applications. Van Nostrand Reinhold, New York (1985)
87. Kendall, S.M., Ord, K.: Time series, 3rd edn. Oxford University Press, New York (1990)
88. Kim, K.J., Han, I.: Genetic algorithms approach to feature discretization in artificial neural networks for the prediction of stock price index. Expert Systems with Applications 19, 125–132 (2000)
89. Kim, K., Lee, W.B.: Stock market prediction using artificial neural networks with optimal feature transformation. Neural Computing and Applications 13, 255–260 (2004)
90. Yoon, K.T., Joo, O.K., Insuk, S., Changha, H.: Usefulness of artificial neural networks for early warning system of economic crisis. Expert Systems with Applications 26, 583–590 (2004)
91. Ko, P.-C., Lin, P.-C.: Resource allocation neural network in portfolio selection. Expert Systems with Applications 35, 330–337 (2008)
92. Kowalczyk, W., Piasta, Z.: Rough set-inspired approach to knowledge discovery in business databases. In: Wu, X., Kotagiri, R., Korb, K.B. (eds.) PAKDD 1998. LNCS, vol. 1394, pp. 186–197. Springer, Heidelberg (1998)
93. Kowalczyk, W., Slisser, F.: Modelling customer retention with rough data models. In: Komorowski, J., Żytkow, J.M. (eds.) PKDD 1997. LNCS, vol. 1263, pp. 4–13. Springer, Heidelberg (1997)
94. Kuchta, D.: Fuzzy capital budgeting. Fuzzy Sets and Systems 111, 367–385 (2000)
95. Kuchta, D.: A generalisation of a solution concept for the linear programming problem with interval coefficients. Operations Research and Decisions 4, 115–123 (2003)
96. Kuo, R.J., Chen, C.H., Hwang, Y.C.: An intelligent stock trading decision support system through integration of genetic algorithm based fuzzy neural network and artificial neural network. Fuzzy Sets and Systems 118, 21–45 (2001)
97. Lam, M.: Neural network techniques for financial performance prediction: integrating fundamental and technical analysis. Decision Support Systems 37, 567–581 (2004)
98. Lee, C.-F., Tzeng, G.-H., Wang, S.-Y.: A new application of fuzzy set theory to the Black-Scholes option pricing model. Expert Systems with Applications 29, 330–342 (2005)
99. Leontief, W.W.: The Structure of the American Economy, 1919-1935. Oxford University Press, London (1949)
100. Leontief, W.: Input-output analysis. In: Input-Output Economics, New York. Oxford University Press, Oxford (1966)
101. Leontief, W., Duchin, F.: The dynamic input-output model. In: The Future Impact of Automation on Workers, New York, pp. 132–138. Oxford University Press, Oxford (1986)
102. Li, D.-F.: Multiattribute decision making models and methods using intuitionistic fuzzy sets. Journal of Computer and System Sciences 70, 73–85 (2005)
103. Li, Q., Sterali, H.D.: An approach for analyzing foreign direct investment projects with application to China's Tumen River Area development. Computers & Operations Research 3, 1467–1485 (2000)

104. Li, Y.P., Huang, G.H., Guo, P., Yang, Z.F., Nie, S.L.: A dual-interval vertex analysis method and its application to environmental decision making under uncertainty. European Journal of Operational Research 200, 536–550 (2010)
105. Liu, F., Geng, H., Zhang, Y.-Q.: Interactive Fuzzy Interval Reasoning for smart Web shopping. Applied Soft Computing 5, 433–439 (2005)
106. Liang, P., Song, F.: Computer-aided risk evaluation system for capital investment. Omega 22, 391–400 (1994)
107. Liang, Y.-H.: Combining seasonal time series ARIMA method and neural networks with genetic algorithms for predicting the production value of the mechanical industry in Taiwan. Neural Computing and Applications 18, 833–841 (2009)
108. Lin, T.Y., Tremba, A.J.: Attribute transformations on numeric databases and its applications to stock market and economic data. In: Terano, T., Chen, A.L.P. (eds.) PAKDD 2000. LNCS, vol. 1805, pp. 181–192. Springer, Heidelberg (2000)
109. Liu, M., Chen, D., Wu, C., Li, H.: Reduction method based on a new fuzzy rough set in fuzzy information system and its applications to scheduling problems. Computers and Mathematics with Applications 51, 1571–1584 (2006)
110. Li, R., Wang, Z.-O.: Mining classification rules using rough sets and neural networks. European Journal of Operational Research 157, 439–448 (2004)
111. Liao, Z., Wang, J.: Forecasting model of global stock index by stochastic time effective neural network. Expert Systems with Applications 37, 834–841 (2010)
112. Lin, C.-T., Tsai, M.-C.: Location choice for direct foreign investment in new hospitals in China by using ANP and TOPSIS. Qual. Quant. 44, 375–390 (2010)
113. Litoiu, M., Tade, R.: Fuzzy scheduling with application to real-time systems. Fuzzy Sets and Systems 121, 523–535 (2001)
114. Liu, S.T.: Computational method for the profit bounds of inventory model with interval demand and unit cost. Applied Mathematics and Computation 183, 499–507 (2006)
115. Lodwick, W.: Interval and Fuzzy Analysis: A Unified Approach. Elsevier, Amsterdam (2007)
116. Mahfoud, S., Mani, G.: Financial forecasting using genetic algorithms. Applications of Artificial Intelligence 10, 543–566 (1996)
117. George, M.: Fuzzy Mathematics Applications In Economics. Campus Books International (2005)
118. McCloskey, A., McIvor, R., Maguire, L., Humphreys, P., O'Donnell, T.: A user-centred corporate acquisition system: a dynamic fuzzy membership functions approach. Decision Support Systems 42, 162–185 (2006)
119. Mehta, K., Bhattacharyy, S.: Adequacy of training data for evolutionary mining of trading rules. Decision Support Systems 37, 461–474 (2004)
120. Mills, D.: Finding the likely buyer using rough sets technology. American Salesman 38, 3–5 (1993)
121. Moore, R.E.: Automatic error analysis in digital computation. Technical Report Space Div. Report LMSD84821, Lockheed Missiles and Space Co. (1959)
122. Moore, R.E.: Interval analysis. Prentice-Hall, Englewood Cliffs (1966)
123. Moore, R.E., Kearfott, R.B., Cloud, M.J.: Introduction to Interval Analysis. SIAM Press, Philadelphia (2009)
124. Moreno, M., Olmeda, I.: Is the predictability of emerging and developed stock markets really exploitable? European Journal of Operational Research 182, 436–454 (2007)
125. Mrozek, A., Skabek, K.: Rough sets in economic applications. In: Polkowski, L., Skowron, A. (eds.) Rough Sets in Knowledge Discovery, vol. 2, pp. 238–271. Physica-Verlag, Heidelberg (1998)

126. Nenortaite, J., Simutis, R.: Development and Evaluation of Decision-Making Model for Stock Markets. Journal of Global Optimization 36, 1–19 (2006)
127. Ng, G.S., Quek, C., Jiang, H.: FCMAC-EWS: A bank failure early warning system based on a novel localized pattern learning and semantically associative fuzzy neural network. Expert Systems with Applications 34, 989–1003 (2008)
128. Ngai, E.W.T., Wat, F.K.T.: Fuzzy decision support system for risk analysis in e-commerce development. Decision Support Systems 40, 235–255 (2005)
129. Nguyen Hanh, H., Chan Christine, W.: Applications of data analysis techniques for oil production prediction. Engineering Applications of Artificial Intelligence 18, 549–558 (2005)
130. Nikolova, M., Nikolov, M., Cornelis, C., Deschrijver, G.: Survey of the research on intuitionistic fuzzy sets. Advanced Studies in Contemporary Mathematics 4, 127–157 (2002)
131. Pardalos, P.M., Siskos, Y., Zopounidis, C.: Advances in Multicriteria Analysis. Kluwer Academic Publishers, Dordrecht (1995)
132. Pasley, A., Austin, J.: Distribution forecasting of high frequency time series. Decision Support Systems 37, 501–513 (2004)
133. Pawlak, Z.: Rough sets. International Journal of Information and Computer Sciences 11, 341–356 (1982)
134. Perez, M.: Artificial neural networks and bankruptcy forecasting: a state of the art. Neural Computing and Applications 15, 154–163 (2006)
135. Perrone, G.: Fuzzy multiple criteria decision model for the evaluation of AMS. Comput. Integrated Manufacturing Systems 7, 228–239 (1994)
136. Poel, D.: Rough sets for database marketing. In: Polkowski, L., Skowron, A. (eds.) Rough Sets in Knowledge Discovery, vol. 2, pp. 324–335. Physica-Verlag, Heidelberg (1998)
137. Poel, D., Piasta, Z.: Purchase prediction in database marketing with the ProbRough System. In: Polkowski, L., Skowron, A. (eds.) RSCTC 1998. LNCS (LNAI), vol. 1424, pp. 593–600. Springer, Heidelberg (1998)
138. Pramanik, S., Roy, T.K.: An intuitionistic fuzzy goal programming approach for a quality control problem: a case study. Tamsui Oxford Journal of Management Sciences 23, 1–18 (2007)
139. Qi, M.: Predicting US recessions with leading indicators via neural network models. International Journal of Forecasting 17, 383–401 (2001)
140. Qian, B., Rasheed, K.: Stock market prediction with multiple classifiers. Applied Intelligence 26, 25–33 (2007)
141. Quek, C., Pasquier, M., Kumar, N.: A novel recurrent neural network-based prediction system for option trading and hedging. Applied Intelligence 29, 138–151 (2008)
142. Raju, K.S., Kumar, D.N., Duckstein, L.: Artificial neural networks and multicriterion analysis for sustainable irrigation planning. Computers and Operations Research 33, 1138–1153 (2006)
143. Roh, T.H.: Forecasting the volatility of stock price index. Expert Systems with Applications 33, 916–922 (2007)
144. Roy, B.: Methodologie Multicriterie dAide a la decision. Economica, Paris (1985)
145. Roy, B., Bouyssou, D.: Aide Multicritere a la Decision: Methodes et Cas. Economica, Paris (1993)
146. Ruggiero, M.: Rules are made to be traded. AI in Finance Fall, 35–40 (1994)
147. Saaty, T.L.: A scaling method for priorities in hierarchical structures. Journal of Mathematical Psychology (15), 59–62 (1977)

148. Scharlig, A.: Pratiquer ELECTRE et PROMETHEE. Presses Polytechniques et Universitaires Romandes, Lausanne (1996)
149. Seçme, N.Y., Bayrakdaroğlu, A., Kahraman, C.: Fuzzy performance evaluation in Turkish Banking Sector using Analytic Hierarchy Process and TOPSIS. Expert Systems with Applications 36, 11699–11709 (2009)
150. Sevastianov, P., Dymova, L.: Synthesis of fuzzy logic and Dempster-Shafer Theory for the simulation of the decision-making process in stock trading systems. Mathematics and Computers in Simulation 80, 506–521 (2009)
151. Sevastianov, P., Sevastianov, D.: Risk and capital budgeting parameters evaluation from the fuzzy sets theory position. Reliable software 1, 10–19 (1997) (in Russian)
152. Sewastjanow, P., Dymowa, L.: On the Fuzzy Internal Rate of Return. In: Cengiz, K. (ed.) Fuzzy Engineering Economics with Applications, pp. 105–128. Springer, Heidelberg (2008)
153. Shapiro Arnold, F.: The merging of neural networks, fuzzy logic, and genetic algorithms. Insurance: Mathematics and Economics 31, 115–131 (2002)
154. Shapiro Arnold, F.: Fuzzy logic in insurance. Insurance: Mathematics and Economics 35, 399–424 (2004)
155. Shen, L., Loh, H.T.: Applying rough sets to market timing decisions. Decision Support Systems 37, 583–597 (2004)
156. Shiue, Y.-R., Su, C.-T.: Attribute Selection for Neural Network-Based Adaptive Scheduling Systems in Flexible Manufacturing Systems. International Journal of Advanced Manufacturing Technology 20, 532–544 (2002)
157. Li, S., Wu, Z., Pang, X.: Job shop scheduling in real-time cases. International Journal of Advanced Manufacturing Technology 26, 870–875 (2005)
158. Shyur, H.-J.: *COTS* evaluation using modified *TOPSIS* and *ANP*. Applied Mathematics and Computation 177, 251–259 (2006)
159. Sikder, I.U., Gangopadhyay, A.: Managing uncertainty in location services using rough set and evidence theory. Expert Systems with Applications 32, 386–396 (2007)
160. Skalko, C.: Rough sets help time the OEX. Journal of Computational Intelligence in Finance 4, 20–27 (1996)
161. Slowinski, R., Zopoundis, C.: Rough-set sorting of firms according to bankruptcy risk. In: Paruccini, M. (ed.) Applying Multiple Criteria Aid for Decision to Environmental Management, pp. 339–357. Kluwer Academic Publishing, Boston (1994)
162. Slowinski, R., Zopounidis, C.: Application of the rough set approach to evaluation of bankruptcy risk. Intelligent Systems in Accounting, Finance and Management 4, 27–41 (1995)
163. Slowinski, R., Zopounidis, C., Dimitras, A.I.: Prediction of company acquisition in Greece by means of the Rough Set approach. European Journal of Operational Research 100, 1–15 (1997)
164. Slowinski, R., Zopounidis, C., Dimitras, A.I., Susmaga, R.: Rough set predictor of business failure. In: Ribeiro, R.A., Zimmermann, H.J., Yager, R.R., Kacprzyk, J. (eds.) Soft Computing in Financial Engineering, pp. 402–424. Physica-Verlag, New York (1999)
165. Steuer, R.E.: Algorithm for linear programming problems with interval objective function co-efficient. Mathematics of Operations Research 6, 333–348 (1981)
166. Steuer, R.E., Na, P.: Multiple criteria decision making combined with finance. A categorical bibliographic study. European Jurnal of Operational Research 150, 496–515 (2003)
167. Susmaga, R., Michalowski, W., Slowinski, R.: Identifying regularities in stock portfolio tilting. Interim Report IR-97-66, International Institute for Applied Systems Analysis (1997)

168. Szmidt, E., Kacprzyk, J.: Atanassov's Intuitionistic Fuzzy Sets as a Promising Tool for Extended Fuzzy Decision Making Models. In: Fuzzy Sets and Their Extensions: Representation, Aggregation and Models, pp. 335–355. Springer, Heidelberg (2008)
169. Tan, A., Quek, C., Yow, K.C.: Maximizing winning trades using a novel RSPOP fuzzy neural network intelligent stock trading system. Applied Intelligence 29, 116–128 (2008)
170. Tanaka, H., Uejima, S., Asia, K.: Linear regression analysis with Fuzzy model. IEEE Transactions on Systems Man and Cybernetics 12, 903–907 (1982)
171. Taskin, A., Guneri, A.F.: Economic analysis of risky projects by ANNs. Applied Mathematics and Computation 175, 171–181 (2006)
172. Tay, E.H., Shen, L.: Economic and financial prediction using rough sets model. European Journal of Operational Research 141, 643–661 (2002)
173. Tong, S.: Interval number and fuzzy number linear programming. Fuzzy Sets and Systems 66, 301–306 (1994)
174. Tsai, C.-F., Wu, J.-W.: Using neural network ensembles for bankruptcy prediction and credit scoring. Expert Systems with Applications 34, 2639–2649 (2008)
175. Tsai, C.-F., Chiou, Y.-J.: Earnings management prediction: A pilot study of combining neural networks and decision trees. Expert Systems with Applications 36, 7183–7191 (2009)
176. Tsou, C.-S.: Multi-objective inventory planning using MOPSO and TOPSIS. Expert Systems with Applications 35, 136–142 (2008)
177. Tsumoto, S., Slowinski, R., Komorowski, H.J., Grzymala-Busse, J.W. (eds.): RSCTC 2004. LNCS (LNAI), vol. 3066. Springer, Heidelberg (2004)
178. Vaidya, O.S., Kumar, S.: Analytic hierarchy process: An overview of applications. European Journal of Operational Research 169, 1–29 (2006)
179. Vanstone, B., Finnie, G.: An empirical methodology for developing stockmarket trading systems using artificial neural networks. Expert Systems with Applications 36, 6668–6680 (2009)
180. Vercher, E., Bermdez, J.D., Segura, J.V.: Fuzzy portfolio optimization under downside risk measures. Fuzzy Sets and Systems 158, 769–782 (2007)
181. Vincke, P.: Multicriteria Decision Aid. Wiley, NewYork (1992)
182. Wang, H., Wei, G.: An Effective Supplier Selection Method with Intuitionistic Fuzzy Information. In: Proc. of 4th International Conference Wireless Communications, Networking and Mobile Computing (WiCOM 2008), pp. 1–4 (2008)
183. Wang, J.-W., Cheng, C.-H., Huang, K.-C.: Fuzzy hierarchical *TOPSIS* for supplier selection. Applied Soft Computing 9, 377–386 (2009)
184. Wang, S.: The Neural Network Approach to Input-Output Analysis for Economic Systems. Neural Computing and Applications 10, 22–28 (2001)
185. Wang, Y.-F.: Mining stock price using fuzzy rough set system. Export Systems with Applications 24, 13–23 (2003)
186. Wang, Y.-M., Yang, J.-B., Xu, D.-L.: Environmental impact assessment using the evidential reasoning approach. European Journal of Operational Research 174, 1885–1913 (2006)
187. Wang, Z., Li, K.W., Wang, W.: An approach to multiattribute decision making with interval-valued intuitionistic fuzzy assessments and incomplete weights. Information Sciences 179, 3026–3040 (2009)
188. Ward, T.L.: Discounted fuzzy cash flow analysis. In: Proceedings of Fall Industrial Engineering Conference, pp. 476–481 (1985)
189. Warmus, M.: Calculus of Approximations. Bull. Acad. Polon. Sci., Cl. III IV, 253–259 (1956)

190. Weck, M., Klocke, F., Schell, H., Ruenauver, E.: Evaluating alternative production cycles using the extended fuzzy AHP method. European Journal of Operational Research 100, 351–366 (1997)
191. Wong Bo, K., Lai Vincent, S., Jolie, L.: A bibliography of neural network business applications research: 1994-1998. Computers and Operations Research 27, 1045–1076 (2000)
192. Wu, H.-C.: Using fuzzy sets theory and Black-Scholes formula to generate pricing boundaries of European options. Applied Mathematics and Computation 185, 136–146 (2007)
193. Yang, J.B., Wang, Y.M., Xu, D.L., Chin, K.S.: The evidential reasoning approach for MADA under both probabilistic and fuzzy uncertainties. European Journal of Operational Research 171, 309–343 (2006)
194. Yang, S., Wang, D., Chai, T., Kendall, G.: An improved constraint satisfaction adaptive neural network for job-shop scheduling. Journal of Scheduling 13, 17–38 (2010)
195. Yong, D.: Plant location selection based on fuzzy TOPSIS. International Journal of Advanced Manufacturing Technology 28, 839–844 (2006)
196. Chen, Y.-S., Cheng, C.-H.: A Delphi-based rough sets fusion model for extracting payment rules of vehicle license tax in the government sector. Expert Systems with Applications 37, 2161–2174 (2010)
197. Yu, L., Wang, S., Lai Kin, K., Wen, F.: A multiscale neural network learning paradigm for financial crisis forecasting. Neurocomputing 73, 716–725 (2010)
198. Yudong, Z., Lenan, W.: Stock market prediction of S&P 500 via combination of improved BCO approach and BP neural network. Expert Systems with Applications 36, 8849–8854 (2009)
199. Zadeh, L.A.: Fuzzy Sets. Information and Control 8, 338–353 (1965)
200. Zeleny, M.: Multiple Criteria Decision Making. Business Venturing (7), 505–518 (1982)
201. Ziarko, W. (ed.): Rough Sets, Fuzzy Sets and Knowledge Discovery. In: Proceedings of International Workshop on Rough Sets and Knowledge Discovery (RSKD 1993). Springer, New York (1994)
202. Zopounidis, C.: Multicriteria decision aid in financial management. European Journal of Operational Research (119), 404–415 (1999)

Chapter 3
The Methods for Uncertainty Modeling

There are many modern methods for uncertainty modeling developed in last decades. Generally, they are not in conflict with the traditional probabilistic approach since they deal with another (non-probabilistic) types of uncertainties. Moreover, in the solution of real-world problems, the probabilistic and the other types of uncertainties often become apparent simultaneously. Therefore, the synthesis of traditional and modern methods for uncertainty modeling usually provides best results in applications. In this chapter, we present an overview of modern methods based on the fuzzy sets theory (including type-2 fuzzy sets and intuitionistic fuzzy sets), interval analysis, and the Dempster-Shafer theory of evidence (*DST*). We show the interrelations between these methods and emphasize some problems that impede their applications. We do not intend to present here a comprehensive overview of all modern methods as now they are well presented in numerous books and handbooks. Therefore, in this chapter we present the modern methods for uncertainty modeling only to the extent needed for understanding the applications presented in the following chapters.

3.1 Fuzzy Set Theory

In this section, we present the fundamentals of fuzzy set theory introduced by Zadeh [154] in 1965 and the generalizations of fuzzy set theory such as type 2 fuzzy sets and intuitionistic fuzzy sets which seem to be the most useful in applications.

3.1.1 Basic Definitions

Classical sets are also called "crisp" sets so as to distinguish them from fuzzy sets. In fact, the crisp sets can be treated as a special case of fuzzy sets. Let A be a crisp set defined in the universe X. Then for any element x in X, either x is a member of A

L. Dymowa: Soft Computing in Economics and Finance, ISRL 6, pp. 41–105.
springerlink.com

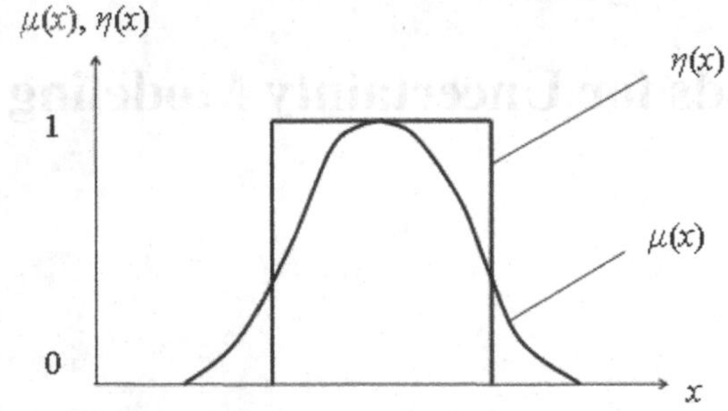

Fig. 3.1 The characteristic function η_X of a crisp set and the membership function $\mu(x)$ of a fuzzy set

or not. In the fuzzy set theory, this property is generalized. Therefore, in a fuzzy set, it is not necessary that x is a full member of a set or not a member. It can be a partial member of a set. The generalization is performed as follows: for any crisp set A, it is possible to define a characteristic function η_X= 0,1, i.e., the characteristic function takes either the values 0 or 1 in a classical set. For a fuzzy set, the characteristic function can take any value between zero and one as it is shown in Fig. 3.1.

Definition 3.1. The membership function $\mu_A(x)$ of a fuzzy set A is μ_A: $X \to [0,1]$. So every element x in X has a membership degree $\mu_A(x) \in [0,1]$. A fuzzy set A is completely determined in the universe X by the set of tuples:

$$A = \{(x, \mu_A(x)), x \in X\}, \forall x \in X. \tag{3.1}$$

The maximal value of $\mu_A(x)$ is equal to 1 for x which completely belongs to A, and the minimal value of $\mu_A(x)$ is equal to 0 when x does not belong to A.

Example 3.1. Suppose one wants to describe the class of cars having the property of being expensive by considering BMW, Rolls Royce, Mercedes, Ferrari, Fiat, Honda and Renault. Some cars like Ferrari and Rolls Royce are definitely expensive and some like Fiat and Renault are not so expensive and do not belong to the set. Then, the fuzzy set of expensive cars can be described as: (Ferrari, 1),(Rolls Royce, 1),(Mercedes, 0.8),(BMW, 0.7), (Honda, 0.4). Obviously, Ferrari and Rolls Royce have the membership value of 1, whereas BMW, which is less expensive, has a membership value of 0.7 and Honda 0.4.

Example 3.2. A set of natural numbers "close to 6" can be defined as a fuzzy set. This can be done, say, buy including all numbers from 3 to 9 as follows: $\tilde{6}$={(3, 0.1), (4, 0.2), (5, 0.5), (6, 1), (7, 0.5), (8, 0.2), (9, 0.1)}.

Zadeh proposed an alternate representation for fuzzy sets, which is more convenient. Suppose the universe of discourse is a finite set with n elements, that is X={x_1, x_2, ..., x_n}, then an alternative representation for A can be presented as follows:

$$A = \sum_{i=1}^{n} \mu_A(x_i)/x_i = \mu_A(x_1)/x_1 + \ldots + \mu_A(x_n)/x_n \tag{3.2}$$

or equivalently

$$A = \sum_{i=1}^{n} \frac{\mu_A(x_i)}{x_i} = \frac{\mu_A(x_1)}{x_1} + ... + \frac{\mu_A(x_n)}{x_n}. \tag{3.3}$$

Here the symbol "+" denotes an enumeration, listing or union rather than addition, the line between the top and bottom entries in the above formulas is just a delimiter or separator.

The fuzzy set of expensive cars (see Example 3.1) can be now written using this notation as: $\left\{\frac{1}{Ferrari} + \frac{1}{\text{RollsRoyce,}} + \frac{0.8}{\text{Mercedes}} + \frac{0.7}{\text{BMW}} + \frac{0.4}{\text{Honda}}\right\}$.

In the case of not finite universe of discourse, a fuzzy set A can be expressed as

$$A = \int_X \mu_A(x)/x \tag{3.4}$$

or

$$A = \int_X \frac{\mu_A(x)}{x}, \tag{3.5}$$

respectively. Symbol $\int$ in the above expressions refers to union rather than to integration. This symbolic notation is used to connect an element and its membership value.

Some other important definitions are presented below.

Definition 3.2. The support of a fuzzy set A, denoted by $supp(A)$, is the set of points in X at which the membership function $\mu_A(x)$ is positive

$$supp(A) = \{x \in X; \mu_A(x) > 0\} \tag{3.6}$$

Definition 3.3. A singleton is a fuzzy set A whose support is a single point x in the universe of discourse X.

Definition 3.4. The core of a fuzzy set A defined in the universe of discourse X, denoted by $core(A)$, also referred to as kernel, is the set of points in X at which the membership function $\mu_A(x)$ equals to 1, that is

$$core(A) = \{x \in X; \mu_A(x) = 1\} \tag{3.7}$$

Definition 3.5. The height of a fuzzy set A defined in the universe of discourse X, denoted by $hgt(A)$, is the maximal value of its membership function $\mu_A(x)$, that is

$$hgt(A) = \sup_{x \in X} \mu_A(x) \tag{3.8}$$

Definition 3.6. A fuzzy set A is empty if for each $x \in X$ the value of membership function $\mu_A(x)$ is equal to zero, that is

$$A = \emptyset \Leftrightarrow \forall x \in X; \mu_A(x) = 0 \tag{3.9}$$

Definition 3.7. A fuzzy set $A \subseteq X$ is included in a fuzzy set $B \subseteq X$, if for any $x \in X$ the value of membership function $\mu_A(x)$ is not greater than the value of membership function $\mu_B(x)$, that is

$$A \subset B \Leftrightarrow \forall x \in X; \mu_A(x) \leq \mu_B(x) \tag{3.10}$$

Definition 3.8. A fuzzy set $A \subseteq X$ is equal to a fuzzy set $B \subseteq X$, if for any $x \in X$ the value of membership function $\mu_A(x)$ is equal to the value of membership function $\mu_B(x)$, that is

$$A = B \Leftrightarrow \forall x \in X; \mu_A(x) = \mu_B(x) \tag{3.11}$$

Definition 3.9. A fuzzy set A defined on X is said to be normalized, if

$$\sup_{x \in X} (\mu_A(x)) = 1 \tag{3.12}$$

Definition 3.10. A fuzzy set A is convex if its membership function is such that:

$$\mu_A(\lambda x_1 + (1-\lambda)x_2) \geq \min\{\mu_A(x_1), \mu_A(x_2)\} \tag{3.13}$$

for any $x_1, x_2 \in X$ and $\lambda \in [0,1]$. Otherwise A is a non-convex fuzzy set.

Definition 3.11. The cardinality of a fuzzy set A is defined as

$$|A| = \int_x \mu_A(x)dx. \tag{3.14}$$

3.1.2 *Operations on Fuzzy Sets*

The fuzzy sets may be treated as the generalization of usual (crisp) sets. Therefore, the complement, union and intersection operators on fuzzy sets have been defined by Zadeh in his early works.

Definition 3.12. The complement of a fuzzy set A, denoted by $\overline{A}$, is defined by its membership function as follows

$$\mu_{\overline{A}}(x) = 1 - \mu_A(x) \tag{3.15}$$

for all $x \in X$.

The value of $\mu_{\overline{A}}(x)$ corresponds to negation of the concept represented by A. For example, if A is a fuzzy set of big houses, the complement of this fuzzy set is the fuzzy set of houses which are not big ones. Fig. 3.2 illustrates the complement of a fuzzy set A.

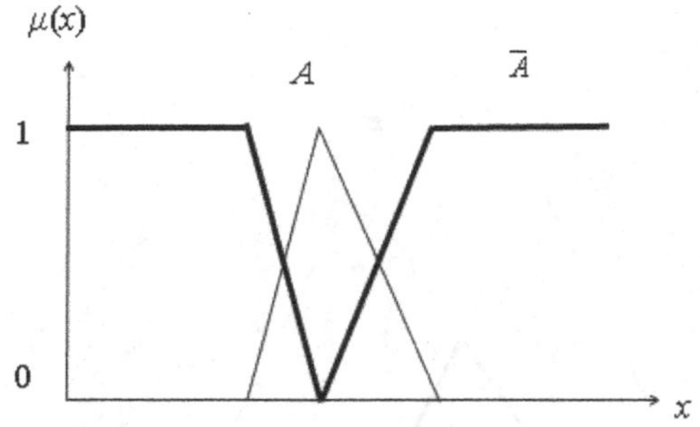

Fig. 3.2 Complement of a fuzzy set

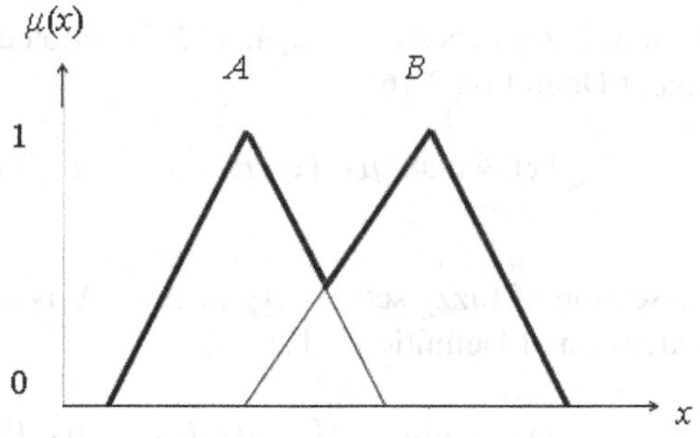

Fig. 3.3 Union of fuzzy sets

Definition 3.13. The union of two fuzzy sets A and B with respective membership functions $\mu_A(x)$ and $\mu_B(x)$ is a fuzzy set denoted by $A \cup B$ whose membership function is defined by

$$\mu_{A\cup B}(x) = \max(\mu_A(x), \mu_B(x)) \tag{3.16}$$

for all $x \in X$.

The graphical illustration of the union of fuzzy sets is presented in Fig. 3.3.

Definition 3.14. The intersection of two fuzzy sets A and B with respective membership functions $\mu_A(x)$ and $\mu_B(x)$ is a fuzzy set denoted by $A \cap B$ whose membership function is defined by

$$\mu_{A\cap B}(x) = \min(\mu_A(x), \mu_B(x)) \tag{3.17}$$

for all $x \in X$.

The illustration is presented in Fig. 3.4.

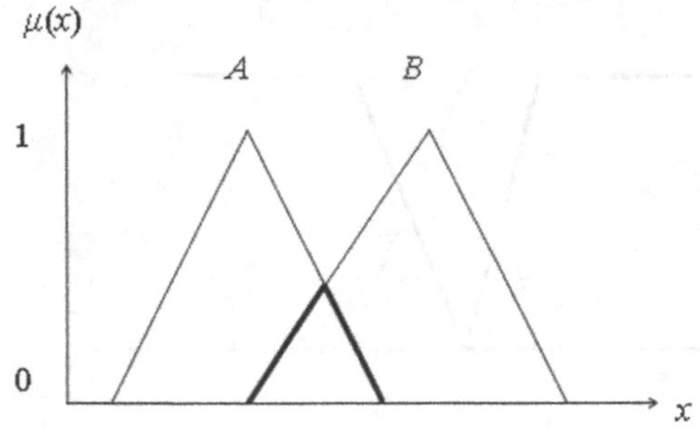

Fig. 3.4 Intersection of fuzzy sets

The union operation of fuzzy sets $A_1, A_2, ..., A_n \subset X$ is denoted by $\bigcup_{i=1}^{n} A_i$ and is given as the extension of Definition 3.16:

$$\mu_{A_1 \cup A_2 \cup ... \cup A_n}(x) = \max(\mu_{A_1}(x), \mu_{A_2}(x), ..., \mu_{A_n}(x)) \quad (3.18)$$

for all $x \in X$.

Similarly, the intersection of fuzzy sets $A_1, A_2, ..., A_n \subset X$ is denoted by $\bigcap_{i=1}^{n} A_i$ and is given as the extension of Definition 3.17:

$$\mu_{A_1 \cap A_2 \cap ... \cap A_n}(x) = \min(\mu_{A_1}(x), \mu_{A_2}(x), ..., \mu_{A_n}(x)) \quad (3.19)$$

for all $x \in X$.

The definitions presented above, constitute the basis for the theory of fuzzy sets formulated by Zadeh [154] and the formulas (3.15), (3.16) and (3.17) are called standard operations. However, these operations are not only possible way to extent classical theory consistently to the fuzzy set theory. Zadeh and other authors have suggested alternative or additional definitions concerning with operations on fuzzy sets.

For example, Yager [145] proposed the more general definitions of union and intersection of fuzzy sets:

$$\mu_{A \cup B}(x)_p = 1 - \min[1, [(\mu_A^p(x) + \mu_B^p(x)]^{1/p}], \quad (3.20)$$

$$\mu_{A \cap B}(x)_p = 1 - \min[1, [(1 - \mu_A(x))^p + (1 - \mu_B(x))^p]^{1/p}], \; p \geq 1. \quad (3.21)$$

For $p \to \infty$ these operations transform to the Zadeh's operations (3.16) and (3.17), respectively.

It is worth noting that operations (3.20) and (3.21) are idempotent ones only in the case of $p = \infty$. Bellman and Zadeh [12] introduced the more "soft" operations of union and intersection of fuzzy sets:

$$\mu_{A\cup B}(x) = \mu_A(x) + \mu_B(x) - \mu_A(x)\mu_B(x), \tag{3.22}$$

$$\mu_{A\cap B}(x) = \mu_A(x)\mu_B(x). \tag{3.23}$$

It is noted in [12] that from practical and theoretical points of view the "hard" operations (3.16) and (3.17) are more preferable than the "soft" operations, but in applications there may be such situations when the "hard" operations do not reflect the sense of union and intersection.

All the definitions presented above are only particular cases of a more wide class of operations of union and intersection.

A general class of intersection operators for fuzzy sets is defined by so-called triangular norms or t-norms, and a general class of union operators is defined, analogously, by s-norms (t-conorms). These operators are extensively applied in fuzzy set theory as logical connective *AND*, which represents the intersection and as logical connective *OR*, which represents the union.

The triangular norms and t-conorms can be defined as follows [38, 42, 99, 138, 158]:

Definition 3.15. A triangular norm t is a function of two arguments:

$$t : [0,1] \times [0,1] \to [0,1], \tag{3.24}$$

which satisfies the following conditions for $a,b,c,d \in [0,1]$

$$\textit{Monotonicity} : t(a,b) \le t(c,d);\ a \le c;\ b \le d \tag{3.25}$$

$$\textit{Commutativity} : t(a,b) = t(b,a) \tag{3.26}$$

$$\textit{Associativity} : t(t(a,b),c) = t(a,t(b,c)) \tag{3.27}$$

$$\textit{Boundary conditions} : t(a,0) = 0;\ t(a,1) = a \tag{3.28}$$

The most frequently used in applications are Zadeh's t-norm: $t(a,b) = \min(a,b)$, Algebraic t-norm: $t(a,b) = a \cdot b$ and Bounded t-norm: $\max(a+b-1,0)$.

Definition 3.16. An s-norm is a function of two arguments:

$$s : [0,1] \times [0,1] \to [0,1], \tag{3.29}$$

which satisfies the following conditions for $a,b,c,d \in [0,1]$

$$\textit{Monotonicity} : s(a,b) \le s(c,d);\ a \le c;\ b \le d \tag{3.30}$$

$$\textit{Commutativity} : s(a,b) = s(b,a) \tag{3.31}$$

$$\textit{Associativity} : s(s(a,b),c) = s(a,s(b,c)) \tag{3.32}$$

$$\textit{Boundary conditions} : s(a,0) = a;\ s(a,1) = 1 \tag{3.33}$$

The most frequently used in applications are Zadeh's s-norm: $s(a,b) = \max(a,b)$, Algebraic s-norm: $s(a,b) = a + b - a \cdot b$ and Bounded s-norm: $\min(a+b,1)$.

The t-norms and s-norms are related in sense of logical duality. Any s-norm can be derived from a t-norm through the following formula [2]:

$$t(a,b) = 1 - s(1-a, 1-b) \tag{3.34}$$

Many other examples of t-norms and s-norms can be found in the literature (see, e.g., [42, 63, 64, 99, 103, 104, 158]).

It is worth noting that there are no good or bad t-norms and s-norms among those proposed in the literature as the choice of appropriate norm for a specific application is a context dependent problem.

3.1.3 Operations on Fuzzy Numbers

There are many definitions of fuzzy number proposed in the literature. One of them, introduced by Dubois and Prade [37], seems to be general and suitable enough for different applications of fuzzy set theory. According to [37], a fuzzy number is a fuzzy subset of the real line whose highest membership values are clustered around a given real number called the mean value; the membership function is monotonic on both sides of this mean value.

The operations on fuzzy numbers are generally based on extension principle introduced by Zadeh [156]. This principle plays an important role in fuzzy set theory, especially in fuzzy arithmetic. It provides a general method for extending crisp mathematical concepts to the fuzzy framework.

Suppose that f is a function that maps points in space X to points in space Y, that is

$$f : X \to Y \tag{3.35}$$

and A is a fuzzy subset of X expressed by Equation (3.3). Then the extension principle asserts that

$$f(A) = \left\{ \frac{\mu_A(x_1)}{f(x_1)} + \ldots + \frac{\mu_A(x_n)}{f(x_n)} \right\} \tag{3.36}$$

If more than one element of X is mapped by f to the same element of $y \in Y$, then the maximum of the membership grades of these elements in the fuzzy set A is chosen as the membership grade for y in Y. If no element $x \in X$ is mapped to y, then the membership grade of y in $f(A)$ is zero.

Fuzzy arithmetic seems to be a highly developed and well-formalized branch of fuzzy sets theory. Nevertheless, there are some problems of the practical implementation of fuzzy arithmetic rules. They are based on the extension principle introduced by Zadeh [154] for arithmetical operations on fuzzy numbers. The general formulation of this principle uses an arbitrary t -norm.

Let A, B, Z be fuzzy values and $@ \in \{+,-,*,/\}$ be an arithmetical operation. Then

$$Z = A@B = \{z = x@y, \mu(z) = \max_z t(\mu(x), \mu(y)), x \in A, y \in B\}. \tag{3.37}$$

As emphasized by Zimmermann and Zysno [159], the choosing of the concrete realization of t-norm is rather an application dependent problem, but three main t-norms are usually used in practice:

$$t(\mu(x), \mu(y)) = min(\mu(x), \mu(y)), \tag{3.38}$$
$$t(\mu(x), \mu(y)) = 0.5(\mu(x) + \mu(y)), \tag{3.39}$$
$$t(\mu(x), \mu(y)) = \mu(x)\mu(y). \tag{3.40}$$

It is worth noting that there are many other t-norms used in fuzzy logic, but expressions (3.38)-(3.40) are most suitable for fuzzy arithmetic applications, although expression (3.39) is not a t-norm at all.

An alternative approach to implementation of fuzzy arithmetic is based on the α-cuts presentation of fuzzy numbers [155, 156].

So, if A is a fuzzy number, then $A = \bigcup_{\alpha} \alpha A_\alpha$, where αA_α is the fuzzy subset: $x \in X$, $\mu_A(x) \geq \alpha$, A_α is the support set of fuzzy subset αA_α and X is the universe of discourse.

To illustrate the representation of fuzzy number with the use α-cuts, consider the following example.

Let $X = 1, 2, ..., 10$ and
$A = 0.1/2 + 0.3/3 + 0.6/4 + 0.8/5 + 1/6 + 0.7/7 + 0.4/8 + 0.2/9.$
Then $A = \sum_{\alpha \in (0,1]} \alpha A_\alpha =$
$0.1(1/2 + 1/3 + 1/4 + 1/5 + 1/6 + 1/7 + 1/8 + 1/9) +$
$+0.2(1/3 + 1/4 + 1/5 + 1/6 + 1/7 + 1/8 + 1/9) +$
$+0.3(1/3 + 1/4 + 1/5 + 1/6 + 1/7 + 1/8) +$
$+0.4(1/4 + 1/5 + 1/6 + 1/7 + 1/8) +$
$+0.6(1/4 + 1/5 + 1/6 + 1/7) +$
$+0.7(1/5 + 1/6 + 1/7) +$
$+0.8(1/5 + 1/6) +$
$+1(1/6) = 0.1/2 + 0.3/3 + 0.6/4 + 0.8/5 + 1/6 + 0.7/7 + 0.4/8 + 0.2/9.$

The graphical interpretation of α-cuts is presented in Fig. 3.5.

It was proved that if A and B are fuzzy numbers, then all the operations on them may be presented as operations on the set of crisp intervals corresponding to their α-cuts:

$$(A@B)_\alpha = A_\alpha @ B_\alpha. \tag{3.41}$$

Of course, the direct α-cut representation of operations on fuzzy numbers seems to be a rough one in comparison with the generalized expression (3.37). But for the practical numerical realization of (3.37) we have to use the discretization of the supports of considered fuzzy numbers A, B and Z if we deal with non-trivial forms

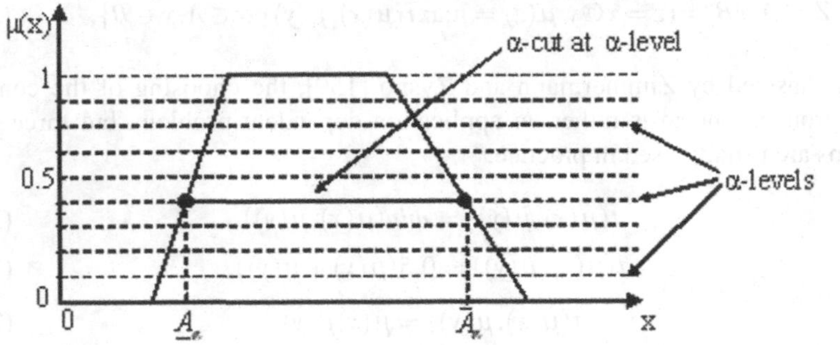

Fig. 3.5 Illustration of α-cuts

of $\mu(x)$ and $\mu(y)$. It is well known for the practicians that any direct discretization of (3.37) leads to unacceptable results [100, 109].

To illustrate, let us consider the example adopted from [100]. Let A = "near 5" and B = "near 7" be fuzzy values represented by corresponding finite fuzzy sets (see Table 3.1).

Table 3.1 Discrete representation of the test fuzzy values A and B

$\mu_A(x_i)$	0	0.33	0.66	1	0.5	0
x_i	2	3	4	5	6	7
$\mu_B(y_i)$	0	0.5	1	0.66	0.33	0
y_i	5	6	7	8	9	10

Then for multiplication $(A \cdot B)$ in the case of $t(\mu(x), \mu(y)) = \mu(x) \cdot \mu(y)$ we get the result shown in Fig. 3.6. We can see that the use of the general definition (3.37) does not provide the convex resulting fuzzy number in the case of multiplied fuzzy numbers, whereas we have no problems when using an approach based on the α-cuts representation of fuzzy numbers. It seems natural that the results obtained with the use of the general definition (3.37) may be improved using a more detailed discretization, but in practice usually it is hard to do this. It should be emphasized here that similar unacceptable results were obtained using all the definitions (3.38)-(3.40) for all arithmetical operations, but there is no problem if we use the α-cut presentation of operations on fuzzy numbers. More details, examples and analysis may be found in [100, 109]. However, for our purposes it is quite enough to treat the result presented above as only the empirical fact, which indicates that there are some difficult problems when using the discretization of general expression (3.37) and that these problems can be easily eliminated using the α-cut representation of fuzzy arithmetic rules. Hence, the α-cut representation of fuzzy numbers and operations on them can be accepted as the basic concept for fuzzy modeling of the real-world processes.

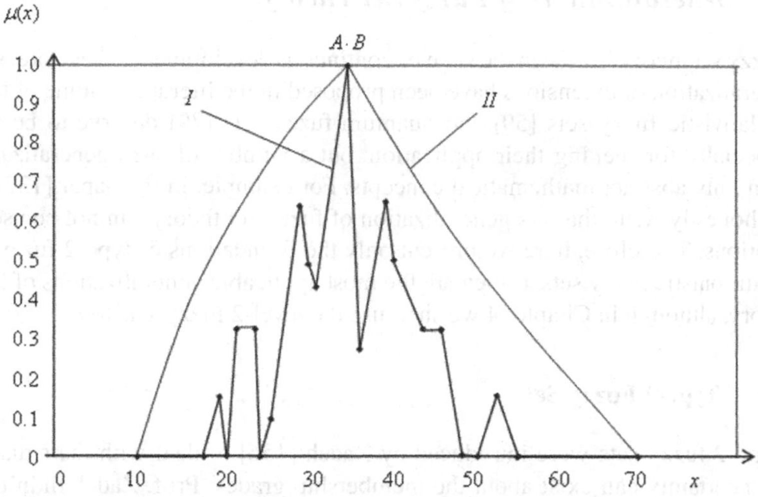

Fig. 3.6 Multiplication of two fuzzy numbers (see, Table 3.1) (I-using the definitions (3.37) and (3.40), II-using α-cuts (3.41)

Since in the case of α-cut presentation, fuzzy arithmetic is based on crisp interval arithmetic rules, the basic definitions of applied interval analysis should be presented too.

There are some definitions of interval arithmetic [57, 89], but in practical applications the so-called "naive" form proved to be the best one. According to it, if $A = [a_1, a_2]$ and $B = [b_1, b_2]$ are crisp intervals and $@ \in \{+, -, *, /\}$, then

$$Z = A@B = \{z = x@y, x \in A, y \in B\}. \tag{3.42}$$

As the direct consequence of the basic definition (3.42) the following expressions were obtained:

$$A + B = [a_1 + b_1, a_2 + b_2], A - B = [a_1 - b_2, a_2 - b_1],$$

$$AB = [min(a_1b_1, a_2b_2, a_1b_2, a_2b_1), max(a_1b_1, a_2b_2, a_1b_2, a_2b_1)],$$

$$A/B = [a_1, a_2][1/b_2, 1/b_1],\ 0 \notin B.$$

Of course, there are many internal problems within applied interval analysis, e.g., the division by zero-containing interval, but in general, it can be considered as a reliable mathematical tool for modeling under conditions of uncertainty.

The methods of interval analysis will be presented in the following section in more detail.

3.1.4 Generalizations of Fuzzy Set Theory

The fuzzy set theory is now in the state of continuous development. Therefore, some its generalization or extensions have been proposed in the literature. Some of them, e.g., relativistic fuzzy sets [59] and quantum fuzzy sets [79] deserve to be studied especially for seeking their application, but a number of such generalizations perform only abstract mathematical concepts. For example, in the paper [137], the author honestly write that his generalization of fuzzy set theory can not be used in applications. Therefore, here we present only the foundations of type-2 fuzzy sets and intuitionistic fuzzy sets as they are the most applicable generalizations of fuzzy set theory, although in Chapter 4 we shell use the level-2 fuzzy sets too.

3.1.4.1 Type-2 Fuzzy Sets

The type-2 fuzzy sets were introduced by Zadeh [156] to deal with the situations when uncertainty can exist about the membership grades. Prof. Zadeh didn't stop with type-2 fuzzy sets, because in [156] he also generalized all of this to type-n fuzzy sets. A classical fuzzy set (type- 1 fuzzy set) is a special case of a type 2 fuzzy set.

Work on type-2 fuzzy sets languished during the 1980's and early-to-mid 1990's, although a small number of papers were published about them. This changed in the latter part of the 1990's as a result of Prof. Jerry Mendel and his student's works on type-2 fuzzy sets and systems (see, e.g., [83]). Since then, more and more researchers around the world are writing papers about type-2 fuzzy sets and systems.

According to [84], there are (at least) four sources of uncertainties in type-1 fuzzy sets: (1) The meanings of the words that are used to define membership grades can be uncertain (words mean different things to different people). (2) The knowledge can be extracted from a group of experts who do not all agree. (3) Measurements that activate a type-1 fuzzy sets may be noisy and therefore uncertain. (4) The data that are used to tune the parameters of a type-1 fuzzy sets may also be noisy. All of these uncertainties translate into uncertainties about fuzzy set membership functions. Type-1 fuzzy sets are not able to directly model such uncertainties because their membership functions are totally crisp. On the other hand, type-2 fuzzy sets are able to model such uncertainties because their membership functions are themselves fuzzy. Membership functions of type-1 fuzzy sets are two-dimensional, whereas membership functions of type-2 fuzzy sets are three-dimensional. It is the new third-dimension of type-2 fuzzy sets that provides additional degrees of freedom that make it possible to directly model uncertainties. Type-2 fuzzy sets have already been used in decision making [146], in economic and financial application such as forecasting of time-series [61, 82] and transport scheduling [58].

To make the formal definition of type-2 fuzzy set more transparent, consider its graphical representation [84].

Imagine blurring the type-1 membership function depicted in Fig. 3.7(a) by shifting the points on the triangle either to the left or to the right and not necessarily by the same amounts, as in Fig. 3.7(b). Then, at a specific value of x, say x', there is no longer a single value for the membership function (u'); instead, the membership function takes on values wherever the vertical line intersects the blur. Those values need not all be weighted the same; hence, we can assign an amplitude distribution to all of those points. Doing this for all $x \in X$, we create a three-dimensional membership function-a type-2 membership function- that characterizes a type-2 fuzzy set.

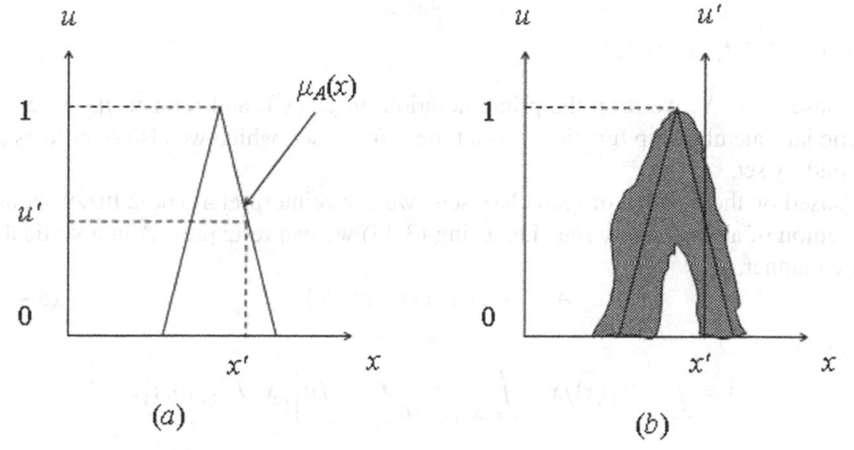

Fig. 3.7 (a) Type-1 membership function and (b) blurred type-1 membership function, including discretization at $x = x'$

Below we present the definitions from [84].

Definition 3.17. A type-2 fuzzy set, denoted $\tilde{A}$, is characterized by a type-2 membership function $\mu_{\tilde{A}}(x,u)$, where $x \in X$ and $u \in J_x \subseteq [0,1]$, i.e.,

$$\tilde{A} = \{((x,u), \mu_{\tilde{A}}(x,u)) \,|\, \forall x \in X, \forall u \in J_x \subseteq [0,1]\} \tag{3.43}$$

in which $0 \leq \mu_{\tilde{A}}(x,u) \leq 1$.

$\tilde{A}$ can also be expressed as

$$\tilde{A} = \int_{x \in X} \int_{u \in J_x} \mu_{\tilde{A}}(x,u) / (x,u), \; J_x \subseteq [0,1], \tag{3.44}$$

where $\int\int$ denotes the union over all admissible x and u. For discrete universes of discourse $\int$ is replaced by $\sum$.

In Definition 3.17, the first restriction that $\forall u \in J_x \subseteq [0,1]$ is consistent with the type-1 constraint $0 \leq \mu_A(x) \leq 1$, i.e., when uncertainties disappear a type-2 membership function must reduce to a type- 1 membership function, in which case the variable

u equals to $\mu_A(x)$ and $0 \le \mu_A(x) \le 1$. The second restriction $0 \le \mu_{\tilde{A}}(x,u) \le 1$ is consistent with the fact that the amplitudes of a membership function should lie between or be equal to 0 and 1.

Definition 3.18. At each value of x, say $x = x'$, the 2-D plane whose axes are u and $\mu_{\tilde{A}}(x',u)$ is called a vertical slice of $\mu_{\tilde{A}}(x,u)$. A secondary membership function is a vertical slice of $\mu_{\tilde{A}}(x,u)$. It is $\mu_{\tilde{A}}(x = x',u) = \mu_{\tilde{A}}(x',u)$ for $x \in X$ and $\forall u \in J_{x'} \subseteq [0,1]$, i.e.,

$$\mu_{\tilde{A}}(x = x',u) \equiv \mu_{\tilde{A}}(x') = \int_{u \in J_{x'}} f_{x'}(u)/u, \ J_{x'} \subseteq [0,1] \tag{3.45}$$

in which $0 \le f_{x'}(u) \le 1$.

Because $\forall x' \in X$, we drop the prime notation on $\mu_{\tilde{A}}(x')$, and refer to $\mu_{\tilde{A}}(x)$ as a secondary membership function; it is a type-1 fuzzy set, which we also refer to as a secondary set.

Based on the concept of secondary sets, we can reinterpret a type-2 fuzzy set as the union of all secondary sets, i.e., using (3.45) we can re-express $\tilde{A}$ in a vertical-slice manner, as

$$\tilde{A} = \{(x, \mu_{\tilde{A}}(x)) \,|\, \forall x \in X\} \tag{3.46}$$

or, as

$$\tilde{A} = \int_{x \in X} \mu_{\tilde{A}}(x)/x = \int_{x \in X} \left[\int_{u \in J_x} f_x(u)/u \right] /x, \ J_x \subseteq [0,1]. \tag{3.47}$$

Definition 3.19. The domain of a secondary membership function is called the primary membership of x. In (3.47), J_x is the primary membership of x, where $J_x \subseteq [0,1]$ for $\forall x \in X$.

Definition 3.20. The amplitude of a secondary membership function is called a secondary grade. In (3.47), $f_x(u)$ is a secondary grade; in (3.45), $\mu_{\tilde{A}}(x',u')$ is a secondary grade.

If X and J_x are both discrete, then the right hand side of (3.47) can be expressed as follows:

$$\begin{aligned} \tilde{A} &= \sum_{x \in X} \left[\sum_{u \in J_x} f_x(u)/u \right] \Big/ x = \sum_{i=1}^{N} \left[\sum_{u \in J_{x_i}} f_{x_i}(u)/u \right] \Big/ x_i = \\ &= \left[\sum_{k=1}^{M_1} f_{x_1}(u_{1k})/u_{1k} \right] \Big/ x_1 + \ldots + \left[\sum_{k=1}^{M_N} f_{x_N}(u_{Nk})/u_{Nk} \right] \Big/ x_N. \end{aligned} \tag{3.48}$$

In this equation, "+" also denotes union. Observe that x has been discretized into N values and at each of these values u has been discretized into M_i values. The discretization along each u_{ik} does not have to be the same, which is why we have shown a different upper sum for each of the bracketed terms. The operations on type-2 fuzzy sets are defined in [84].

One of the useful simplification (type reduction) of a type-2 fuzzy set is an interval type-2 fuzzy set. Interval type-2 fuzzy sets have received the most attention because the mathematics that is needed for such sets-primarily interval arithmetic-is much simpler than the mathematics that is needed for general type-2 fuzzy sets. So, the literature about interval type-2 fuzzy sets is large, whereas the literature about general type-2 fuzzy sets is much smaller. Both kinds of fuzzy sets are being actively researched by an ever-growing number of researchers around the world.

More information about type-2 fuzzy sets, interval type-2 fuzzy sets and their applications can be found in [73, 76, 77, 81, 85, 86, 87].

3.1.4.2 Intuitionistic Fuzzy Sets

Intuitionistic fuzzy set proposed by Atanassov [3], is one of the possible generalizations of fuzzy sets theory and appears to be relevant and useful in some applications. As the so-called "intuitionistic fuzzy set theory" was independently introduced by Takeuti and Titani [129], there are some terminological difficulties in fuzzy set theory. Dubois et al. [36] noted that "Takeuti-Titani's intuitionitic fuzzy logic is simply an extension of intuitionistic logic [27], i.e., all formulas provable in the intuitionistic logic are provable in their logic. Intuitionistic fuzzy set theory by Takeuti and Titani is an absolutely legitimate approach, in the scope of intuitionistic logic, but it has nothing to do with Atanassov's intuitionistic fuzzy sets." Therefore, to avoid a misunderstanding, in this book, the Atanassov's intuitionistic fuzzy sets is abbreviated as $A\text{-}IFS$. Generally, the Atanassov's model ($A\text{-}IFS$) may be treated as a classification model subject to a valuation space with three classes and defining certain structure [88].

The concept of $A\text{-}IFS$ is based on the simultaneous consideration of membership μ and non-membership ν of an element of a set in the set itself [3]. It is postulated that $0 \leq \mu + \nu \leq 1$. The similar approach, the so-called vague sets, proposed by Gau and Buehrer in [44] is proved to be equivalent to the $A\text{-}IFS$ (see [19]). Since vague sets were proposed later than $A\text{-}IFS$, in this book, we shall always speak of $A\text{-}IFS$.

To make the basic definition of $A\text{-}IFS$ more clear and transparent, consider an example from [5].

Let E be the set of all countries with elective governments. Assume that we know for every country $x \in E$ the percentage of the electorate that have voted for the corresponding government. Denote it by $M(x)$ and let $\mu(x) = \frac{M(x)}{100}$ (degree of membership, validity, etc.). Let $\nu(x) = 1 - \mu(x)$. This number corresponds to the part of electorate who have not voted for the government. By fuzzy set theory alone we cannot consider this value in more detail. However, if we define $\nu(x)$ (degree of non-membership, non-validity, etc.) as the number of votes given to parties or persons outside the government, then we can show the part of electorate who have not voted at all or who have given bad voting-paper and the corresponding number will be $\pi(x) = 1 - \mu(x) - \nu(x)$ (degree of indeterminacy, uncertainty, etc.). Thus we can construct the set $\{\langle x, \mu(x), \nu(x)\rangle \,|x \in E\}$ and obviously, $0 \leq \mu(x) + \nu(x) \leq 1$.

Obviously, for every ordinary fuzzy set $\pi_A(x) = 0$ for each $x \in E$ and these sets have the form $\{\langle x, \mu_A(x), 1 - \mu_A(x)\rangle \,|x \in E\}$.

In [3], Atanassov defined $A\text{-}IFS$ as follows.

Definition 3.21. Let $X = \{x_1, x_2, \ldots, x_n\}$ be a finite universal set. An intuitionistic fuzzy set A in X is an object having the following form: $A = \{< x_j, \mu_A(x_j), \nu_A(x_j) > |x_j \in X\}$, where the functions $\mu_A : X \to [0,1], x_j \in X \to \mu_A(x_j) \in [0,1]$ and $\nu_A : X \to [0,1], x_j \in X \to \nu_A(x_j) \in [0,1]$ define the degree of membership and degree of non-membership of the element $x_j \in X$ to the set $A \subseteq X$, respectively, and for every $x_j \in X, 0 \leq \mu_A(x_j) + \nu_A(x_j) \leq 1$.

Following to [3], we call $\pi_A(x_j) = 1 - \mu_A(x_j) - \nu_A(x_j)$ the intuitionistic index (or the hesitation degree) of the element x_j in the set A. It is obvious that for every $x_j \in X$ we have $0 \leq \pi_A(x_j) \leq 1$.

As we noted above, $A\text{-}IFS$ is an extension of the standard fuzzy set. All results which are typical for ordinary fuzzy sets theory can be transformed here, too. Also, any research based on fuzzy sets can be described in terms of $A\text{-}IFS$. On the other hand, there have been denned over $A\text{-}IFS$ not only operations similar to the ordinary fuzzy set ones, but also operators that cannot be denned in case of fuzzy sets. $A\text{-}IFS$ have geometrical interpretations (see Fig. 3.8 and Fig. 3.9).

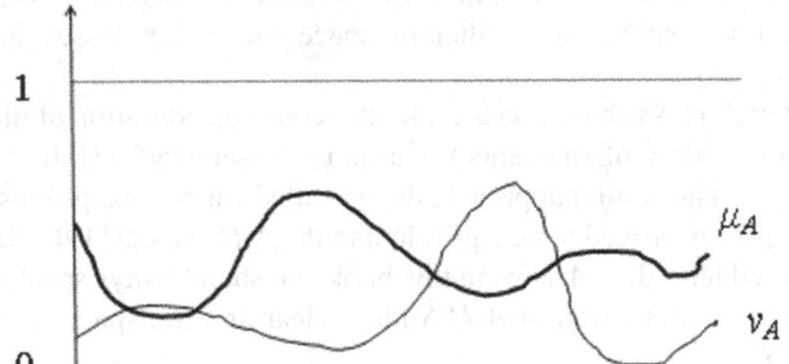

Fig. 3.8 Illustration of μ_A and ν_A

Similarly to the fuzzy set theory, a large number of relations and operations over $A\text{-}IFS$ are denned, but more interesting are the modal operators that can be defined over the $A\text{-}IFS$. They do not have analogues in the ordinary fuzzy set theory.

Let A be an $A\text{-}IFS$ and t $\alpha, \beta \in [0,1]$. The simplest operators are

$$\Omega A = \{\langle x, \mu_A(x), 1 - \mu_A(x)\rangle \,|x \in E\}, \tag{3.49}$$

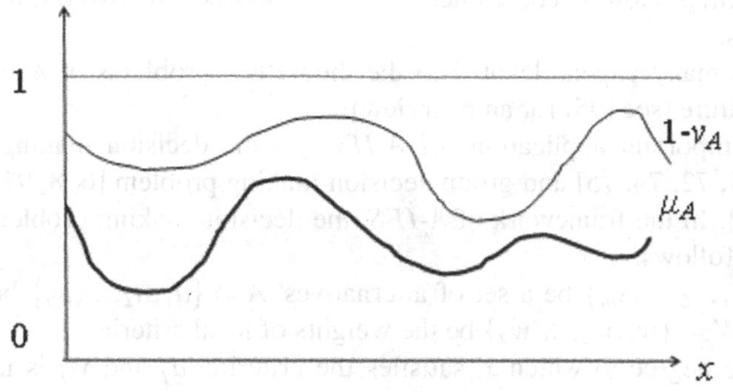

Fig. 3.9 Illustration of μ_A and $1-\nu_A$

$$\Delta A = \{\langle x, 1-\nu_A(x), \nu_A(x)\rangle \,|x \in E\}. \tag{3.50}$$

They are analogous of the modal logic operators "necessity" and "possibility". In the frameworks of the *A-IFS* theory, we can extend these operators, defining the following ones [7]:

$$D_\alpha(A) = \{\langle x, \mu_A(x) + \alpha \cdot \pi_A(x), \nu_A(x) + (1-\alpha)\cdot \pi_A(x)\rangle \,|x \in E\}, \tag{3.51}$$

$$F_{\alpha,\beta}(A) = \{\langle x, \mu_A(x) + \alpha \cdot \pi_A(x), \nu_A(x) + \beta \cdot \pi_A(x)\rangle \,|x \in E\}, \tag{3.52}$$

where $\alpha + \beta \le 1$,

$$G_{\alpha,\beta}(A) = \{\langle x, \alpha \cdot \mu_A(x), \beta \cdot \nu_A(x)\rangle \,|x \in E\}, \tag{3.53}$$

$$H_{\alpha,\beta}(A) = \{\langle x, \alpha \cdot \mu_A(x), \nu_A(x) + \beta \cdot \pi_A(x)\rangle \,|x \in E\}, \tag{3.54}$$

$$H^*_{\alpha,\beta}(A) = \{\langle x, \alpha \cdot \mu_A(x), \nu_A(x) + \beta \cdot (1 - \alpha \cdot \mu_A(x) - \nu_A(x))\rangle \,|x \in E\}, \tag{3.55}$$

$$J_{\alpha,\beta}(A) = \{\langle x, \mu_A(x) + \alpha \cdot \pi_A(x), \beta \cdot \nu_A(x)\rangle \,|x \in E\}, \tag{3.56}$$

$$J^*_{\alpha,\beta}(A) = \{\langle x, \mu_A(x) + \alpha \cdot (1 - \mu_A(x) - \beta \cdot \nu_A(x)), \beta \cdot \nu_A(x)\rangle \,|x \in E\}. \tag{3.57}$$

If we have an ordinary fuzzy set A, then $\Omega A = A = \Delta A$, while for a proper *A-IFS* A: $\Omega A \subset A \subset \Delta A$ and $\Omega A \ne A \ne \Delta A$.

Also the following equalities are valid for each *A-IFS* A: $\Omega \bar{A} = \overline{\Delta A}$, $\Delta \bar{A} = \overline{\Omega A}$.

In modal logic, both operators Ω and Δ are related to the last two connections, but no other connection between them is observed. In the *A-IFS*, we can see that operators D_α and $F_{\alpha,\beta}$ ($\alpha, \beta \in [0,1]$) and $\alpha + \beta \le 1$) are their direct extensions, because $\Omega A = D_0(A) = F_{0,1}(A)$, $\Delta A = D_1(A) = F_{0,1}(A)$.

These equalities show a deeper interconnection between the two ordinary modal logic operators.

There were many papers devoted to the theoretical problems of $A\text{-}IFS$ in the scientific literature (see [95] for an overview).

The most important applications of $A\text{-}IFS$ are the decision making problem [25, 53, 70, 71, 72, 74, 75] and group decision making problem [6, 8, 97, 98, 123, 124, 125, 126]. In the framework of $A\text{-}IFS$, the decision making problem may be formulated as follows.

Let $X = \{x_1, x_2, ..., x_m\}$ be a set of alternatives, $A = \{a_1, a_2, ..., a_n\}$ be a set of local criteria, $W = \{w_1, w_2, ..., w_n\}$ be the weights of local criteria.

If μ_{ij} is the degree to which x_i satisfies the criterion a_j and ν_{ij} is the degree to which x_i does not satisfy the criterion a_j and $0 \leq \mu_{ij} + \nu_{ij} \leq 1$ then alternative x_i may be presented by its characteristics as follows: $x_i = \{(w_1, <\mu_{i1}, \nu_{i1}>), (w_2, <\mu_{i2}, \nu_{i2}>), ..., (w_n, <\mu_{in}, \nu_{in}>)\}, i = 1, ..., m.$

It seems quite natural that if alternative's attributes are intuitionistic fuzzy values ($IFVs$) then the resulting alternative's evaluation should be an IFV too. To avoid this problem, the different real valued score functions based on μ_{ij} and ν_{ij} are usually used. Chen and Tan [25] proposed the score function $S(x_i) = \mu(x_i) - \nu(x_i)$, where $\mu(x_i) = \min(\mu_{i1}, \mu_{i2}, ..., \mu_{in})$, $\nu(x_i) = \max(\nu_{i1}, \nu_{i2}, ..., \nu_{in})$. To take into account the weights of local criteria, Chen and Tan [25] proposed the weighted score function: $WS(x_i) = \sum\limits_{j=1}^{n} w_j(\mu_{ij} - \nu_{ij})$. Hong and Choi [53] in addition introduced the so-called accuracy function $H(x_i) = \mu(x_i) + \nu(x_i)$ and the weighted accuracy function $T(x_i) = \sum\limits_{j=1}^{n} w_j(\mu_{ij} + \nu_{ij})$. Two similar real valued score functions that jointly serve as a degree to which an alternative satisfies the decision-maker's requirements have been introduced in [70, 71]. Liu and Wang [75] proposed score functions based on the intuitionistic fuzzy point operators originating from IF triangular norm and conorm [18, 34]. Xu [141] proposed a method for ranking alternatives based on the Hamming distances $d_H(\alpha, \beta) = \frac{1}{2}(|\mu_\alpha - \mu_\beta| + |\nu_\alpha - \nu_\beta|)$ and normalized Euclidean distances $d_E(\alpha, \beta) = \sqrt{\frac{1}{2}(|\mu_\alpha - \mu_\beta|^2 + |\nu_\alpha - \nu_\beta|^2)}$ introduced by Burillo and Bustince [17]. In [127], a new similarity measure for intuitionistic fuzzy sets is proposed. This measure is based on the normalized Hamming distance $l_{IFS}(A, B) = \frac{1}{2n}\sum\limits_{i=1}^{n} (|\mu_A(x_i) - \mu_B(x_i)| + |\nu_A(x_i) - \nu_B(x_i)| + |\pi_A(x_i) - \pi_B(x_i)|)$.

The measures obtained in [127] were used in [128] to rank alternatives expressed via $A\text{-}IFS$.

Different aggregation operators are usually used to obtain a final real valued evaluation of an alternative on the base of score functions.

Li [72] proposed the most complicated formulation of $A\text{-}IFS$ decision making problem for the case when the weights of local criteria are IFV too. To avoid the above mentioned problem of final alternative's evaluation in the form of IFV, Li [72] proposed its reduction to the linear programming task with real valued parameters. Lin et al. [74] developed a method based on the aggregation of the weighted score and accuracy functions. Using this more simple method the same final results as in [72] were obtained.

It is well known [3] that the couple $< \mu(x), \nu(x) >$ can be mapped bijectively onto regular interval $[\mu(x), 1-\nu(x)]$. This interval can be presented in the equivalent form $[\mu(x), \mu(x)+\pi(x)]$, where $\pi(x) = 1-\mu(x)-\nu(x)$ is the so-called intuitionistic fuzzy index or hesitation degree. In [53], the evaluation of the alternative $x_j \in X$ with respect to the criterion $a_i \in A$ is represented by interval $[\mu_{ij}, 1-\nu_{ij}]$, where μ_{ij} denotes the degree to which x_j satisfies the criterion a_i, ν_{ij} denotes the degree to which x_j does not satisfy a_i. For this purpose, Li [72] proposed to use intervals $[\mu_{ij}^l, \mu_{ij}^u] = [\mu_{ij}, \mu_{ij}+\pi_{ij}]$. These intervals were considered in [72] only as the constraints in the linear programming task. It is important to note that in [53, 72], the semantics of the right bound μ_{ij}^u is not clarified.

It is very important for us that *A-IFS* is not isolated theory. There are different links between *A-IFS* and some other theories modeling uncertainty and imprecision. For example, Deschrijver and Kerre [35] established some interrelations between *A-IFS* and such theories as interval valued fuzzy sets, type 2 fuzzy sets and soft sets. Grzegorzewski and Mrowka [46] analyzed the semantic aspects of such interrelations.

In this book, we show that there exists also a strong link between *A-IFS* and the Dempster-Shafer theory of evidence (*DST*). We use this link to provide a transparent and fruitful semantics for interval $[\mu_{ij}^l, \mu_{ij}^u]$ in terms of *DST*.

Another link between *A-IFS* and *DST* was shown in [51, 52] where the authors developed a theory of mass assignment as a variant of *DST* linked with *A-IFS* including inconsistent and contradictory evidence.

In our opinion, the main problem of *IF* decision making is that, generally, the resulting alternative's evaluation should be presented in the form of *IFV*, whereas usually different real valued score functions are used. Of course, such approaches provide useful, but only approximate results since any intermediate defuzzification in the solution procedure leads inevitably to information losses. Therefore, in the papers [140, 143, 144] proposed a set of *MCDM* models in *A-IFS* setting based on some operations on *IFVs* defined in [4, 29]. In [143], the intuitionistic fuzzy weighted averaging ($IFWA_w$) operator has been constructed as follows (similarly, the intuitionistic fuzzy ordered weighted averaging (*IFOWA*) and intuitionistic fuzzy hybrid averaging (*IFHA*) operators were introduced too).

Let $\alpha_i = < \mu_i, 1-\nu_i >$, i=1 to n, be a set of *IFVs* and $w = (w_i, w_2, ..., w_n)$, be the weight vector. Then $IFWA_w = w_1\alpha_1 \oplus w_2\alpha_2 \oplus ... \oplus w_n\alpha_n$ is defined as

$$IFWA_w = \left[1-\prod_{i=1}^{n}(1-\mu_i)^{w_i}, 1-\prod_{i=1}^{n}\nu_i^{w_i}\right]. \tag{3.58}$$

This operator is not idempotent. For example, if $\alpha_1 = \alpha_2 = < 0.5, 0.7 >$ and $w_1 = w_2 = 0.5$ then $IFWA_w = < 0.5, 0.3 >$. Among basic properties of aggregation operations (boundary conditions, monotonicity, continuity, idempotency, symmetry, associativity and some others) defined in [20], idempotence seems to be especially important in the decision making. In the framework of *MCDM* problems, the idempotency of an aggregation operator means that if all criteria are satisfied in the same degree x, then the global score should also be equal to x. Although this property sometimes is supposed to be a genuine one in *MCDM* [43], we prefer to treat it as a desirable, but not obligatory property, as in practice, it is not always possible to construct an idempotency aggregation operator which reflects well the decision maker's reasoning.

Xu and Yager [144] proposed also (in context of dynamic intuitionistic fuzzy multi-attribute decision making and interval valued *A-IFS*) the other $IFWA_w$ operator based on the modified operations on $IFVs$:

$$IFWA_w = \left[1 - \prod_{i=1}^{n}(1-\mu_i)^{w_i}, \prod_{i=1}^{n}\nu_i^{w_i}\right] \tag{3.59}$$

Unlike the operator (3.58), this operator is idempotent. It has been proved in [143, 144] that introduced $IFWA_w$ operator provides $IFVs$. On the other hand, the problem with these operators (3.58) and (3.59) is that contrary to the initial intentions they do not seem like weighted averaging operators (look at the weighting procedure), but rather as specific forms of a geometric aggregation operator.

Xu and Yager [140] developed some geometric aggregation operators using the modified operations on $IFVs$.

The intuitionistic fuzzy weighted geometric operator $IFWG_w = \alpha_1^{w_1} \otimes \alpha_2^{w_2} \otimes \ldots \otimes \alpha_n^{w_n}$ is defined in [140] as follows:

$$IFWG_w = \left[\prod_{i=1}^{n}\mu_i^{w_i}, \prod_{i=1}^{n}(1-\nu_i)^{w_i}\right]. \tag{3.60}$$

This operator results in IFV, but it is not idempotent: if $\alpha_1 = \alpha_2 =< 0.5, 0.7 >$ and $w_1 = w_2 = 0.5$ then $IFWA_w =< 0.5, 0.3 >$. The merit of this operator is that it remains the structure of usual weighted geometric aggregation operator, whereas the initial weighted averaging structure is violated in (3.58) and (3.59). The common limitation of the aggregation operators proposed in [140, 143, 144], is that the weights w_i, i=1 to n, are supposed to be real values, although, in general, they may be presented by $IFVs$ too.

In Subsection 3.3.4, we present a new approach based on DST, which makes it possible to aggregate local criteria without this limitation.

If the final scores of alternatives are presented by $IFVs$, the problem of comparison of such values arises. Therefore, the specific methods were developed to compare $IFVs$. For this purpose, Chen and Tan [25] proposed to use the score function $S(x_j) = \mu(x_j) - \nu(x_j)$. It is intuitively obvious that if $S(x_k) > S(x_l)$ then x_k should be greater (better) than x_l, but if $S(x_k) = S(x_l)$ this does not always mean that x_k is equal to x_l. Therefore, Hong and Choi [53] in addition to the above score function introduced the so-called accuracy function $H(x_j) = \mu(x_j) + \nu(x_j)$ and showed that the relation between functions S and H is similar to the relation between mean and variance in statistics. Xu [142] used the functions S and H to construct order relations between any pair of intuitionistic fuzzy values as follows:

If $S(x_l) > S(x_k)$, then x_k is smaller than x_l ;
If $S(x_l) = S(x_k)$, then
(1) If $H(x_l) = H(x_k)$, then $x_l = x_k$;
(2) If $H(x_l) < H(x_k)$, then x_l is smaller than x_k.

Basing on these relations, Xu [142] introduced the concepts of intuitionistic preference relation, consistent intuitionistic preference relation, incomplete intuitionistic preference relation and acceptable intuitionistic preference relation.

The method for *IFVs* comparison based on the functions S and H seems to be intuitively obvious and this is its undeniable merit. On the other hand, as two different functions S and H are needed to compare *IFVs*, this method generally does not provide an appropriate technique for the estimation of an extent to which one *IFV* is grater/smaller than the other, whereas such information is usually important for a decision maker. This problem was discussed in [53], where a heuristic method to the aggregation of functions S and H has been developed. In this book, we show that it is possible to get over this difficulty using the *DST* semantics for *A-IFS*. In this *DST/IFS* approach, the problem of *IFVs* comparison reduces to the comparison of belief intervals and is solved using the methods of interval analysis. Hence, there is no need for the methods proposed in [25, 53, 142] in the framework of this approach.

Summarizing, we note that there exist two important problems in *MCDM* in the intuitionistic fuzzy setting: aggregation of local criteria without intermediate defuzzification in the case when criteria and their weights are *IFVs*; comparison of *IF* valued scores of alternatives basing on the degree to which one *IFV* is grater/smaller than the other. In Subsection 3.3.4, we propose an approach to the solution of these problems based on the interpretation of *A-IFS* in the framework of *DST*.

3.2 Interval Arithmetic

This section presents the basics of interval arithmetic, operations on intervals and some important problems concerned with the interval extension of real valued functions and equations.

Interval arithmetic is not a completely new phenomenon in mathematics; it has appeared several times under different names in the course of history. For example, Archimedes calculated lower and upper bounds $223/71 < \pi < 22/7$ in the 3rd century BC. Actual calculation with intervals has neither been as popular as other numerical techniques, nor been completely forgotten. Rules for calculating with intervals and other subsets of the real numbers were published in a 1931 work by Rosalind Cicely Young, a doctoral candidate at the University of Cambridge.

The birth of modern interval arithmetic was marked by the appearance of the book Interval Analysis by Ramon E. Moore in 1966 [89]. He had the idea in Spring 1958, and a year later he published a report [90] on how interval arithmetic could be implemented on a computer. Independently in 1956, Mieczyslaw Warmus suggested formulae for calculations with intervals [136].

Now the modern interval analysis is well presented in the book [91] and some important applications of interval analysis can be found in [57]. During the 1990s, interval analysis has recruited a large community. It now has its own journal Interval Computations, created in 1991 and renamed Reliable Computing in 1995, and several regular international conferences.

The main focus in the interval arithmetic is on the simplest way to calculate upper and lower endpoints for the range of values of a function of one or more variables. These barriers need not be necessarily the supremum or infimum, since the precise calculation of those values are often too difficult; it can be shown that this task is, in general, *NP*-hard.

An interval $[x] = [\underline{x}, \overline{x}]$, where $\underline{x}$ and $\overline{x}$ are its lover and upper bounds respectively, is a subset of a real line $\Re$.

The treatment is typically limited to real intervals, so quantities of form

$$[a,b] = \{x \in \Re | a \leq x \leq b\},$$

where $a = -\infty$ and $b = \infty$ are allowed; with one of them infinite we would have an unbounded interval, while with both infinite we would have the whole real number line.

One of the most undesirable negative features of interval arithmetic is the fast increasing of width of intervals obtained as the results of interval calculations (excess width effect). To reduce this undesirable effect, several different modifications of interval arithmetic were proposed. The most known are: Non- standard interval arithmetic [80] based on the special form of interval subtraction and division, Generalized interval arithmetic [48], Segment interval analysis [105], MV-form [21]. All of these approaches provide good results only in specific conditions. On the other hand, in practice the so-called "naive" form proposed by Moore [90], is proved to be the best one.

Definition 3.22. If $@ \in \{+,-,*,/\}$ and X, Y, Z are intervals, then

$$Z = X@Y = \{z = x@y, x \in X \wedge y \in Y\}. \tag{3.61}$$

The rule (3.61) was first presented in the context of bounded and closed intervals by Moore [90] and then extended to open-ended unbounded intervals [28, 49, 60]. The operations (3.61) can be redefined in the context of closed intervals as operations on their bounds: the bounds of the result of an interval operation are expressed as functions of the bounds of its interval arguments.

Definition 3.23

$$Z = X@Y = \left[\min\left\{\underline{x}@\underline{y}, \underline{x}@\overline{y}, \overline{x}@\underline{y}, \overline{x}@\overline{y}\right\}, \max\left\{\underline{x}@\underline{y}, \underline{x}@\overline{y}, \overline{x}@\underline{y}, \overline{x}@\overline{y}\right\}\right]. \tag{3.62}$$

This definition makes it possible to provide the following operation on intervals:

$$X + Y = \left[\underline{x} + \underline{y}, \overline{x} + \overline{y}\right], \tag{3.63}$$

$$X - Y = \left[\underline{x} - \overline{y}, \overline{x} - \underline{y}\right], \tag{3.64}$$

$$X \cdot Y = \left[\min\left\{\underline{x}\cdot\underline{y}, \underline{x}\cdot\overline{y}, \overline{x}\cdot\underline{y}, \overline{x}\cdot\overline{y}\right\}, \max\left\{\underline{x}\cdot\underline{y}, \underline{x}\cdot\overline{y}, \overline{x}\cdot\underline{y}, \overline{x}\cdot\overline{y}\right\}\right], \tag{3.65}$$

$$X/Y = X \cdot ([1/\underline{y}, 1/\overline{y}], (0 \notin Y). \tag{3.66}$$

if Y is zero containing interval ($0 \in Y$), then

$$\begin{cases} [\overline{x}/\underline{y}, \infty] \text{ if } \overline{x} \le 0 \text{ and } \overline{y} = 0 \\ [-\infty, \overline{x}/\overline{y}] \cup [\overline{x}/\underline{y}, \infty] \text{ if } \overline{x} \le 0 \text{ and } \underline{y} < 0 < \overline{y} \\ [-\infty, \overline{x}/\overline{y}] \text{ if } \overline{x} \le 0 \text{ and } \underline{y} = 0 \\ [-\infty, \infty] \text{ if } \underline{x} < 0 < \overline{x} \\ [-\infty, \underline{x}/\underline{y}] \text{ if } \underline{x} \ge 0 \text{ and } \overline{y} = 0 \\ [-\infty, \underline{x}/\underline{y}] \cup [\underline{x}/\overline{y}, \infty] \text{ if } \underline{x} \ge 0 \text{ and } \underline{y} < 0 < \overline{y} \\ [\underline{x}/\overline{y}, \infty] \text{ if } \underline{x} \ge 0 \text{ and } \underline{y} = 0 \end{cases} \tag{3.67}$$

The properties of the basic operators for intervals generally differ from operations in $\mathfrak{R}$. If A, B and C are intervals then:

$A + (B + C) = (A + B) + C$, $A + B = B + A$, $A \cdot (B \cdot C) = (A \cdot B) \cdot C$, $A \cdot B = B \cdot A$, $A + 0 = 0 + A = A$, $A \cdot 1 = 1 \cdot A = A$.

If at least one of operands of operation is an interval, then the result is an interval too. An exclusion is multiplication by $0 = [0,0]$. If $A + B = 0$, $B \cdot C = 1$, then A, B and C are degenerated intervals (real numbers).

Addition is not operation opposite to subtraction and multiplication is not operation opposite to division. Therefore, $A - A \neq 0$, $A/A \neq 1$, when $\omega(A) > 0$ ($\omega(A) = \overline{a} - \underline{a}$ is the width of interval), but $0 \in A - A$ and $1 \in A/A$.

We can see that addition and multiplication remain associative and commutative, but multiplication is no longer distributive with respect to addition.

Instead, $A(B + C) \subset A \cdot B + A \cdot C$, a property known as subdistributivity takes place, which is a direct consequence of the dependency effect, as A appears only once on the left-hand side, but twice on the right-hand side.

There exists another explanation. If $d \in A \cdot (B + C)$, then $d = a \cdot (b + c)$, where $a \in A$, $b \in B$, $c \in C$. Since $a \cdot b \in A \cdot B$ and $a \cdot c \in A \cdot C$, we obtain $d = a \cdot b + a \cdot c \in A \cdot B + A \cdot C$. It is important to show the situations when distributivity takes place. We shall call interval A zero containing if $\underline{a} < 0 < \overline{a}$.

Let us introduce the function

$$sign(A) = \begin{cases} 1, & A > 0, \\ 0, & 0 \in A, \\ -1, & A < 0. \end{cases} \tag{3.68}$$

It is easy to show that
if $A = [0,0]$ or $B = [0,0]$ or $C = [0,0]$, then $A \cdot (B + C) = A \cdot B + A \cdot C$;
if A is zero containing interval, then $A \cdot (B + C) = A \cdot B + A \cdot C$ when $sign(B) = sign(C)$;
if $\underline{d} \ge 0$ for $D = [\underline{d}, \overline{d}] = B \cdot C$, then $A \cdot (B + C) = A \cdot B + A \cdot C$,
if B and C are symmetrical intervals, then $A \cdot (B + C) = A \cdot B + A \cdot C$.
Interval arithmetic is monotonic with respect to inclusion.

Therefore, if $A \in C$, $B \in D$, then $A+B \in C+D$, $A-B \in C-D$, $A \cdot B \in C \cdot D$, $A/B \in C/D$ (if $0 \notin B$ and $0 \notin D$).

Using the above properties, the following theorem was proved by Moore [89].

Theorem 3.1. *If $F(X_1,X_2,,X_n)$ is a function of interval arguments $X_1,X_2,,X_n$, which is a finite combination of $X_1,X_2,,X_n$ and a finite set of operations on intervals, then inclusions $X_i^{(1)} \subset X_i^{(2)}$, $i = 1,2,...,n$, lead to*

$$F\left(X_1^{(1)},X_2^{(1)},...,X_n^{(1)}\right) \subset \left(X_1^{(2)},X_2^{(2)},...,X_n^{(2)}\right). \tag{3.69}$$

This theorem is usually called the main theorem of interval arithmetic and a function $F(X_1,X_2,,X_n)$ is called natural interval extension of real valued function $f(x_1,x_2,,x_n)$.

Let f is a usual or interval function of real valued arguments $x_1,x_2,,x_n$. An interval extension of function f is then the interval function F of interval arguments $X_1,X_2,,X_n$, such that for real valued arguments $f(x_1,x_2,,x_n) = F(x_1,x_2,,x_n)$.

The following theorem was proved in [89].

Theorem 3.2. *If F is an interval extension of function f, then*

$$f(X_1,X_2,,X_n) \subset F(X_1,X_2,,X_n). \tag{3.70}$$

Thus, if we have to make interval extension of real valued function, all argument of this function should be replaced with corresponding intervals and all operations should be replaced with corresponding operations on intervals. Such approach to the interval extension seems to be justified enough and intuitively clear. Nevertheless, the so-called dependency problem is a major obstacle to the application of extension principle in interval arithmetic. Although interval methods can determine the range of elementary arithmetic operations and functions very accurately, this is not always true with more complicated functions. If an interval occurs several times in a calculation, and each occurrence is taken independently then this can lead to an unwanted expansion of the resulting intervals.

The accuracy of resulting interval strongly depends on the expression of f, as illustrated by the following example [57].

Example 3.3. Consider the following four formal expressions of the same function $f(x)$:

$$f_1(x) = x \cdot (x+1), \tag{3.71}$$

$$f_2(x) = x \cdot x + x, \tag{3.72}$$

$$f_3(x) = x^2 + x, \tag{3.73}$$

$$f_4(x) = (x+\frac{1}{2})^2 - \frac{1}{4}. \tag{3.74}$$

Evaluate their natural extensions for $[x] = [-1, 1]$:

$$[f_1]([x]) = [x] \cdot ([x]+1) = [-2, 2], \tag{3.75}$$

$$[f_2]([x]) = [x] \cdot [x] + [x] = [-2, 2], \tag{3.76}$$

$$[f_3]([x]) = [x]^2 + [x] = [-1, 2], \tag{3.77}$$

$$[f_4]([x]) = ([x] + \frac{1}{2})^2 - \frac{1}{4} = [-\frac{1}{2}, 2]. \tag{3.78}$$

We can see that the accuracy of interval result depends on the formal expression of f. Since $[x]$ occurs only ones in f_4 and f_4 is continuous, f_4 produces the narrower resulting interval. We can see that the difference between results providing by f_1, f_2, f_3 and f_4 is great enough to be taken into account. Obviously, such results do not promote the popularity of interval methods especially in applications. The use of natural interval extensions of functions is not always to be recommended, however. Their efficiency depends strongly on the number of occurrences of each variable, which is often difficult to reduce. An important field of investigation in interval analysis is then the seeking for the other types of extension that would provide less pessimistic results [57, 101].

One of the important problems of interval analysis and fuzzy set theory is the comparison of interval and fuzzy numbers. There are many different methods for interval and fuzzy number comparison proposed in the literature which provide the results of comparison in the form of a real or Boolean value. Nevertheless, there is one important problem of rather methodological nature. It is well known that all arithmetic operations on intervals or fuzzy numbers result in intervals or fuzzy numbers as well. From this point of view, it seems intriguing enough that interval and fuzzy number relations have resulted in real numbers or Boolean values only. So the aspiration to express the results of interval and fuzzy number comparison in terms of intervals and fuzzy numbers looks as quite natural. Of course, this assertion may be considered as disputable. Nevertheless, whatever the case, it is interesting and useful to embody this idea in some mathematical form. In [108], using the Dempster-Shafer theory of evidence (*DST*) with its probabilistic interpretation, a new method which provides the result of comparison in the form of an interval or a fuzzy number is developed. The complete and consistent set of expressions for inequality and equality relations between intervals obtained in the framework of probabilistic approach is presented in [108]. These relations make it possible to compare intervals with real values as well. It is shown that these relations may be considered as an asymptotic limit of the results obtained using *DST*. A natural fuzzy extension of proposed approach is considered and discussed using some illustrative examples.

Since the general approach to interval and fuzzy nubmers comparison proposed in [108] is based on the *DST*, we present it with the review of other methods for interval and fuzzy number comparison in the following section after the presentation of the basics of *DST*.

3.3 Dempster-Shafer Theory of Evidence

In this section, we present the basics of Dempster-Shafer theory of evidence (*DST*), its application to the solution of the problem of interval and fuzzy numbers comparison and the link between *DST* and intuitionistic fuzzy sets with its useful application to the solution of the multiple criteria decision making problems.

3.3.1 Basic Definitions

The origins of *DST* go back to the work by A.P. Dempster [30, 31] who developed a system of upper and lower probabilities. Following this work his student G. Shafer [117] included in his 1976 book "A Mathematical Theory of Evidence" a more thorough explanation of belief functions. In [153], the authors provide a collection of articles by some of the leading researchers in this field. The close connection between *DS* structure and random sets is discussed in [45]. In the following, we provide a brief introduction to basic ideas of *DST*.

Let us start from informal analysis of some problems of the classical Bayesian theory of probability presented in [15]. As it was pointed out in [15], the Bayesian technique is not without its critics, including among others Walley [132], as well as Caselton and Luo [22] who discussed the difficulty arising when conventional Bayesian analysis is presented only with weak information sources. In such cases, we have the "Bayesian dogma of precision", whereby the information concerning uncertain statistical parameters, no matter how vague, must be represented by conventional, exactly specified, probability distributions. Some of the difficulties can be understood through the "principle of insufficient reason", as illustrated by Wilson [139].

Suppose we are given a random device that randomly generates integer numbers between 1 and 6 (its "frame of discernment"), but with unknown chances. What is our belief in "1" being the next number? A Bayesian will use a symmetry argument, or the principle of insufficient reason to say that the Bayesian belief in a "1" being the next number, say $P(1)$ should be $1/6$. In general in a situation of ignorance a Bayesian is forced to use this principle to evenly allocate subjective (additive) probabilities over the frame of discernment. To further understand the Bayesian approach, especially with the regard to representation of ignorance, consider the following example adopted from [15], similar to that in [139].

Let a be a proposition that "I live in Kings Road, Cardi". How could one construct $P(a)$, a Bayesian belief in a? Firstly we must choose a frame of discernment, denoted by Θ and a subset A of Θ representing the proposition a; then would need to use the principle of insufficient reason to arrive at a Bayesian belief. The problem is there are a number of possible frames of discernment Θ that we could choose, depending effectively on how many Cardi roads can be enumerated. If only two such streets are identifiable, then $\Theta=\{x_1,x_2\}$, $A=x_1$. The principle of insufficient reason then gives $P(A)$, to be 0.5, through evenly allocating subjective probabilities over the frame of discernment. If it is estimated that there are about 1000 roads in

Cardi, then Θ=$\{x_1, x_2, ..., x_{1000}\}$ with again $A = \{x_1\}$ and the other x_i's representing the other roads. In this case the theory of insufficient reason gives $P(A) = 0.001$.

Either of these frames may be reasonable, but the probability assigned to A is crucially dependent upon the frame chosen. Hence one's Bayesian belief is a function not only of the information given and one's background knowledge, but also of a sometimes arbitrary choice of frame of discernment. To put the point another way, we need to distinguish between uncertainty and ignorance. Similar arguments hold where we are discussing not probabilities per se but weights which measure subjective assessments of relative importance.

In summary, *DST* is a numerical method for evidential reasoning (a term often used to denote the body of techniques) specifically designed for manipulation of reasoning from evidence, based upon the *DST* of belief functions (see [78]).

Following on from the above example concerning Cardi roads, one of the primary features of the *DST* model is that we are relieved of the need to force our probability or belief measures to sum to unity. There is no requirement that belief not committed to a given proposition should be committed to its negation. As can be seen in the further analysis below, this allows us to construct and analyze our "frame of discernment" in a more flexible way. The total allocation of belief can vary to suit the extent of our knowledge.

The second basic idea of *DST* is that numerical measures of uncertainty may be assigned to overlapping sets and subsets of hypotheses, events or propositions as well as to individual hypothesis.

To illustrate, consider the following expression of knowledge concerning murderer identification adapted from [96]. Mr. Jones has been murdered, and we know that the murderer was one of three notorious assassins, Peter, Paul and Mary, so we have a set of hypotheses, i.e., frame of discernment, Θ=Peter, Paul, Mary. The only evidence we have is that the person who saw the killer leaving is 80 percent sure that it was a man, i.e., $P(man) = 0.8$. The measures of uncertainty, taken collectively are known in *DST* terminology as a "basic probability assignment" (*bpa*). Hence we have a *bpa*, say m_1 of 0.8 given to the focal element Peter, Paul, i.e., m_1({Peter, Paul })=0.8, since we know nothing about the remaining probability it is allocated to the whole of the frame of the discernment, i.e., m_1({Peter,Paul, Mary })=0.2.

The key point to note is that assignments to "singleton" sets may operate at the same time as assignments to sets made up of a number of propositions. Such a situation is simply not permitted in a conventional Bayesian framework, although it is possible to have a Bayesian assignment of prior probabilities for groups of propositions (since conventional probability theory can cope with joint probabilities).

As pointed out by Schubert [119], *DST* is in this sense a generalization of the Bayesian theory. It avoids the problem of having to assign non-available prior probabilities and makes no assumptions about non-available probabilities.

The next question to consider is what can then be done with the *bpa*'s. The answer in the first instance is twofold. We may firstly collect *bpa*'s together in such a way as to express our overall belief. This is examined below. More importantly, we may then combine *bpa*'s from different sources. Continuing our example, if we gained further evidence that it was reported with confidence 0.6 that Peter was leaving on a jet plane when the murder occurred, in this case we have *bpa*, say m_2 ({Paul, Mary })=0.6. Since we know nothing about the remaining probability it is allocated to the whole of the frame of the discernment, i.e., m_2({Peter, Paul,Mary })=0.4.

Combination of conventional probabilities is achieved through the familiar multiplication rule. In the more generalized *DST* approach a more complex multiplication rule is required, which combines two pieces of evidence. We can see how this works in the context of the example.

Put simply, the result of combining two assignments is that for any intersecting sets A and B, where A has mass M_1 from assignment m_1 (i.e., $m_1(\{A\}) = M_1$) and B has mass M_2 from assignment m_2, the belief accruing to their intersection is the product of M_1 and M_2. For example, m_3({Paul,Mary })=m_1({Peter, Paul, Mary })$\cdot m_2$({Paul, Mary })= $0.2 \cdot 0.6$= 0.12.

The new piece of evidence has a more spread-out allocation of probabilities to varying subsets of the frame of discernment. We can bring together this evidence to find some level of belief: the belief in any set is the sum of all the probabilities of all the subsets of that set. Hence, for example: *Bel*({Peter, Paul })=m_3({Peter })+m_3({Paul })+ m_3({Paul, Peter })= $0 + 0.48 + 0.32$=0.8.

The above loose definitions and reasoning may help to understand better the following formal definitions.

Assume V is a variable whose domain is the set X. It is important to note that variable V may be treated also as a question or proposition and X as a set of propositions or mutually exclusive hypotheses or answers [130, 150].

A *DS* belief structure has associated with it a mapping m, called basic assignment function (*bpa*), from subsets of X into the unit interval, $m : 2^X \to [0,1]$ such that $m(\emptyset) = 0, \sum\limits_{A \subset X} m(A) = 1$.

The subsets of X for which the mapping does not assume a zero value are called the focal elements. We shall denote these as A_i, for $i = 1$ to n. We note that the null set is never a focal element. In [117], Shafer introduced a number of measures associated with this structure.

The measure of belief is a mapping $Bel : 2^X \to [0,1]$ such that for any subset B of X

$$Bel(B) = \sum_{i=1}^{n} m(A_i), \;\; A_i \subseteq B, \; i = 1 \; to \; n. \tag{3.79}$$

With V a variable taking its value in the set X under the semantics provided by Shafer [117], *Bel(B)* is our degree of belief that the value of V lies in the set B. It can be easily seen that *Bel* is a kind of fuzzy measure [150]:

$$Bel(\emptyset) = 0, Bel(X) = 1, \text{if } B_1 \subset B_2 \text{ then } Bel(B_2) \geq Bel(B_1).$$

In [117], it is shown that m can be uniquely recovered from *Bel*.

A second measure introduced by Shafer [117] is the measure of plausibility. The measure of plausibility associated with m is a mapping $Pl : 2^X \to [0,1]$ such that for any subset B of X

$$Pl(B) = \sum_{i=1}^{n} m(A_i), \; A_i \cap B \neq \emptyset, \; i = 1 \; to \; n. \tag{3.80}$$

The semantics associated with this measure is that *Pl(B)* is the degree of plausibility that the value of *V* lies in the set *B*. It can be easily seen that plausibility is also a fuzzy measure
$Pl(\emptyset) = 1, Pl(X) = 1$, if $B_1 \subset B_2$ then $Pl(B2) \geq Pl(B1)$.

It is easy to see that $Bel(B) \leq Pl(B)$.
Shafer denoted the doubt that *V* was contained in the set *B* t as $Dou(B)$.
It can be shown that

$$Pl(B) = 1 - Dou(B)$$

and hence

$$Pl(B) = 1 - Bel(\overline{B})$$

and equivalently

$$Bel(B) = 1 - Pl(\overline{B}).$$

DS provides an explicit measure of ignorance about an event *B* and its complementary $\overline{B}$ as the length of the interval $[Bel(B), Pl(B)]$ called belief interval (*BI*). It can also be interpreted as imprecision on the "true probability" of *B*.

A couple of special *DS* belief structures are worth pointing out. One special case is what Shafer [117] calls a Bayesian belief structure.

A belief structure is called a Bayesian belief structure if the focal elements are singletons, that is each A_i consists of exactly one element. In this case it can be shown that for every *B*, *Pl(B)=Bel(B)* , the plausibility and belief of a subset are the same. Furthermore, it can be shown that when $A \cap B = \emptyset$ we get:

$$Bel(A \bigcup B) = Bel(A) + Bel(B).$$

Essentially this is a probability distribution where

$A_i = \{x_i\}$ and $m(A_i) = Prob(x_i)$.

In this case, the measures *Bel* and *Pl* are probability measures.

Another special case of *DS* belief structure is what Shafer calls consonant belief structures. A belief structure is called a consonant belief structure if the focal elements can be indexed such that $A_1 \subset A_2 \subset ... \subset A_n$.

It was shown [117] that if our structure is consonant then for all *A* and *B*

$$Bel(A \cap B) = \min[Bel(A), Bel(B)],$$

$$Pl(A \cup B) = \max[Pl(A), Pl(B)].$$

In this case $Pl(B)$ is a possibility measure [150]. In this consonant case there exists some function $\Pi : X \to [0,1]$, called a possibility distribution such that if $\Pi(x) = Pl(x)$ then:

$Pl(A) = \max[\Pi(x)]$ over all $x \in A$.

Thus, *DST* may be treated as the generalization of the probability theory as well as the possibility theory.

To illustrate DST, a simple example is adopted from [13].
A certain politician has been accused of committing an illegal act. The politician either knew that he was committing an illegal act or he was naive in his judgment. At an inquiry into the case two witnesses gave testimonies and estimated the following basic probability assignments (*bpa*) over the power set of the politician's act, as given in Table 3.2.

Table 3.2 Breakdown of evidence for politician

	0	Politician knew (K)	Politician naive (N)	Either K,N
Witness X	0	0.3	0.1	0.6
Witness Y	0	0.4	0.0	0.6

Hence in our notation, the *bpa* for witness *X* is as follows:

$$m_X(\{K\}) = 0.3, \;\; m_X(\{N\}) = 0.1, \;\; m_X(\{K,N\}) = 0.6.$$

A similar *bpa* can be constructed for witness *Y*.
It follows, for witness *X*:

$$Bel(\{K\}) = m_X(\{K\}) = 0.3, \;\; Pl(\{K\}) = m_X(\{K\}) + m_X(\{K,N\}) =$$
$$= 0.3 + 0.6 = 0.9.$$

3.3.2 Combination of Evidence in the Dempster-Shafer Theory

The core of the evidence theory is the Dempster's rule of combination of evidence from different sources [117]. The rule assumes that information sources are independent and uses the so-called orthogonal sum to combine multiple belief structures: $m = m_1 \oplus m_2 \oplus ... \oplus m_k$, where $\oplus$ represents the operator of combination. With two belief structures m_1, m_2, the Dempster's rule of combination is defined as

$$m_{12}(A) = \frac{\sum_{B \cap C = A} m_1(B) m_2(C)}{1-K}, \; A \neq \emptyset, \; m_{12}(\emptyset) = 0, \qquad (3.81)$$

where $K = \sum_{B \cap C = \emptyset} m_1(B)m_2(C)$. The denominator $1-K$ is called the normalization factor, K is called the degree of conflict which measures the conflict between the pieces of evidence and the process of dividing by $1-K$ is called normalization.

For example, using Dempster's rule of combination on evidence from witnesses X and Y (see Table 3.2) gives the following new *bpa*:

$$m_{XY}\{K\}) = 0.5625,\ m_{XY}(\{N\}) = 0.0625,\ m_{XY}(\{K,N\}) = 0.3750.$$

A considerable body of literature has been devoted to the discussion of the appropriateness of this fusion operation.

In [157], Zadeh has underlined that this normalization involves counter-intuitive behaviors in the case of considerable conflict.

Zadeh provides a compelling example of erroneous results.

Suppose that a patient is seen by two physicians regarding the patients neurological symptoms. The first doctor believes that the patient has either meningitis with a probability of 0.99 or a brain tumor, with a probability of 0.01. The second physician believes the patient actually suffers from a concussion with a probability of 0.99 but admits the possibility of a brain tumor with a probability of 0.01. Using the values to calculate the $m\{brain\ \ tumor\}$ with Dempster's rule, we find that $m\{brain\ \ tumor\} = Bel\{brain\ \ tumor\} = 1$. Clearly, this rule of combination yields a result that implies complete support for a diagnosis that both physicians considered to be very unlikely.

Therefore, a number of other approaches to the combination of evidence were proposed in the literature. Smets [120] proposed the version of Dempster's rule introduced in the transferable belief model usually referred to as the *TBM* conjunctive rule. The Smet's rule of combination is nothing but the non-normalized version of the conjunctive consensus (equivalent to the non-normalized version of Dempster's rule). It is commutative and associative and allows positive mass on the null/empty set $\emptyset$ (the so-called open-world assumption). Smet's rule of combination of two independent (equally reliable) sources of evidence (denoted here by index S) is given by:

$$m_S(\emptyset) = \sum_{X_1 \cap X_2 = \emptyset} m_1(X_1)m_2(X_2) \tag{3.82}$$

and for $(\forall X \neq \emptyset) \in 2^{\Theta}$, where Θ is the so-called frame of discernment consisting in a finite set of exclusive and exhaustive hypotheses, by

$$m_S(X) = \sum_{X_1 \cap X_2 = X} m_1(X_1)m_2(X_2). \tag{3.83}$$

The main limitations of the Dempster's rule and the *TBM* conjunctive rule seem to be their lack of robustness with respect to conflicting evidence (a criticism which mainly applies to the Dempster's rule), and the requirement that the items of evidence combined be distinct [33].

The Yager's rule of combination [148] admits that in the case of conflict the result is not reliable, so that $K = \sum_{B \cap C = \emptyset} m_1(B)m_2(C)$ plays the role of an absolute discounting term added to the weight of ignorance. This commutative, but not associative rule, denoted here by index Y is given by $m_Y(\emptyset) = 0$ and $\forall X \in 2^{\Theta}, X \neq \Theta$ by

$$m_Y(X) = \sum_{X_1 \cap X_2 = X} m_1(X_1)m_2(X_2) \tag{3.84}$$

and when $X = \Theta$ by

$$m_Y(\Theta) = m_1(\Theta)m_2(\Theta) + \sum_{X_1 \cap X_2 = \emptyset} m_1(X_1)m_2(X_2) \tag{3.85}$$

The Dubois and Prade's rule of combination [40] admits that the two sources are reliable when they are not in conflict, but one of them is right when a conflict occurs. Then if one observes a value in set X_1 while the other observes this value in a set X_2, the truth lies in $X_1 \cap X_2$ as long $X_1 \cap X_2 \neq \emptyset$. If $X_1 \cap X_2 = \emptyset$ then the truth lies in $X_1 \cup X_2$. According to this principle, the commutative (but not associative) Dubois and Prade's hybrid rule of combination, denoted here by index DP, which is a reasonable trade-off between precision and reliability, is defined by $m_{DP}(\emptyset) = 0$ and $(\forall X \neq \emptyset) \in 2^{\Theta}, X \neq 0$ by

$$m_{DP}(X) = \sum_{\substack{X_1 \cap X_2 = X \\ X_1 \cap X_2 \neq \emptyset}} m_1(X_1)m_2(X_2) + \sum_{\substack{X_1 \cup X_2 = X \\ X_1 \cap X_2 = \emptyset}} m_1(X_1)m_2(X_2). \tag{3.86}$$

Recently Murphy [92] have proposed other combination rules. It is shown in [107, 121] that, in general, they (and some other too) are not associative. On the other hand, the problems of conflict management with Dempster's rule (and, to a lesser extent, with the *TBM* conjunctive rule) are often due to incorrect or incomplete modelisation of the problem at hand, and these rules often yield reasonable results when they are properly applied [47].

Since we shall use the Demster's rule for the aggregation of local criteria in decision making problems, it is important to note that the Dempster's rule and the *TBM* conjunctive rule are commutative and associative, but not idempotency operators.

Nevertheless, in spite of the lack of idempotency, the Dempster's rule is successfully used in different real-world applications. Beynon et al. [15] proposed to use the *DST* as an alternative approach to multiple criteria decision modeling with the use of the Dempster's rule for aggregation of local criteria. This approach has been developed and studied in the framework of analytic hierarchy process in [13, 14, 16], where its advantages are exposed. Hua et al. [55] extended this approach to the case of incomplete information. It is important to note that in this approach the weights of local criteria are supposed to be only real values. In Subsection 3.3.4, we present a new approach without this limitation.

In his recent paper [33], Denoeux introduced two new commutative, associative and idempotent combination operators for belief functions. Contrary to the *TBM* conjunctive and disjunctive rules, these operators (cautious conjunctive rule and bold disjunctive rule) do not require the assumption of independence or distinctness of the information sources from which basic assignment functions are derived. These operators can be useful in *MCDM* in the case of dependent (correlated) local criteria or/and their weights. The problem of dependent local criteria is not widely discussed in the literature. The probable cause of this is well described by Denoeux [33]: "However, it is often the case that, although two items of evidence (such as, e.g., opinions expressed by two experts sharing some experiences, or observations of correlated random quantities) can clearly not be regarded as distinct, the interaction between them is ill known and, in many cases, almost impossible to describe."

Finally, Denoeux wrote "As expected, the *TBM* conjunctive rule achieves higher performance in the case of independent features. However, it is outperformed by the cautious rule when features are no longer independent" [33]. Thus, the use of the combination rules proposed by Denoeux is justified in the case when interdependence between criteria is evident and important in context of considered decision problem.

Nevertheless, taking into account that this is not usually the case, in this book, we shall use only the classical Dempster's rule of combination (3.81) as it is more popular in applications than *TBM* conjunctive rule and other rules.

3.3.3 The Methods for Interval and Fuzzy Numbers Comparison Based on the Probabilistic Approach and Dempster-Shafer Theory

The problem of interval and fuzzy number comparison is of perennial interest, because of its direct relevance in modeling and optimization of the real-world processes. In this subsection, we present an approach developed in [108] for interval and fuzzy number comparison based on the probabilistic approach and *DST*.

The merit of this approach is that it makes it possible to obtain the result of interval comparison in form of belief interval and to get the complete and consistent set of expressions for inequality and equality relations between intervals and fuzzy numbers. Since this approach and based on it methods were successfully used for the solution of the decision making and optimization problems presented in the following chapters, here we describe this approach with all details including the critical review of other methods for interval and fuzzy number comparison presented in [108].

Nowadays, many scientists make a distinction between fuzzy intervals and fuzzy numbers depending on the multiplicity or uniqueness of modal values [39], i.e., the real values at which a membership degree is equal to 1. Therefore, a trapezoidal fuzzy quantity (multiplicity of modal values) is treated as a fuzzy interval, whereas triangular fuzzy quantity (singular modal value) is considered as a fuzzy number. We shall use the term "fuzzy number" in its most general sense in this book. A fuzzy number may be viewed as an elastic constraint acting on a certain variable which is only known to lay "around" a certain value. This generalizes both the concepts of a real number and a closed interval [32].

Theoretically, intervals and fuzzy numbers can only be partially ordered and hence cannot be compared in usual sense. However, if interval or fuzzy numbers are used in practical applications, the comparison of fuzzy numbers becomes necessary. There exist numerous definitions of ordering relations over fuzzy numbers (as well as intervals) in the literature [9, 10, 23, 39, 41, 50, 56, 66, 89, 102, 147, 149, 151].

Usually, the authors use some quantitative indices. The values of such indices present a degree to which one interval or fuzzy number is greater/less than another one. Existing approaches to interval and fuzzy number comparison may be clustered into three groups: the methods for only qualitative ordering [9, 23, 56, 66, 89], the methods permitting quantitative ordering with the use of some indices obtained from the basic definitions of fuzzy sets theory [1, 39, 50] and the methods based on the representation of fuzzy numbers using α-cuts [102, 147, 149, 151].

The widest review of fuzzy numbers comparison problem based on more than 35 literature indices has been presented in [134], where the authors proposed a new interesting classification of the methods for fuzzy numbers comparison. Nevertheless, this problem is still open. In [135], it is noted that the most of proposed interval comparison methods are "totally based on the midpoints of interval numbers". The authors of [135] write "Our experience told us that the use of midpoints to compare or rank interval numbers was sometimes inconvincible and not easy to be accepted". Therefore, the authors developed a simple heuristic method which makes no use of the midpoints of intervals. In [135], the degree of preference of interval $A = [a_1, a_2]$ over $B = [b_1, b_2]$ (or $A > B$) is defined as

$$P(A > B) = \frac{\max(0, a_2 - b_1) - \max(0, a_1 - b_2)}{(a_2 - a_1) + (b_2 - b_1)}.$$

The degree of preference of B over A is defined in the same way:

$$P(B > A) = \frac{\max(0, b_2 - a_1) - \max(0, b_1 - a_2)}{(a_2 - a_1) + (b_2 - b_1)}.$$

It is obvious that $P(A > B) + P(B > A) = 1$ and $P(A > B) = P(B > A) \equiv 0.5$ when $A = B$, i.e., $a_1 = b_1, a_2 = b_2$. The main limitation of this approach is the lack of interval equality relation. Furthermore, the midpoint of interval numbers still plays a key role in this approach since for all nested intervals with the same midpoints we have $P(A > B) = P(B > A) \equiv 0.5$. Therefore, if a midpoint of A is greater than that of B then $P(A > B) > P(B > A)$.

In [62], a possibilistic approach to fuzzy numbers comparison originally proposed in [39] is developed in context of sequencing problems.The authors of [54, 69] had shown the need for separate inequality ($<, >$) and equality ($=$) interval and fuzzy number relations. In [118], a method based on the comparison of means and variances of intervals and fuzzy numbers is developed in the framework of financial profitability analysis. It is interesting that using such approach we implicitly deal with two sometimes convincing local criteria: the interval mean and variance.

This problem is exposed explicitly in [108]. In this paper, the author presents the generalization of such methods. The proposed approach is based on the α-cut representation of fuzzy numbers. The author uses the probability approach and *DST* to get a numerical evaluation of fact that a certain interval is greater than or equal to another interval.

The idea to use the probability interpretation of intervals is not novel. Nevertheless, now only a few works based on it can be cited [24, 65, 67, 68, 93, 94, 109, 110, 111, 112, 113, 114, 115, 116, 131, 152]. The main advantage of the probabilistic approach is the ability to infer a complete set of interval and fuzzy number relations as well as their comparison with real numbers using only one major assumption: an interval is a support of uniformly distributed random value. The probabilistic approach to interval and fuzzy number comparison has proved to be a powerful practical tool for modeling and optimization in the interval and fuzzy setting [111, 113, 114].

Nevertheless, there are some methodological problems within this approach and as a consequence the different expressions for probabilities estimation were obtained in [24, 65, 67, 68, 93, 94, 109, 110, 111, 112, 113, 114, 115, 116, 131, 152].

In [131], the set of expressions for the probabilities $P(A < B)$ and $P(A > B)$ has been obtained. These were the same expressions as those obtained later in [109, 110, 111, 112, 113, 114, 115, 116] using other assumptions. On the other hand, the probability $P(A = B)$ in [131] has been presented (in our notation) as $P(A = B) = \varepsilon^2/(W(A)W(B))$, where $W(A)$, $W(B)$ are the lengths of compared intervals and ε is an arbitrary small number.

The authors of [131] assumed that $\varepsilon \approx 0$. In other words, in [131] it is implicitly assumed that in any case $P(A = B) = 0$.

Nevertheless, in practice there may be situations when intervals , e.g., such as $A = [0, 1000.1]$ and $B = [0, 1000.2]$, from common sense, should be considered rather as the equal ones. Obviously, in such situations even intuitively, we feel that $P(A = B) > P(A < B)$ and requirement $P(A = B) = 0$ for all cases seems to be too much restrictive one.

There are no comparisons of a real number with interval, nor the fuzzy interval relations in [131]. The similar problems may be found in [65, 67, 68]. In [152], the expression for $P(A \leq B)$ for overlapping case (see Fig. 3.10) has been obtained in the form which is equivalent to the expression for $P(A < B)$ proposed in [109, 110, 111]. It has been shown in [109, 110, 111] that different expressions for calculation of $P(A < B)$ and $P(A = B)$ should be used in both overlapping and inclusion cases. For the inclusion case (see Fig. 3.10), the authors of [152] proposed (in our notation $A = [a_1, a_2], B = [b_1, b_2]$) $P(A \leq B) = (2b_2 - a_2 - a_1)/(b_2 - b_1)/2$. It is easy to see that in the asymptotic case when $a_1 \to b_1$, $a_2 \to b_2$, i.e., $A \to B$, we get $P(A \leq B) \to 0.5$. The same asymptotic result $P(A \leq B) \to 0.5$ if $A \to B$ was obtained in [94]. These results can be qualified as the discussable ones. So some additional comments are needed [108].

According to the classical interpretation [89], any interval A is completely defined by its bounds ($A = [a_1, a_2]$). Obviously, if we consider an interval in a probabilistic sense, i.e., as an interval of uniformly distributed random value, it is completely defined by its bounds too. In other words, we can treat an interval A as the mathematical object defined by pair $[a_1,a_2]$. Therefore, if we meet two such objects A and B with equal bounds ($a_1=b_1,a_2=b_2$) we can say that they are equal objects. Let us introduce a measure $m(A,B)\in[0,1]$ of such objects equality/inequality. The natural properties of $m(A,B)$ should be such that $m(A=B)=1$, $m(A>B)=m(A<B)=0$ for the equal A and B and $m(A=B)=0$ for the completely different A and B when their intersection is empty. That is why, if the probability P is treated as a measure or degree of equality/inequality of the intervals then the only reasonable result in the case of $A\rightarrow B$ should be $P(A<B)\rightarrow 0$, $P(A>B)\rightarrow 0$, $P(A=B)\rightarrow 1$.

Nevertheless, the discussed results ($P(A=B)\rightarrow 0.5$ when $a_1\rightarrow b_1,a_2\rightarrow b_2$) obtained in [94] and some other papers make a sense if we look at the problem from another point of view. In [94], the authors as the conceptual model of the proposed approach used the next clear example: "... Control rule bases often include rule like "if temperature a is higher than the temperature b, then open valve 1, else open valve 2. In practice, after measurements, we only have intervals A and B of possible values of a and b. If the corresponding two intervals intersect, then none of the temperatures is guaranteed to be higher than another. A natural idea is therefore to choose an interval for which the probability that $a\geq b$ is greater than the probability that $a\leq b$". As it is proved in [94], such a reasoning leads to the expression (in our notation):

$$P(A\geq B)=\frac{a_2-b_1}{a_2-b_1-a_1+b_2}\ \text{if}\ a_2\geq b_1\ \text{and}\ 0\ \text{else}.$$

Although this expression leads to $P(A=B)\rightarrow 0.5$ when $a_1\rightarrow b_1,a_2\rightarrow b_2$, the result is completely justified in context of considered example. It is easy to see that only possible real valued temperatures ranging in corresponding intervals are compared, not intervals. It is important that there are no comparisons between real number and interval in [94]. It is shown in [94] that there are many real-life situations in practice when such reasoning is valid. We think that observed variety of proposed methods reflects the next fact: the interval and fuzzy number comparison is a context dependent problem.

Therefore, to avoid a possible misunderstanding, it is emphasized in [108] that the main author's aspiration is to develop a method for comparison of intervals treated as the mathematical objects defined completely by their bounds.

It is worth noting that such approach to the interval comparison is not a purely mathematical conception, it is originated from real-world problems of simulation and optimization in the interval and fuzzy setting [111, 112, 113, 114].

It is asserted in [108] that described differences observed in probabilistic methods for interval comparison are caused by the limited ability of a purely probability approach to deal with such objects as intervals or fuzzy numbers. The problem is that the probability theory allows us to represent only uncertainty, whereas interval and fuzzy objects in addition are inherently characterized by imprecision and ambiguity.

Consider another important problem of rather methodological nature. It is well known that all arithmetic operations on intervals or fuzzy numbers result in intervals or fuzzy numbers as well. From this point of view, it seems intriguing enough that interval and fuzzy number relations have resulted in real numbers or Boolean values only. So the aspiration to express the results of interval and fuzzy number comparison in terms of intervals and fuzzy numbers looks as quite natural. Of course, this assertion may be considered as disputable. Nevertheless, whatever the case, it is interesting and useful to embody this idea in some mathematical form.

3.3.3.1 Interval Comparison Based on the Probabilistic Approach

Since the method proposed in [108] is based on the α-cut representation of fuzzy numbers, the main problem is to compare intervals. There are only two nontrivial cases of interval locations deserve to be considered which were called in [108] overlapping and inclusion cases (see Fig. 3.10).

Let $A = [a_1, a_2]$ and $B = [b_1, b_2]$ be independent intervals and $a \in [a_1, a_2]$, $b \in [b_1, b_2]$ be random variables distributed on these intervals. As we are dealing with non-fuzzy intervals, it is natural to assume that values of random variables a and b are uniformly distributed. This assumption is in accordance with the Bayesian "principle of insufficient reason" [11]: if there is no reliable information about probabilities of events, they are treated as equally probable ones. Moreover, when dealing with intervals representing α-cuts of fuzzy number, the uniform distribution is the only reasonable choice.

There are some subintervals which play an important role in further analysis. For example, in the overlapping case (see Fig. 3.10), falling of random variables $a \in [a_1, a_2]$, $b \in [b_1, b_2]$ into subintervals $[a_1, b_1]$, $[b_1, a_2]$, $[a_2, b_2]$ may be treated as a set of independent random events.

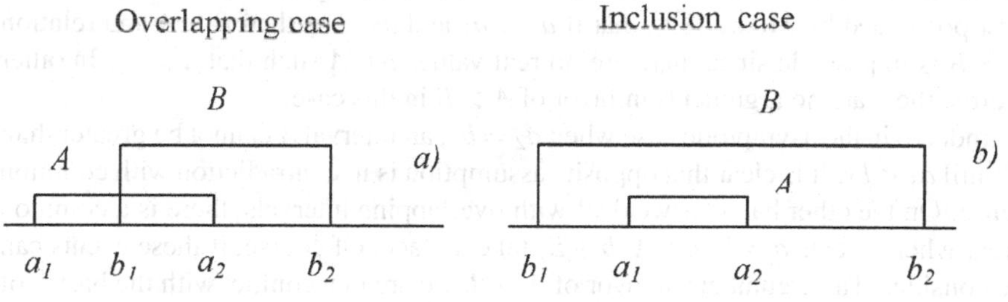

Fig. 3.10 Examples of interval relations

Let us define the events $H_k : a \in A_i, b \in B_j$, for $k = 1$ to n, where A_i and B_j are the subintervals formed by the boundaries of compared intervals A and B such that $A = \bigcup_i A_i$, $B = \bigcup_j B_j$. It is easy to see that the events H_k form a complete group of events which represents all the cases of random values a and b falling into various subintervals A_i and B_j, respectively. Let $P(H_k)$ be the probability of event H_k, and

$P(B > A/H_k)$ be a conditional probability of $B > A$ given H_k. Hence, the composite probability may be expressed as follows:

$$P(B > A) = \sum_{k=1}^{n} P(H_k)P(B > A/H_k). \tag{3.87}$$

Let us consider the case of overlapping intervals.

Since we are dealing with the uniform distributions of random variables a and b in given subintervals, the probabilities $P(H_k)$ can be easily obtained geometrically.

In the overlapping case (see Fig. 3.10), there is a set of four events:

$$\begin{aligned} &H_1 : a \in [a_1, b_1] \wedge b \in [b_1, a_2], H_2 : a \in [a_1, b_1] \wedge b \in [a_2, b_2], \\ &H_3 : a \in [b_1, a_2] \wedge b \in [b_1, a_2], H_4 : a \in [b_1, a_2] \wedge b \in [a_2, b_2]. \end{aligned} \tag{3.88}$$

Since the events $a \in [a_1, b_1], b \in [b_1, a_2], \ldots$ are independent, we obtain the following probabilities:

$$\begin{aligned} P(H_1) &= \frac{b_1 - a_1}{a_2 - a_1} \frac{a_2 - b_1}{b_2 - b_1}, P(H_2) = \frac{b_1 - a_1}{a_2 - a_1} \frac{b_2 - a_2}{b_2 - b_1}, \\ P(H_3) &= \frac{a_2 - b_1}{a_2 - a_1} \frac{a_2 - b_1}{b_2 - b_1}, P(H_4) = \frac{a_2 - b_1}{a_2 - a_1} \frac{b_2 - a_2}{b_2 - b_1}. \end{aligned} \tag{3.89}$$

Obviously, the events H_1 , H_2 and H_4 may be considered as the evidences of the event $B > A$ only, but the event H_3 can be treated from two different points of view since it is simultaneously an evidence of the events $a \in [b_1, a_2]$ and $b \in [b_1, a_2]$. So some comments are needed.

An important point is that in the framework of the conventional interval analysis, the relation $A > B$ for overlapping intervals shown in Fig. 3.10 is senseless. It was first postulated by Moore [89], that if $a_1 < b_1$ and $a_2 < b_2$ then $B > A$ and relation $A > B$ is impossible since there are no real values $a \in A$ such that $a > b_2$. In other words, there are no arguments in favor of $A > B$ in this case.

Indeed, in the asymptotic case when $a_2 = b_2$, an interval A cannot be greater than B until $a_1 < b_1$. It is clear that opposite assumption is in contradiction with common sense. On the other hand, as we deal with overlapping intervals, there is a common area, where events $a > b$ ($a \in A$, $b \in B$) take a place. Of course, if these events can be considered as arguments in favor of $A > B$, we are in a conflict with the basics of interval analysis and common sense.

The source of this contradiction is an assumption (not obvious) that relations between particular $a \in [b_1, a_2]$ and $b \in [b_1, a_2]$ may be used to compare intervals as a whole. Since this assumption leads to the absurd conclusion (we can not ignore common sense) it is wrong in context of interval analysis. Thus, the relation $A > B$ for overlapping intervals shown in Fig. 3.10 is senseless, but there are no such strong reasons to exclude the possibility (in some extent) of events $A = B$ and $A < B$.

There are two different possible assumptions concerned with conditional probabilities in observed situations defined in [108] as "weak" and "strong" relations.

Let us consider the weak relations.

Since H_3 is the evidence of events $a \in [b_1, a_2]$ and $b \in [b_1, a_2]$ simultaneously, it may be treated as an evidence of events $A < B$ and $A = B$ only. According to the Bayesian principle of insufficient reason [11], we assume that there are equal chances for $A = B$ and $B > A$ when the event H_3 occurs, i.e., $P(B > A/H_3) = P(A = B/H_3) = \frac{1}{2}$. The same assumption as a natural one is used in [152] without additional clarifications.

Thus, for the conditional probabilities we get:

$$P(B > A/H_1) = 1,\ P(B > A/H_2) = 1,$$
$$P(B > A/H_3) = \frac{1}{2},\ P(B > A/H_4) = 1. \tag{3.90}$$

From (3.89), (3.90) and (3.87) we have:

$$P(B > A) = 1 - \frac{1}{2}\frac{(a_2 - b_1)^2}{(a_2 - a_1)(b_2 - b_1)}.$$

In a similar way, we obtain:

$$P(B = A) = \frac{1}{2}\frac{(a_2 - b_1)^2}{(a_2 - a_1)(b_2 - b_1)}.$$

Of course, we have $P(B > A) + P(A = B) = 1$. It is easy to see that in asymptotic conditions, where $a_1 = b_1$ and $a_2 = b_2$, i.e., $A = B$, we get $P(B > A) = P(B = A) = \frac{1}{2}$. This seems as rather wrong result if intervals are treated as the mathematical objects defined only by their bounds. On the other hand, common sense is not always the best adviser in mathematically complicated situations. For example, we can say that in our case we deal with intervals and therefore the fact $A = B$ may be an evidence of $B > A$ and in the same degree of $B = A$. The similar reasoning was used in [94].

Let us consider the strong relations.

In this case, we assert that the event H_3 is not a strong evidence of $A < B$, but is a satisfactory evidence of $A = B$, i.e., $P(B > A/H_3) = 0$ and $P(A = B/H_3) = 1$. Thus, for the conditional probabilities we get:

$$P(B > A/H_1) = 1, P(B > A/H_2) = 1,$$
$$P(B > A/H_3) = 0, P(B > A/H_4) = 1. \tag{3.91}$$

From (3.87), (3.89) and (3.91) we obtain:

$$P(B > A) = 1 - \frac{(a_2 - b_1)^2}{(a_2 - a_1)(b_2 - b_1)}. \tag{3.92}$$

In a similar way, we get:

$$P(B = A) = \frac{(a_2 - b_1)^2}{(a_2 - a_1)(b_2 - b_1)}. \tag{3.93}$$

Obviously, $P(B > A) + P(B = A) = 1$. In the case of $A = B$, from Eqs.(3.92) and (3.93) we get $P(B > A) = 0$, $P(B = A) = 1$ and there are no problems with interpretation of these results. To simplify our further analysis, consider another simple but exact method for inferring the probabilities $P(B > A)$, $P(B = A)$. It easy to prove that in our case

$$P(H_1) + P(H_2) + P(H_3) + P(H_4) = 1. \tag{3.94}$$

Since in the case of "weak" relations we have $P(B > A/H_1) = 1$, $P(B > A/H_2) = 1$, $P(B > A/H_3) = \frac{1}{2}$, $P(B > A/H_4) = 1$, for the compound probability from Eq.(3.94) we get:

$$\begin{aligned} P(B > A) &= P(H_1) + P(H_2) + \frac{1}{2}P(H_3) + P(H_4) = \\ &= 1 - \frac{1}{2}\frac{(a_2 - b_1)^2}{(a_2 - a_1)(b_2 - b_1)}. \end{aligned} \tag{3.95}$$

It is easy to see that the same expression have been obtained above using different reasoning. In our further analysis, we shell use the similar argumentations when inferring the expressions for the calculation of probabilities.

Let us consider the case of inclusion.

There are three possible events in this case:

$$\begin{aligned} H_1: a \in [a_1, a_2] \wedge b \in [b_1, a_1], H_2: a \in [a_1, a_2] \wedge b \in [a_1, a_2], \\ H_3: a \in [a_1, a_2] \wedge b \in [a_2, b_2]. \end{aligned} \tag{3.96}$$

The corresponding probabilities are:

$$P(H_1) = \frac{a_1 - b_1}{b_2 - b_1}, P(H_2) = \frac{a_2 - a_1}{b_2 - b_1}, P(H_3) = \frac{b_2 - a_2}{b_2 - b_1}. \tag{3.97}$$

Since $b_1 \leq a_1$, in this case the relation $A > B$ may be true. For instance, there are no doubts that $A > B$ if $b_1 < a_1$ and $b_2 = a_2$.

Consider the weak relations.

Let us assume that there are equal chances for $A < B$, $A = B$ and $A > B$ when the event H_2 takes place. This means that

$$P(A < B/H_2) = P(A = B/H_2) = P(A > B/H_2) = \frac{1}{3}.$$

As a consequence, for the compound probabilities we get:

$$P(A < B) = \frac{1}{3}P(H_2) + P(H_3) = \frac{1}{3}\frac{a_2 - a_1}{b_2 - b_1} + \frac{b_2 - a_2}{b_2 - b_1}, \tag{3.98}$$

$$P(A = B) = \frac{1}{3}P(H_2) = \frac{1}{3}\frac{a_2 - a_1}{b_2 - b_1}, \tag{3.99}$$

$$P(A > B) = \frac{1}{3}P(H_2) + P(H_1) = \frac{1}{3}\frac{a_2 - a_1}{b_2 - b_1} + \frac{a_1 - b_1}{b_2 - b_1}. \tag{3.100}$$

In the case of $A = B$, we get $P(A < B) = P(A = B) = P(A > B) = \frac{1}{3}$. In a case of the degenerate interval A, i.e., $a_1 = a_2 = a$, from Eqs. (3.98)–(3.100) we infer $P(A < B) = \frac{b_2 - a}{b_2 - b_1}$, $P(A > B) = \frac{a - b_1}{b_2 - b_1}$ and $P(A = B) = 0$. Some remarks are needed to clarify obtained results. Of course, if B is an interval, and a is a real number then equality relation $B = a$ is senseless since the simultaneous fulfillment of the conditions $b_1 = a$ and $b_2 = a$ is impossible. Thus, in such cases we have $P(B = a) = 0$. On the other hand, inequality relation $B < a$ may be used in analysis since in the case, for example, $b_2 < a$, there is no doubt that $P(B < a) = 1$. It is clear that in the case of $b_1 \leq a \leq b_2$, the probability $P(a < B)$ makes sense. An interesting situation we have in the case of estimation of $P(A = B)$, where A and B are intervals. The simplest decision is to introduce a "strong" rule: $A = B$ only if $a_1 = b_1$ and $a_2 = b_2$. On the other hand, when dealing with optimization problems we often use the equality type restrictions. Obviously, the interval or fuzzy extension of such tasks inevitably leads to the extension of corresponding equality type restrictions. It is clear that fulfillment of "strong" equality rules in these cases, especially when using numerical optimization methods, is rather impossible.

Therefore, in the framework of proposed in [108] probabilistic approach the "weak" equality rules have been developed. So, if $a_1 \approx b_1$ and $a_2 \approx b_2$, then $P(A = B) \neq 0$.

Consider the strong relations.

We assert that only H_2 is an evidence of $A = B$, only H_1 is a witness of $A > B$ and only H_3 may confirm $A < B$. Hence

$$P(A < B) = P(H_3) = \frac{b_2 - a_1}{b_2 - a_2}, P(A = B) = P(H_2) = \frac{a_2 - a_1}{b_2 - b_1},$$
$$P(A > B) = P(H_1) = \frac{a_1 - b_1}{b_2 - b_1}. \tag{3.101}$$

For $A = B$, from Eqs.(3.101) we get $P(A < B) = P(A > B) = 0$ and $P(A = B) = 1$. For a degenerated A, i.e., $a_1 = a_2 = a$, from Eqs.(3.101) we get the same expressions as in the "weak" relation case: $P(A < B) = \frac{b_2 - a}{b_2 - b_1}$, $P(A > B) = \frac{a - b_1}{b_2 - b_1}$ and $P(A = B) = 0$.

Thus, the complete set of expressions for "weak" and "strong" interval relations is inferred. An interesting and useful task is an analysis of the transitivity properties of interval relations (see [67]), but it is out of scope of this book.

The question arises: which approach - "strong" or "weak" - is the best one? To answer, let us consider them in the case of $A = B$. In the "weak" case, for overlapping intervals we have $P(B > A/H_3) = P(B = A/H_3) = \frac{1}{2}$. In the inclusion case, the "weak" approach leads to the assumption $P(A > B/H_2) = P(A < B/H_2) = P(A = B/H_2) = \frac{1}{3}$ when the event H_2 occurs. In the case of $A = B$, we get $P(A > B) = P(A < B) = P(A = B) = \frac{1}{3}$. Obviously, it is not easy to explain such a result, but when using the "strong" approach we have no problems in the interpretation of the results obtained for the case of $A = B$.

Let us compare obtained expressions for the probabilistic interval relations with those proposed by other authors. The main merit of the approach proposed in [108] is an explicit introduction of interval equality relation as an essential constituent of complete set of interval relations such that in all cases the fundamental property $P(A<B)+P(A=B)+P(A>B)=1$ is verified. It is important to note that in the frameworks of most of known approaches - not only probabilistic ones - interval equality is usually considered as an impossible relation [131], identity [89] or only in conjunction with inequality relation [152]. It is shown in [108] that these treatments lead to some theoretical problems.

We think that treating an interval equality as identity ($A=B$ only if $a_1=b_1$, $a_2=b_2$) can not provide good solutions of some practical problems, e.g., when dealing with interval extension of optimization task under equality type restrictions. In approach proposed in [108], an equality is not equivalent to identity since $P(A=B)\leq 1$. Let's consider the example of comparing the intervals $A=[1,10]$ and $B=[2,11]$. Using the method proposed [108], e.g., strong relations, we get $P(A<B)=0.21$, $P(A=B)=0.79$, $P(A>B)=0$. An explanation of such results is obvious, especially taking into account the large common area of compared intervals, which is an argument for equality relation's dominance in the considered example.

There is no doubt that if we deal with the probability approach, the fundamental condition $P(A<B)+P(A=B)+P(A>B)=1$ should be automatically verified. It will be shown below that above consideration remains valid in a fuzzy setting as well. An interesting point is that the introduced set of interval relations makes it possible to look at the general problems of comparison from some debatable, but new point of view [108]. If we compare two alternatives A and B then using their real valued estimations a and b we usually introduce some value ε as a measure of indiscernibility. If $|\mathrm{a}-\mathrm{b}|<\varepsilon$ then considered alternatives are treated as indiscernible. It is well known that the value of ε is usually depends on inaccuracy of measurement, subjective factors and other sources of imprecision and uncertainty. Nevertheless, when thinking in interval or fuzzy spirit we can say that such estimation is pushing the problem of uncertainty assessment from the stage of alternative evaluation to the stage of their comparison. Suppose we have alternative estimations in the form of intervals A and B. Then using our approach we can conclude that these alternatives are indiscernible if $P(A=B)>max(P(A<B),P(A>B))$ or A is greater (better, preferable and so on) than B if $P(A>B)>max(P(A<B),P(A=B))$. It easy to see that there is no need for ε in such approach. Of course, we do not insist here that proposed in [108] approach is a universal remedy for all comparison problems, but what we really see is an absence of additional and - in our opinion - artificial measure of indiscernibility, ε.

Recently, a numerical algorithm for minimization of interval and fuzzy valued cost functions based on the described probabilistic approach to interval comparison has been developed in [111, 113, 114]. In the fuzzy case, the decomposition of a fuzzy number into a set of intervals representing the α-cuts of fuzzy number has been used. Proposed numerical methods have been successfully used for modeling and optimization of real-world processes [111, 113, 114].

Nevertheless, one can say that an existence of two quite different results gained from mutually exclusive assumptions ("weak"and "strong") may be considered as a weakness or incompleteness of proposed approach. It is noted in [111, 113, 114] that all observed ambiguities may be treated as a consequence of the limited ability of a purely probabilistic approach to deal with such objects as intervals or fuzzy numbers. The problem is that the probability theory allows us to represent only uncertainty. Nevertheless, interval and fuzzy objects are inherently characterized by imprecision and ambiguity. To solve this problem, the *DST* is used in [108].

Among numerous approaches to interval comparison, there is a group of rather qualitative methods which deserve to be mentioned because they have served as a starting point for the development of interval relations using *DST*.

This group of methods has a long history. They can be traced back to the fundamental work by Moore [89], and now appear in software [133] which embodies most of the modern concepts in the field of interval arithmetic. The main idea behind these methods is contained in the following definition: "An interval is less than another interval if it contains some value(s) that are less than some value(s) in another interval". As this definition is not broad enough to capture all the cases of interval location and intersection, three main classes of interval relation are used: certainty, possibility, set. To implement these relations, a set of relation functions was proposed in [26]. It is important to note that these functions return only Boolean values. In fact, these methods provide only qualitative results, expressed in verbal form. Unfortunately, the linguistic term "Possibility" has in practice a wide, but finite (as a consequence of the limited nature of any natural language) set of nuances (senses). In turn, each "nuance" reflects a certain group of qualitatively equivalent interval relations, which, on the other hand, differ quantitatively.

Indeed, all the arithmetic operations on intervals provide us intervals as well, and only interval comparisons have resulted in real numbers (different indices or probabilities) or Boolean values. In order to catch not only an uncertainty of initial information available, but an ambiguity and imprecision as well, it seems natural to make attempt to build an approach which can provide the results of comparisons in the interval form, for instance, as a probability interval. For this purpose, the *DST* may be successfully applied [108].

3.3.3.2 Interval Comparison Based on the Dempster-Shafer Theory

In this subsection, we present the method developed in [108].

Let us consider a case of overlapping intervals (Fig. 3.10). Assume they are independent ones and $a \in [a_1, a_2], b \in [b_1, b_2]$ be random values distributed on these intervals.

Only four mutually exclusive events $H_i, i = 1$ to 4, may take place in considered situation. For probabilities (see previous subsection) we get :

$$P(H_1) + P(H_2) + P(H_3) + P(H_4) = 1.$$

Thus, in the spirit of *DSTthe* probabilities $P(H_i), i = 1$ to 4, can be used to construct a basic assignment function (*bpa*). In a case of overlapping intervals $a_1 \leq b_1$ and $a_2 \leq b_2$, only two interval relations make sense: $A < B, A = B$ (see previous subsection).

It is easy to see that events H_1, H_2 and H_4 may be considered as "strong" evidences of $A < B$, otherwise H_3 can be treated as only a "weak" evidence of $A < B$ since it is simultaneously evidence of $A = B$.

Using *DST* notation we can represent the above conclusions as $H_1 \subseteq (A < B), H_2 \subseteq (A < B), H_4 \subseteq (A < B)$. Since the events H_1, H_2, H_4 are independent, a sum of their probabilities may be treated as an argument (evidence) in favor of $A < B$. So we have the first focal element of *bpa*: $m(\{A < B\}) = P(H_1) + P(H_2) + P(H_4)$. Event H_3 is an evidence of events $A < B$ and $A = B$ simultaneously. Since they are not mutually exclusive, i.e., $(A < B) \cap (A = B) \neq 0$, the probability $P(H_3)$ may be treated as an evidence of composed event $(A < B, A = B)$ and the next focal element of *bpa* may be represented as $m(\{A < B, A = B\}) = P(H_3)$.

Thus, obtained *bpa* consists of only two focal elements. Using the basic definitions (3.79), (3.80) and taking into account Eqs.(3.88) and (3.94) we get

$$
\begin{aligned}
Bel(A < B) &= m(\{A < B\}) = P(H_1) + P(H_2) + P(H_4) = \\
&= 1 - P(H_3) = 1 - \frac{(a_2 - b_1)^2}{(a_2 - a_1)(b_2 - b_1)},
\end{aligned}
\tag{3.102}
$$

$$
\begin{aligned}
Pl(A < B) &= m(\{A < B\}) + m(\{A < B, A = B\} = \\
&= P(H_1) + P(H_2) + P(H_3) + P(H_4) = 1.
\end{aligned}
\tag{3.103}
$$

In a similar way, for $A = B$ we obtain:

$$
Bel(A = B) = 0,
\tag{3.104}
$$

$$
Pl(A = B) = P(H_3) = \frac{(a_2 - b_1)^2}{(a_2 - a_1)(b_2 - b_1)}.
\tag{3.105}
$$

Denoting all the probabilities we have inferred in previous subsection for the "strong" case as $P_s(\bullet)$ and for the "weak" case as $P_w(\bullet)$, we get the belief intervals *BI* as follows:

$$
\begin{aligned}
BI(A < B) &= [Bel(A < B), Pl(A < B)] = [P_s(A < B), 1] \\
&= [1 - P_s(A = B), 1],
\end{aligned}
\tag{3.106}
$$

$$
BI(A = B) = [Bel(A = B), Pl(A = B)] = [0, P_s(A = B)].
\tag{3.107}
$$

So, using an approach based on *DST* we get interval estimations for degrees of interval inequality and equality.

An important property of $BI(A < B)$ and $BI(A = B)$ relations for overlapping A and B is $BI(A = B) < BI(A < B)$.

It is worth noting that the last inequality is "strong" only if $P_s \leq 0.5$ (see Fig. 3.11).

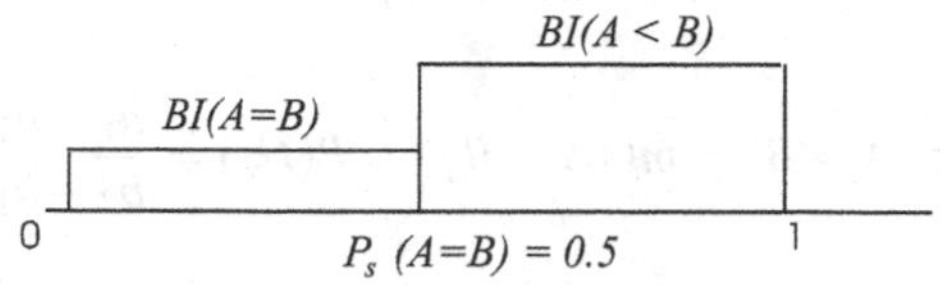

Fig. 3.11 Relation between $BI(A < B)$ and $BI(A = B)$

In the case of $a_1 = b_1, a_2 = b_2$, i.e., $A \equiv B$ from (3.106) and (3.107) we get: $BI(A < B) = BI(A = B) = [0, 1]$. This result seems quite natural in the spirit of *DST*.

Introduced interval estimations can be considered as an embodiment of usually not explicitly expressed, but pivotal requirement of interval arithmetic: the result of interval operation should be an interval as well.

In [108], the degree of imprecision or ambiguity *ID* of interval relations *rel* as a whole was proposed:

$$ID(rel) = BI(A < B) + BI(A = B) = [1 - P_s(A = B), 1 + P_s(A = B)] =$$
$$= [1 - \frac{(a_2 - b_1)^2}{(a_2 - a_1)(b_2 - b_1)}, 1 + \frac{(a_2 - b_1)^2}{(a_2 - a_1)(b_2 - b_1)}].$$

It easy to see that the length of *ID(rel)* may be considered as a natural real valued estimation of interval relation imprecision which decreases with lowering the length of overlapping region.

Consider the case of inclusion.

In this case, we have three possible events $H_i, i = 1$ to 3, with corresponding probabilities (see previous subsection). The elementary evidences of events $A < B, A = B, A > B$ were considered in [108] and they were taken then into account to construct the *bpa* using nearly the same reasoning as in the case of overlapping intervals. Finally, the following expressions were obtained in [108]:

$$m(\{A < B\}) = P(H_3), m(\{A > B\}) = P(H_1),$$
$$m(\{A < B, A = B, A > B\}) = P(H_2),$$

$$Bel(A < B) = m(\{A < B\}) = P(H_3) = \frac{b_2 - a_2}{b_2 - b_1}, \tag{3.108}$$

$$\begin{aligned} Pl(A < B) &= m(\{A < B\}) + m(\{A < B, A = B, A > B\}) = \\ &= P(H_3) + P(H_2) = \frac{a_2 - a_1}{b_2 - b_1} + \frac{b_2 - a_1}{b_2 - b_1} = \\ &= P_s(A < B) + P_s(A = B), \end{aligned} \tag{3.109}$$

$$Bel(A = B) = 0, \tag{3.110}$$

$$Pl(A = B) = m(\{A < B, A = B, A > B\}) = P(H_2) = \frac{a_2 - a_1}{b_2 - b_1}, \tag{3.111}$$

$$Bel(A > B) = m(\{A > B\}) = P(H_1) = \frac{a_1 - b_1}{b_2 - b_1}, \tag{3.112}$$

$$\begin{aligned} Pl(A > B) &= m(\{A > B\}) + m(\{A < B, A = B, A > B\}) = \\ &= P(H_1)) + P(H_2) = \frac{a_1 - b_1}{b_2 - b_1} + \frac{a_2 - a_1}{b_2 - b_1} = \\ &= \frac{a_2 - b_1}{b_2 - b_1} = P_s(A > B) + P_s(A = B). \end{aligned} \tag{3.113}$$

Let us consider some asymptotic cases.
In the case of $A \equiv B$ we get:

$$Bel(A < B) = Bel(A > B) = Bel(A = B) = 0, \tag{3.114}$$

$$Pl(A < B) = Pl(A > B) = Pl(A = B) = 1. \tag{3.115}$$

In the case of $a_2 = b_2, a_1 > b_1$ we have:

$$Bel(A < B) = 0, Pl(A < B) = 1, \tag{3.116}$$

$$Bel(A = B) = 0, Pl(A = B) = \frac{b_2 - a_1}{b_2 - b_1}, \tag{3.117}$$

$$Bel(A > B) = \frac{a_1 - b_1}{b_2 - b_1}, Pl(A > B) = 1. \tag{3.118}$$

In the case of $a_1 = b_1, b_2 > a_2$ we obtain:

$$Bel(A < B) = \frac{b_2 - a_2}{b_2 - b_1}, Pl(A < B) = 1, \tag{3.119}$$

$$Bel(A = B) = 0, Pl(A = B) = \frac{a_2 - b_1}{b_2 - b_1}, \tag{3.120}$$

$$Bel(A > B) = 0, Pl(A > B) = \frac{a_2 - b_1}{b_2 - b_1}. \tag{3.121}$$

These results are in good agreement with common sense. Indeed, if A is a degenerate interval, i.e., $a_1 = a_2 = a$ then from (3.108)-(3.113) we get:

$$Bel(a < B) = Pl(a < B) = \frac{b_2 - a}{b_2 - b_1}, \tag{3.122}$$

$$Bel(a = B) = Pl(a = B) = 0, \tag{3.123}$$

$$Bel(a > B) = Pl(a > B) = \frac{a - b_1}{b_2 - b_1}. \tag{3.124}$$

It is important that we have real valued resulting estimations only in the case of comparison of interval with real value.

For the belief intervals BI we get:

$$\begin{aligned} BI(A < B) &= [Bel(A < B), Pl(A < B)] = [\frac{b_2 - a_2}{b_2 - b_1}, \frac{b_2 - a_1}{b_2 - b_1}] \\ &= [P_s(A < B), P_s(A < B) + P_s(A = B)], \end{aligned} \tag{3.125}$$

$$BI(A = B) = [Bel(A = B), Pl(A = B)] = [0, \frac{a_2 - a_1}{b_2 - b_1}] = [0, P_s(A = B)], \tag{3.126}$$

$$\begin{aligned} BI(A > B) &= [Bel(A > B), Pl(A > B)] = [\frac{a_1 - b_1}{b_2 - b_1}, \frac{a_2 - b_1}{b_2 - b_1}] \\ &= [P_s(A > B), P_s(A > B) + P_s(A = B)]. \end{aligned} \tag{3.127}$$

Observe that in the inclusion case using (3.125)-(3.127) we have

$$BI(A = B) < BI(A > B), BI(A < B),$$

but only in a "weak" sense, since $(BI(A = B) \cap BI(A > B) \neq \emptyset$ and/or $BI(A = B) \cap BI(A < B) \neq \emptyset$.

Another important feature is that if A and B have a common center, i.e., $(a_1 + a_2) = (b_1 + b_2)$ we always have

$$BI(A = B) < BI(A > B) = BI(A < B)$$

and this inequality relation is a "strong" one only if $b_2 - a_2 > a_2 - a_1$. For instance, if $A = [1, 7], B = [0, 8]$ we get the result shown in Fig. 3.12.

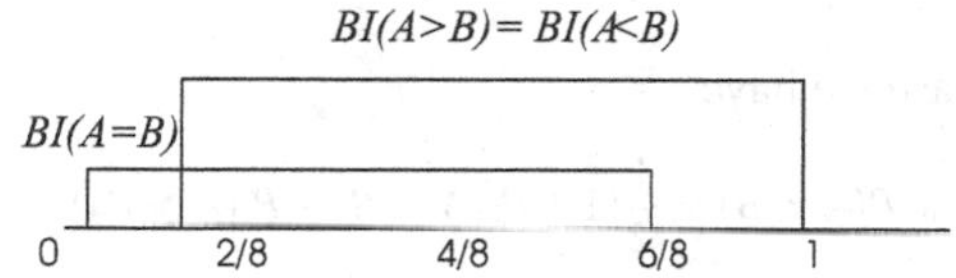

Fig. 3.12 The case when A and B have a common center

In the inclusion case (see Fig. 3.10), we introduce an overall degree of imprecision or ambiguity of interval relations *rel* as

$$\begin{aligned} ID(rel) &= BI(A < B) + BI(A = B) + BI(A > B) \\ &= [1 - P_S(A = B), 1 + 2P_S(A = B)] = \\ &= [1 - \frac{a_2 - a_1}{b_2 - b_1}, 1 + 2\frac{a_2 - a_1}{b_2 - b_1}]. \end{aligned} \tag{3.128}$$

In contrast to the overlapping case we get resulting interval for degree of imprecision or ambiguity, which is asymmetrical in relation to 1. However, its length decreases with lowering the length of included interval A. This seems quite natural, since for the degenerate A there is no any ambiguity or imprecision, i.e., if $rel \in \{<,>,=\}$, then $Bel(A\ rel\ B) = Pl(A\ rel\ B)$ (see Eqs.(3.122)–(3.124)). Thus, in the inclusion case the real valued estimation of ambiguity or imprecision of interval relation is determined by the relation of lengths of included and enveloping intervals.

Finally, several real valued criteria may be applied in order to make a reasonable choice when comparing intervals. Non-exhaustively, we can distinguish:

- strong preference:

$$B > A \text{ if } Bel(B > A) > Pl(A < B),$$

- weak preference:

$$B > A \text{ if } Bel(B > A) > Bel(B < A),$$

- mixed preference:

$$B > A \text{ if } MP(B > A) > MP(B < A),$$

where $MP(\bullet) = \alpha Bel(\bullet) + (1-\alpha)Pl(\bullet)$ with $0 \leq \alpha \leq 1$ (the value of α reflects the risk aversion of decision maker).

Obviously, the mixed preference is a more flexible criterion. In the simplest case ($\alpha = 0.5$) for overlapping intervals we get:

$$MP(A < B) = \frac{1}{2}(Bel(A < B) + Pl(A < B)) = P_w(A < B), \tag{3.129}$$

$$MP(A = B) = \frac{1}{2}(Bel(A = B) + Pl(A = B)) = P_w(A = B). \tag{3.130}$$

It is interesting that

$$MP(A < B) + MP(A = B) = 1. \tag{3.131}$$

In the inclusion case we have

$$MP(A < B) = \frac{1}{2}(1 + P_s(A < B) - P_s(A > B)), \tag{3.132}$$

$$MP(A = B) = \frac{1}{2}P_s(A = B). \tag{3.133}$$

It is easy to see that in this case

$$MP(A < B) + MP(A > B) + MP(A = B) = 1 + \frac{1}{2}P_s(A = B). \tag{3.134}$$

Comparing expressions (3.131) and (3.134), we can see that interval relations in the inclusion case are more doubtful and imprecise than those in the overlapping case.

Expressions (3.129)–(3.134) expose the similarity between the DST and probability approaches to interval the comparison.

When observing belief intervals BI in the overlapping and inclusion cases, we can see that only "strong" interval comparison probability estimations (see previous subsection) and their sums can be considered as natural bounds of corresponding belief intervals. It is important that probabilities $P_s(A > B), P_s(A < B)$ in all cases represent the left bounds of belief intervals, i.e.,

$$Bel(A > B) = P_s(A > B), Bel(A < B) = P_s(A < B),$$

whereas the probabilities $P_s(A = B)$ occur only in representations of right bounds of BI.

Since DST may be treated as a generalization of probability theory, we can conclude that only the set of "strong" interval probability relations introduced in [108] (see previous subsection) should be used in analysis as a direct asymptotic limit of DST.

Consider the fuzzy number relations based on DST.

Let $\tilde{A}$ and $\tilde{B}$ be fuzzy numbers on X with corresponding membership functions $\mu_A(x), \mu_B(x) : X \to [0,1]$. They can be represented by the sets of α-cuts: $\tilde{A} = \bigcup_\alpha A_\alpha$, $\tilde{B} = \bigcup_\alpha B_\alpha$, where $A_\alpha = \{x \in X : \mu_A(x) \geq \alpha\}$, $B_\alpha = \{x \in X : \mu_B(x) \geq \alpha\}$ are intervals. Then all fuzzy number relations $\tilde{A}$ rel $\tilde{B}$, $rel \in \{<,=,>\}$ may be presented by the sets of α-cut relations

$$\tilde{A} \; rel \; \tilde{B} = \bigcup_\alpha A_\alpha \; rel \; B_\alpha. \tag{3.135}$$

Since in the framework of DST all interval relations A_α rel B_α provide the results in form of corresponding intervals $BI(A_\alpha \; rel \; B_\alpha)$ we conclude that in the left hand side of (3.135) we have a fuzzy number. More strictly,

$$\tilde{A} \; rel \; \tilde{B} = \bigcup_\alpha A_\alpha \; rel \; B_\alpha = \bigcup_\alpha BI(A_\alpha \; rel \; B_\alpha), \tag{3.136}$$

where

$$BI(A_\alpha \; rel \; B_\alpha) = [Bel(A_\alpha \; rel \; B_\alpha), Pl(A_\alpha \; rel \; B_\alpha)] \tag{3.137}$$

are belief intervals corresponding to the interval relations A_α rel B_α on α-cuts.

Using mathematical tools presented in previous subsection, we can calculate all the values $Bel(A_\alpha \; rel \; B_\alpha), Pl(A_\alpha \; rel \; B_\alpha)$ needed to determine fuzzy numbers representing the result of fuzzy number relation. So, the fuzzy numbers

$$BI(\tilde{A} \; rel \; \tilde{B}) = \bigcup_\alpha BI(A_\alpha \; rel \; B_\alpha)$$

can be considered as results of fuzzy number comparisons.

The resulting fuzzy estimations $BI(\tilde{A} \; rel \; \tilde{B})$ can be used directly. For instance, let $\tilde{A}$, $\tilde{B}$, $\tilde{C}$ be fuzzy numbers and $BI(\tilde{A} \; rel \; \tilde{B})$, $BI(\tilde{A} \; rel \; \tilde{C})$ be fuzzy estimations of fuzzy relations $\tilde{A} > \tilde{B}$ and $\tilde{A} > \tilde{C}$, respectively. Then estimation $BI(BI(\tilde{A} > \tilde{B}) > BI(\tilde{A} > \tilde{C}))$ is a fuzzy number as well. Such fuzzy calculations may be useful at intermediate stages of analysis since they preserve fuzzy information available.

For practical purposes, it is useful to introduce some indices obtained as the result of defuzzification. A simple, but perhaps the most useful one is:

$$BI_{Df}(\tilde{A}\ rel\ \tilde{B}) = 2\int_{0}^{1} \alpha BI_{Df}(A_{\alpha}\ rel\ B_{\alpha})d\alpha. \tag{3.138}$$

Last expression is a form of defuzzification (or type reduction) and the result is an interval. Formula (3.138) emphasizes that contribution of α-cut to an overall estimation rises with increasing of its number. Of course, the set of complementary parameterized functions of α can be used in (3.138) instead of α as it is proposed in [151], but for the sake of simplicity only expression (3.138) has been used to obtain the results presented below. Some typical cases of fuzzy number comparison are represented in Fig. 3.13 and 3.14.

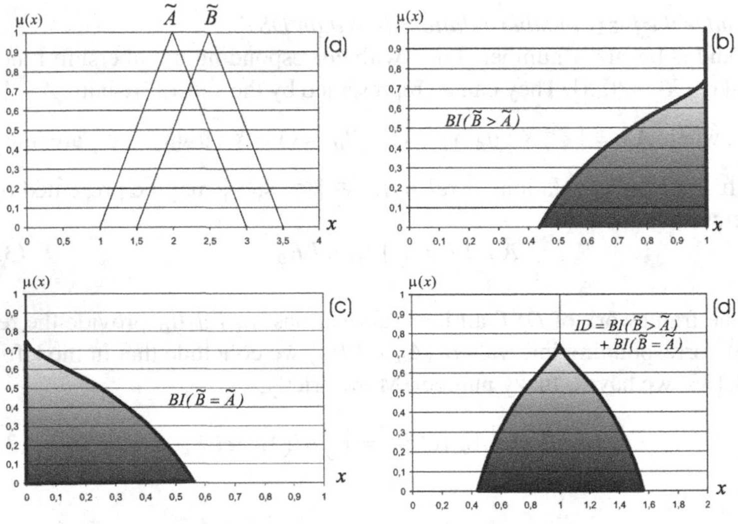

Fig. 3.13 Special case of fuzzy number comparison: $BI_{Df}(\tilde{B}>\tilde{A}) = [0.86,1], BI_{Df}(\tilde{B}=\tilde{A}) = [0,0.13]$, $ID_{Df} = (BI(\tilde{B}>\tilde{A})+BI(\tilde{B}=\tilde{A}))_{Df} = [0.87,1.12]$.

It is easy to see when analyzing the case presented in Fig. 3.13 that imprecision degree of fuzzy relations $ID = BI(\tilde{B} > \tilde{A}) + BI(\tilde{B} = \tilde{A})$ may be treated as a fuzzy number with linguistic interpretation ID="near 1", where "near 1" is a fuzzy number symmetrical in respect to 1. It worth noting here that in the discussed case the main properties of probability are remained, but in the fuzzy sense. On the other hand, the degree of fuzziness of the fuzzy numbers "near 1" provides us additional information that may be useful for the estimation of overall uncertainty of fuzzy

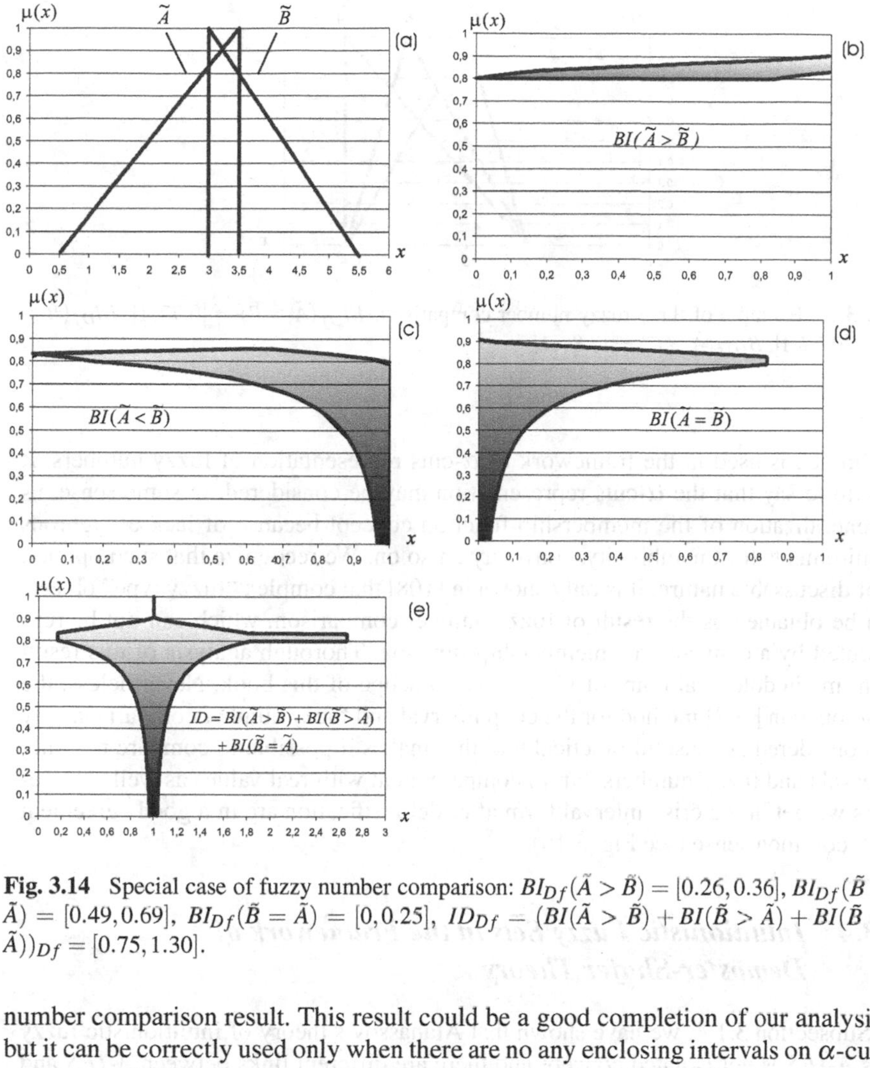

Fig. 3.14 Special case of fuzzy number comparison: $BI_{Df}(\tilde{A} > \tilde{B}) = [0.26, 0.36]$, $BI_{Df}(\tilde{B} > \tilde{A}) = [0.49, 0.69]$, $BI_{Df}(\tilde{B} = \tilde{A}) = [0, 0.25]$, $ID_{Df} = (BI(\tilde{A} > \tilde{B}) + BI(\tilde{B} > \tilde{A}) + BI(\tilde{B} = \tilde{A}))_{Df} = [0.75, 1.30]$.

number comparison result. This result could be a good completion of our analysis, but it can be correctly used only when there are no any enclosing intervals on α-cuts of compared fuzzy numbers. The situation when we have inclusion of intervals on some a-cuts is presented in Fig. 3.14. It is easy to scc that asymmetric fuzzy *ID* is obtained and that the height of asymmetric part of *ID* is proportional to a number of α-cut with included intervals. It was shown above that in inclusion case of interval comparison, an interval *ID* is asymmetrical in relation to 1 (see expression (3.128)). When looking at Fig. 3.14, the natural question arises: Is it possible to interpret the objects $BI(\tilde{A} > \tilde{B})$, $BI(\tilde{A} < \tilde{B})$, $BI(\tilde{A} = \tilde{B})$ (obtained results of fuzzy number comparison) as fuzzy numbers? Of course, it is impossible from a conventional point of view, since ambiguous membership functions are needed to represent such objects. On the other hand, we can use (for fuzzy arithmetic only) a less rigorous definition of fuzzy, perhaps, "fuzzy type" object as a "set of closed α-cuts with $\alpha \in [0, 1]$". Indeed, there is no need for concept of membership function when fuzzy

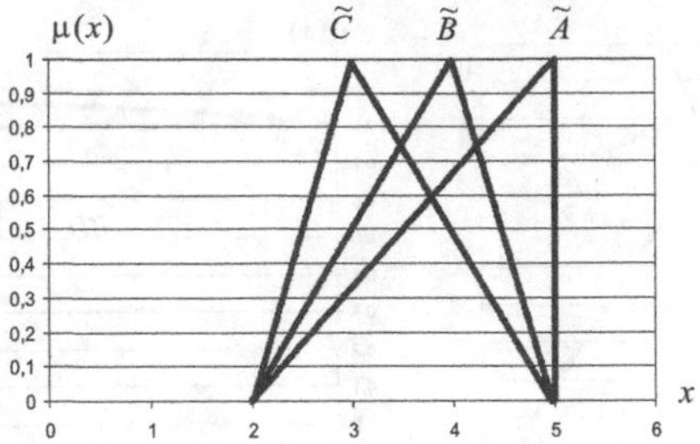

Fig. 3.15 Example of three fuzzy number comparison: $BI_{Df}(\tilde{A} > \tilde{B}) = [0.77, 1]$, $BI_{Df}(\tilde{B} > \tilde{C}) = [0.77, 1]$, $BI_{Df}(\tilde{A} > \tilde{C}) = [0.89, 1]$

arithmetic is used in the framework of α-cuts representation of fuzzy numbers. It is safe to say that the α-cuts representation may be considered, in some sense, as a generalization of the membership function concept because of lack of rigorous requirements for unambiguity, convexity an so on. We recognize that such opinion is of discussable nature. It is only shown in [108] that complex "fuzzy type" objects can be obtained as the result of fuzzy number comparison, which can not be represented by a conventional membership function. Thorough analysis of this result from methodological point of view is out of scope of this book. Nevertheless, the developed in [108] method for the crisp interval and fuzzy number comparison may be considered as a useful practical tool that makes it possible to compare not only intervals and fuzzy numbers, but to compare them with real values as well. The results we get in the crisp interval form after defuzzification are in a good agreement with common sense (see Fig. 3.15).

3.3.4 Intuitionistic Fuzzy Sets in the Framework of Dempster-Shafer Theory

In Subsection 3.1.4, we have shown that Atanassov's theory of intuitionistic fuzzy sets *A-IFS* is not isolated concept and there are different links between *A-IFS* and some other theories modeling imprecision such as interval valued fuzzy sets, type 2 fuzzy sets and soft sets. Here we show that there exists also a strong link between *A-IFS* and *DST*. We use this link to provide a transparent and fruitful semantics for *A-IFS* in terms of *DST*.

We have noted also that there exist two important problems in multiple criteria decision making *MCDM* in the intuitionistic fuzzy setting: aggregation of local criteria without intermediate defuzzification in the case when criteria and their weights are *IFVs*; comparison of *IF* valued scores of alternatives basing on the degree to which one *IFV* is grater/smaller than the other.

Here we propose an approach to the solution of these problems based on the interpretation of *A-IFS* in the framework of the *DST*.

The concept of *A-IFS* is based on the simultaneous consideration of membership μ and non-membership ν of an element of a set to the set itself such that $0 \leq \mu + \nu \leq 1$. For the reader's convenience, we repeat here the Definition 3.21.

Let $X = \{x_1, x_2, ..., x_n\}$ be a finite universal set. An intuitionistic fuzzy set A in X is an object of the following form: $A = \{< x_j, \mu_A(x_j), \nu_A(x_j) > | x_j \in X\}$, where the functions $\mu_A : X \to [0,1], x_j \in X \to \mu_A(x_j) \in [0,1]$ and $\nu_A : X \to [0,1]$, $x_j \in X \to \nu_A(x_j) \in [0,1]$ define the degree of membership and degree of non-membership of the element $x_j \in X$ to the set $A \subseteq X$, respectively, and for every $x_j \in X, 0 \leq \mu_A(x_j) + \nu_A(x_j) \leq 1$.

Following to [3], we call $\pi_A(x_j) = 1 - \mu_A(x_j) - \nu_A(x_j)$ the intuitionistic index (or the hesitation degree) of the element x_j in the set A. It is obvious that for every $x_j \in X$ we have $0 \leq \pi_A(x_j) \leq 1$.

Hong and Choi [53] proposed to use the interval representation $[\mu_A(x_j), 1 - \nu_A(x_j)]$ of intuitionistic fuzzy set A in X instead of pair $< \mu_A(x_j), \nu_A(x_j) >$ in context of *MCDM* problem. The first obvious advantage of such approach is that expression $[\mu_A(x_j), 1 - \nu_A(x_j)]$ represents a regular interval as its right bound always is not smaller than its left bound (this is a consequence of the condition $0 \leq \mu_A(x_j) + \nu_A(x_j) \leq 1$. Obviously, this approach is equivalent to the interval valued fuzzy sets interpretation of *A-IFS*. The second advantage is the possibility to redefine the basics of *A-IFS* in terms of *DST*. Here we show that convenient in the practical applications methods for *MCDM* can be developed using *DST* semantics for *A-IFS*.

Firstly, we show that in the framework of *DST* the triplet $(\mu_A(x_j), \nu_A(x_j), \pi_A(x_j))$ represents the basic assignment function (*bpa*).

Really, when analyzing any situation in context of *A-IFS*, we implicitly deal with the following three hypotheses: *Yes*: $x_j \in A$, *No*: $x_j \notin A$, (*Yes*,*No*): both the hypotheses $x_j \in A$ and $x_j \notin A$ can not be rejected (the case of hesitation).

In this context, $\mu_A(x_j)$ may be treated as the probability or evidence of $x_j \in A$, i.e., as the focal element of the basic assignment function: $m(Yes)$=$\mu_A(x_j)$. Similarly, we can assume that $m(No)$=$\nu_A(x_j)$. Since $\pi_A(x_j)$ is usually treated as the hesitation degree, a natural assumption is $m(Yes,No)$=$\pi_A(x_j)$. Taking into account that $\mu_A(x_j) + \nu_A(x_j) + \pi_A(x_j) = 1$ we come to the conclusion that triplet $(\mu_A(x_j), \nu_A(x_j), \pi_A(x_j))$ represents a correct basic assignment function. According to the *DST* formalism we get $Bel_A(x_j)$=$m(Yes)$=$\mu_A(x_j)$ and $Pl_A(x_j)$=$m(Yes)$+$m(Yes,No)$= $\mu_A(x_j) + \pi_A(x_j)$=$1 - \nu_A(x_j)$.

Therefore, the following definition can be introduced:

Definition 3.24. Let $X = \{x_1, x_2, ..., x_n\}$ be a finite universal set and x_j is an object in X represented by the functions $\mu_A(x_j), \nu_A(x_j)$ which represent the degree of membership and degree of non-membership of $x_j \in X$ to the set $A \subseteq X$ such that $\mu_A : X \to [0,1], x_j \in X \to \mu_A(x_j) \in [0,1]$ and $\nu_A : X \to [0,1], x_j \in X \to \nu_A(x_j) \in [0,1]$ and for every $x_j \in X, 0 \leq \mu_A(x_j) + \nu_A(x_j) \leq 1$. An intuitionistic fuzzy set A in X is an object having the following form: $A = \{< x_j, BI_A(x_j) > | x_j \in X\}$, where $BI_A(x_j) = [Bel_A(x_j), Pl_A(x_j)]$ is the belief interval, $Bel_A(x_j) = \mu_A(x_j)$ and $Pl_A(x_j) = 1 - \nu_A(x_j)$ are the measures of belief and plausibility that $x_j \in X$ belongs to the set $A \subseteq X$.

At first glance, this definition seems as a simple redefinition of *A-IFS* in terms of interval valued fuzzy sets, but we show that using the *DSF* semantics it is possible to enhance the performance of *A-IFS* when dealing with *MCDM* problems. Particularly, this approach allows us to use directly the Dempster's rule of combination to aggregate local criteria presented by *IFVs* and develop a method for *MCDM* without intermediate defuzzification when local criteria and their weights are *IFVs*. As the result, we get final alternative's evaluations in the form of belief interval. Hence, an appropriate method for such intervals comparison is needed. In the previous subsection, we have presented a new method (based on the *DST*) providing the results of interval comparison in the form of belief interval, i.e., without loss of information caused by intermediate type reductions.

Let us consider *MCDM* problems in the framework of Intuitionistic/*DST* approach.

To make our consideration more transparent and comparable with the results obtained earlier by the other authors, we shall use here the example analyzed in [72] since only in this paper the *MCDM* problem is considered in the case when not only local criteria, but also their weights are *IFVs*: "Consider an air-condition system selection problem. Suppose there exist three air-condition systems x_1, x_2 and x_3. Suppose three criteria a_1 (economical), a_2 (function) and a_3 (being operative) are taken into consideration." The degrees μ_{ij} and ν_{ij} for the alternatives with respect to the criteria representing the fuzzy concept "excellence" were presented in [72] as follows:

$$((\mu_{ij},\nu_{ij}))_{3\times 3} = \begin{matrix} \\ a_1 \\ a_2 \\ a_3 \end{matrix} \begin{pmatrix} x_1 & x_2 & x_3 \\ (0.75,0.10) & (0.80,0.15) & (0.40,0.45) \\ (0.60,0.25) & (0.68,0.20) & (0.75,0.05) \\ (0.80,0.20) & (0.45,0.50) & (0.60,0.30) \end{pmatrix}. \quad (3.139)$$

The degrees ρ_i of membership and the degrees τ_i of non-membership representing the fuzzy concept "importance" were presented in [72] as follows:

$$((\rho_i,\tau_i))_{1\times 3} = \begin{pmatrix} a_1 & a_2 & a_3 \\ (0.25,0.25) & (0.35,0.40) & (0.30,0.65) \end{pmatrix}. \quad (3.140)$$

To get the final alternative's evaluations $FAE(x_i)$ on the base of data from the structures (3.130), (3.140) we use the *DST* interpretation of *A-IFS*.

For each pair x_i, a_j there are two sources of information concerned with x_i goodness: the degree of the local criterion satisfaction and the weight (importance) of this criterion. Let us consider the first of them. It is easy to see that in this case we deal with three hypotheses: *Yes*: the alternative x_i is good as it satisfies the local criterion a_j; *No*: the alternative x_i is rather bad (not good) as it does not satisfies a_j; (Yes,No): the compound hypothesis (we hesitate over a choice of *Yes* or *No*). The degree of local criterion satisfaction can be treated as the first source of evidence for estimation of x_i goodness. Therefore, for the pair x_i, a_j it can be presented by the basic assignment function as follows: $m_1^{ij}(Yes)$, $m_1^{ij}(No)$, $m_1^{ij}(Yes,No)$, where $m_1^{ij}(Yes) = \mu_{ij}$, $m_1^{ij}(No) = \nu_{ij}$, $m_1^{ij}(Yes,No) = 1 - \mu_{ij} - \nu_{ij} = \pi_{ij}$.

The other source of evidence of x_i goodness is the relative importance (weight) of the local criterion.

So three hypothesis should be considered: *Yes*: the criterion a_j is important; *No*: the criterion a_j is not important; (Yes, No): the compound hypothesis (we hesitate over a choice of *Yes* or *No*). Then for the local criterion a_j the basis assignment function corresponding to its importance can be presented as follows: $m_2^j(Yes) = \rho_j$, $m_2^j(No) = \tau_j$, $m_2^j(Yes,No) = 1 - \rho_j - \tau_j$. To obtain the combined basic assignment function m_{com} based on these sources of evidence presented by particular assignment functions $m_1^{ij}(Yes)$, $m_1^{ij}(No)$, $m_1^{ij}(Yes,No)$ and $m_2^j(Yes)$, $m_2^j(No)$, $m_2^j(Yes,No)$, we have used the Dempster's combination rule (3.81):

$$m_{com}^{ij}(A) = \frac{\sum\limits_{B \cap C = A} m_1^{ij}(B) m_2^j(C)}{1-K}, \tag{3.141}$$

where $K = \sum\limits_{B \cap C = \emptyset} m_1^{ij}(B) m_2^j(C)$, $A, B, C \in \{Yes, No, (Yes, No)\}$. As the result, for each pair x_i, a_j the basic assignment function $m_{com}^{ij}(Yes)$, $m_{com}^{ij}(No)$, $m_{com}^{ij}(Yes,No)$ can be calculated.

Consequently, in the spirit of *DST*, the local criteria a_j can be treated as the particular sources of information (evidence) for the generalized estimation of x_i goodness. There are three local criteria in our example. Hence, each alternative x_i can be presented by the structure M_i as follows:

$$M_i = \begin{pmatrix} m_{com}^{i1}(Yes) & m_{com}^{i1}(No) & m_{com}^{i1}(Yes,No) \\ m_{com}^{i2}(Yes) & m_{com}^{i2}(No) & m_{com}^{i2}(Yes,No) \\ m_{com}^{i3}(Yes) & m_{com}^{i3}(No) & m_{com}^{i3}(Yes,No) \end{pmatrix}. \tag{3.142}$$

The final basic assignment function based on the particular evidences presented by the local criteria combined with their importances can be obtained using Dempster's rule (3.81). For a *MCDM* problem with more than two local criteria, we can first obtain the combination of focal elements of two assignment functions using the Dempster's rule and combine the obtained result with the third assignment function and so on. It is easy to show that in our case of three local criteria, this process leads to following expression:

$$m_{com}^{i}(A) = \frac{\sum\limits_{B \cap C \cap D = A} m_{com}^{i1}(B) m_{com}^{i2}(C) m_{com}^{i3}(D)}{1-K}, \tag{3.143}$$

where $K = \sum\limits_{B \cap C \cap D = \emptyset} m_{com}^{i1}(B) m_{com}^{i2}(C) m_{com}^{i3}(D); A, B, C, D \in Yes, No, (Yes, No)$.

From this expression for each alternative x_i we obtain the final basic assignment function $m_{com}^{i}(Yes)$, $m_{com}^{i}(No)$, $m_{com}^{i}(Yes,No)$ and the bounds of believe interval $Bel(x_i) = m_{com}^{i}(Yes)$, $Pl(x_i) = m_{com}^{i}(Yes) + m_{com}^{i}(Yes,No)$. If it is needed, this result may be represented in terms of *A-IFS* theory since according to the Definition 3.24 $\mu^i = m^i_{com}(Yes)$, $\nu^i = m^i_{com}(No)$ and $\pi^i = m^i_{com}(Yes,No)$.

The final alternative's evaluation $FAE(x_i)$ can be presented both by the final basic assignment function m^i_{com} and by the belief interval $[Bel(x_i), Pl(x_i)]$. On the other hand, the last presentation seems to be more convenient as it allows us to compare the final alternative's evaluations $FAE(x_i)$=$[Bel(x_i), Pl(x_i)]$ using the method for interval comparison based on *DST*, presented in previous subsection.

The local criteria aggregation on the base of Dempster's combination rule is only one of the methods we can use for *MCDM* in the framework of *A-IFS*/*DST* approach. Therefore we shall denote the corresponding final alternative's evaluation as $FAE_{com}(x_i)$.

Using above approach, the following results for the considered example (3.139), (3.140) have been obtained:

$$FAE_{com}(x_1) = [Bel(x_1), Pl(x_1)] = [0.9257, 0.9257],$$

$$FAE_{com}(x_2) = [Bel(x_2), Pl(x_2)] = [0.8167, 0.8168],$$
$$FAE_{com}(x_3) = [Bel(x_3), Pl(x_3)] = [0.7377, 0.7379].$$

Obviously, to select the best alternative, their final evaluations $FAE_{com}(x_i)$ presented by corresponding intervals should be compared. To obtain the final ranking on the set of comparing alternatives, the real valued criteria introduced in [108] and presented in previous subsection (strong, weak and mixed preferences) could be used, but in the considered example we get the results of such comparison in the form of degenerated belief intervals:

$$BI_{comb}(x_1 > x_2) = 1, BI_{comb}(x_2 > x_3) = 1, BI_{comb}(x_1 > x_3) = 1,$$

$$BI_{comb}(x_1 < x_2) = 0, BI_{comb}(x_2 < x_3) = 0, BI_{comb}(x_1 < x_3) = 0,$$
$$BI_{comb}(x_1 = x_2) = 0, BI_{comb}(x_2 = x_3) = 0, BI_{comb}(x_1 = x_3) = 0.$$

Since in our example only non interval results (0 or 1) of $FAE_{com}(x_i)$ comparison have been obtained, it is easy to see that the final alternative's ranking is $x_3 \prec x_2 \prec x_1$. It is worth noting that using the same example, the substantially different result $x_2 \prec x_3 \prec x_1$ has been obtained in [72].

We can explain such divergence of the results only by the absence of any intermediate type reduction in our method for *MCDM* in *A-IFS* setting that makes it possible to avoid the loss of important information. Summarizing, we can say that *DST* may serve as a good methodological base for interpretation of *A-IFS*. The use of *DSF* semantics makes it possible to enhance the performance of *A-IFS* when dealing with the *MCDM* problems. Particularly, when solving *MCDM* problems, the proposed approach allows us to use the Dempster's rule of combination directly to aggregate the local criteria presented by *IFV*s when their weights are *IFV*s too without intermediate defuzzification.

3.4 Summary and Discussion

In this chapter, we have presented an overview of modern methods for uncertainty modeling based on fuzzy sets and level 2 fuzzy sets, intuitionistic fuzzy sets, interval analysis and Dempster-Shafer theory of evidence. We have shown the interrelations between these methods and emphasize some problems that impede their applications.

It is shown that one of the most undesirable negative features of interval arithmetic is the fast increasing of width of intervals obtained as the results of interval calculations (excess width effect).

Another important problem of interval analysis is the so-called natural interval extension. If we have to make interval extension of real valued function, all argument of this function should be replaced with corresponding intervals and all operations should be replaced with corresponding operations on intervals. Such approach to interval extension seems to be justified enough and intuitively clear. Nevertheless, the so-called dependency problem is a major obstacle to the application of extension principle in interval arithmetic. Although interval methods can determine the range of elementary arithmetic operations and functions very accurately, this is not always true with more complicated functions. If an interval occurs several times in a calculation, and each occurrence is taken independently then this can lead to an unwanted expansion of the resulting intervals.

An important problem of interval extension is also that the accuracy of resulting interval strongly depends on the algebraic form of function chosen for interval extension.

It is worth noting that as the fuzzy arithmetic operations are usually based on the α -cut representation of fuzzy numbers, the above mentioned problems of interval analysis are the problems of fuzzy arithmetic as well.

It is noted that the most important applications of Atanassov's intuitionistic fuzzy sets (*A-IFS*) are the multiple criteria decision making problems (*MCDM*).

It is shown that there exist two important problems in *MCDM* in the intuitionistic fuzzy setting: aggregation of local criteria without intermediate defuzzification in the case when criteria and their weights are *IFVs*; comparison of *IF* valued scores of alternatives basing on the degree to which one *IFV* is grater/smaller than the other. In this chapter we have shown that there exist a strong link between *DST* and *A-IFS* which makes it possible to reformulate the basic definitions of *A-IFS* in terms of *DST*. We show that using the *DST* semantics it is possible to enhance the performance of *A-IFS* when dealing with MCDM problems. Particularly, this approach allows us to use directly the Dempster's rule of combination to aggregate local criteria presented by *IFVs* and develop a method for *MCDM* without intermediate defuzzification when local criteria and their weights are *IFVs*. As the result we get final alternative's evaluations in the form of belief interval. Hence, an appropriate method for such intervals comparison is needed.

Therefore, we present a method for interval and fuzzy numbers comparison based on *DSF* which provides the results of comparison in the form of belief intervals.

It is noted that now besides the classical Dempster's rule of combination a number of other methods for combination of evidence are proposed in the literature. All of them have own merits and drawbacks and the problem of choosing the best method is now open.

References

1. Abbasbandy, S., Asady, B.: Ranking of fuzzy numbers by sing distance. Information Sciences 176, 2405–2416 (2006)
2. Alsina, C.: On a family of connectives for fuzzy sets. Fuzzy Sets and Systems 16, 231–235 (1985)
3. Atanassov, K.T.: Intuitionistic fuzzy sets. Fuzzy Sets and Systems 20, 87–96 (1986)
4. Atanassov, K.: New operations defined over the intuitionistic fuzzy sets. Fuzzy Sets and Systems 61, 137–142 (1994)
5. Atanassov, K.: Intuitionistic Fuzzy Sets. Springer Physica-Verlag, Berlin (1999)
6. Atanassov, K., Pasi, G., Yager, R.: Intuitionistic fuzzy interpretations of multi-person multicriteria decision making. In: Proc. of 2002 First International IEEE Symposium Intelligent Systems, vol. 1, pp. 115–119 (2002)
7. Atanassov, K.: Intuitionistic fuzzy sets, Past, present, future. In: Proc. Third European Conf. on Fuzzy Logic and Technol (Eusflat 2003), Zittau, Germany, pp. 12–19 (2003)
8. Atanassov, K., Pasi, G., Yager, R., Atanassova, V.: Intuitionistic fuzzy group interpretations of multi-person multi-criteria decision making. In: Proc. of the Third Conference of the European Society for Fuzzy Logic and Technology, EUSFLAT 2003, Zittau, September 10-12, pp. 177–182 (2003)
9. Baas, S.M., Kwakernaak, H.: Rating and ranking multiple-aspect alternatives using fuzzy sets. Automatica 13, 47–58 (1977)
10. Bartolan, G., Degani, R.: A review of some methods for ranking fuzzy subsets. Fuzzy sets and Systems 15, 1–19 (1985)
11. Bayes, T.: An assay toward solving a problem in the doctrine of chances. Phil. Trans. Roy. Soc. (London) 53, 370–418 (1763)
12. Bellman, R., Zadeh, L.: Decision-making in fuzzy environment. Management Science 17, 141–164 (1970)
13. Beynon, M.: DS/AHP method: A mathematical analysis, including an understanding of uncertainty. European Journal of Operational Research 140, 148–164 (2002)
14. Beynon, M.: Understanding local ignorance and non-specificity within the DS/AHP method of multi-criteria decision making. European Journal of Operational Research 163, 403–417 (2005)
15. Beynon, M., Curry, B., Morgan, P.: The Dempster-Shafer theory of evidence: an alternative approach to multicriteria decision modeling. Omega 28, 37–50 (2000)
16. Beynon, M., Cosker, D., Marshall, D.: An expert system for multi-criteria decision making using Dempster Shafer theory. Expert Systems with Applications 20, 357–367 (2001)
17. Burillo, P., Bustince, H.: Entropy on intuitionistic fuzzy sets and interval-valued fuzzy sets. Fuzzy Sets and Systems 78, 305–316 (1996)
18. Burillo, C., Deschrijver, G., Kerre, E.: Implication in intuitionistic fuzzy and interval-valued fuzzy set theory: Construction, classification, application. International Journal of Approximate Reasoning 35, 55–95 (2004)

19. Bustince, H., Burillo, P.: Vague sets are intuitionistic fuzzy sets. Fuzzy Sets and Systems 79, 403–405 (1996)
20. Calvo, T., Kolesarova, A., Komornikova, M., Mesiar, R.: Aggregation operators: properties, classes and construction methods. In: Calvo, T., Mayor, G., Mesiar, R. (eds.) Aggregation Operators New Trends and Applications, pp. 3–104. Physica-Verlag, Heidelberg (2002)
21. Caprani, O., Madsen, K.: Mean value forms in interval analysis. Computing 25, 147–154 (1980)
22. Caselton, W.F., Luo, W.: Decision making with imprecise probabilities: Dempster-Shafer theory and applications. Water Resources Research 28, 3071–3083 (1992)
23. Chanas, S., Kuchta, D.: Multi-objective Programming in optimization of the Interval Objective Functions - a generalized approach. European Journal of Operational Research 94, 594–598 (1996)
24. Chanas, S., Zielinski, P.: Ranking fuzzy numbers in the settung of random sets-further results. Information Sciences 117, 191–200 (1999)
25. Chen, S.M., Tan, J.M.: Handling multicriteria fuzzy decision-making problems based on vague set theory. Fuzzy Sets and Systems 67, 163–172 (1994)
26. C++ Interval Arithmetic Library Reference, `http://docs.sun.com/htmlcollcoll.693/iso-8859-1/CPPARIT.../iapgrefman.htm`
27. Van Dalen, D.: Intuitionistic logic. In: Handbook of Philosophical Logic, 2nd edn., vol. 5, pp. 1–115. Kluwer, Dordrecht (2002)
28. Davis, E.: Constraint propagation with interval interval lebels. Artificial Intelligence 32, 353–366 (1987)
29. De, S.K., Biswas, R., Roy, A.R.: Some operations on intuitionistic fuzzy sets. Fuzzy Sets and Systems 114, 477–484 (2000)
30. Dempster, A.P.: Upper and lower probabilities induced by a muilti-valued mapping. Ann. Math. Stat. 38, 325–339 (1967)
31. Dempster, A.P.: A generalization of Bayesian inference (with discussion). J. Roy. Stat. Soc., Series B. 30, 208–247 (1968)
32. Deneux, T.: Modeling vague beliefs using fuzzy-valued belief structures. Fuzzy Sets and Systems 116, 167–199 (2000)
33. Denoeux, T.: Conjunctive and disjunctive combination of belief functions induced by nondistinct bodies of evidence. Artificial Intelligence 172, 234–264 (2008)
34. Deschrijver, G., Cornelis, C., Kerre, E.: On the representation of intuitionistic fuzzy t-norms and t-conorms. IEEE Transactions on Fuzzy Systems 12, 45–61 (2004)
35. Deschrijver, G., Kerre, E.E.: On the position of intuitionistic fuzzy set theory in the framework of theories modelling imprecision. Information Sciences 177, 1860–1866 (2007)
36. Dubois, D., Gottwald, S., Hajek, P., Kacprzyk, J., Prade, H.: Terminological difficulties in fuzzy set theory-The case of "Intuitionistic Fuzzy Sets". Fuzzy Sets and Systems 156, 485–491 (2005)
37. Dubois, D., Prade, H.: Operations on fuzzy numbers. International Journal of Systems Science 9, 613–626 (1978)
38. Dubois, D., Prade, H.: Fuzzy Sets and Systems: Theory and Applications. Academic Press, London (1980)
39. Dubois, D., Prade, H.: Ranking of fuzzy numbers in the setting of possibility theory. Information Sciences 30, 183–224 (1983)
40. Dubois, D., Prade, H.: Representation and combination of uncertainty with belief functions and possibility measures. Computational Intelligence 4, 244–264 (1998)

41. Facchinetti, G., Ricci, R.G., Muzzioli, S.: Note on ranking fuzzy triangular numbers. International Journal of Intelligent Systems 13, 613–622 (1998)
42. Fuller, R.: Introduction to Neuro-Fuzzy Systems, Advances in Soft Computing. Physica-Verlag, A Springer-Verlag Company, Heidelberg, New York (2000)
43. Fodor, J.C., Roubens, M.: Fuzzy Preference Modelling and Multicriteria Decision Support. Kluwer Academic Publishers, Dordrecht (1994)
44. Gau, W.L., Buehrer, D.J.: Vague sets. IEEE Trans. Systems Man Cybernet 23, 610–614 (1993)
45. Goodman, I.R., Nguyen, H.T.: Uncertainty Models for Knowledge-Based System. North-Holand, Amsterdam (1985)
46. Grzegorzewski, P., Mrowka, E.: Some notes on (Atanassovs) intuitionistic fuzzy sets. Fuzzy Sets and Systems 156, 492–495 (2005)
47. Haenni, R.: Are alternatives to Dempster's rule of combination real alternatives?: Comments on "about the belief function combination and the conflict management problem"-Lefevre et al. Information Fusion 3, 237–239 (2002)
48. Hansen, E.: A generalized interval arithmetic. In: Nickel, K. (ed.) Interval Mathematics. LNCS, vol. 29, pp. 7–18. Springer, Heidelberg (1975)
49. Hanson, R.J.: Interval arithmetic as a closed arithmetic system on a komputer. Technical Memorandum 197, Jet Propulsion Laboratory, Section 314, Kalifornia Instytut of Technology, Pasadena, CA (1968)
50. Hielpern, S.: Representation and application of fuzzy numbers. Fuzzy sets and Systems 91, 259–268 (1997)
51. Hinde, C.J., Patching, R.S., McCoy, S.A.: Inconsistent Intuitionistic Fuzzy Sets and Mass Assignment. EXIT (2007)
52. Hinde, C.J., Patching, R.S., McCoy, S.A.: Semantic transfer and contradictory evidence in intuitionistic fuzzy sets. In: Proc. of 2008 IEEE International Conference on Fuzzy Systems, pp. 2095–2102 (2008)
53. Hong, D.H., Choi, C.-H.: Multicriteria fuzzy decision-making problems based on vague set theory. Fuzzy Sets and Systems 114, 103–113 (2000)
54. Hong, D.H., Lee, S.: Some properties and a distance measure for interval-valued fuzzy numbers. Information Sciences 148, 1–10 (2002)
55. Hua, Z., Gong, B., Xu, X.: A DS-AHP approach for multi-attribute decision making problem with incomplete information. Expert Systems with Applications 34, 2221–2227 (2008)
56. Ishihashi, H., Tanaka, M.: Multiobjective programming in optimization of the Interval Objective Function. European Journal of Operational Research 48, 219–225 (1990)
57. Jaulin, L., Kieffer, M., Didrit, O., Walter, E.: Applied interval analysis. Springer, London (2001)
58. John, R.I.: Type-2 inferencing and community transport scheduling. In: Proc. 4th Euro. Congress Intelligent Techniques Soft Computing, Aachen, Germany, pp. 1369–1372 (1996)
59. Jumarie, G.: Relativistic fuzzy sets. Toward a new approach to subjectivity in human systems. Mathématiques et Scienses Humainies 18, 39–75 (1980)
60. Kahan, W.: A more complete interval arithmetic. Lectures notes for a summer course. University of Toronto, Canada (1968)
61. Karnik, N.N., Mendel, J.M.: Applications of type-2 fuzzy logic systems to forecasting of timeseries. Information Sciences 120, 89–111 (1999)
62. Kasperski, A.: A possibilistic approach to sequencing problems with fuzzy parameters. Fuzzy Sets and Systems 150, 77–86 (2005)

63. Klement, E.P., Mesiar, R., Pap, E.: Triangular Norms. Kluwer Academic Publishers, Dordrecht (2000)
64. Klir, G.J., Yuan, B.: Fuzzy Sets and Fuzzy Logic: Theory and Applications. Prentice Hall, Upper Saddle River (1995)
65. Krishnapuram, R., Keller, J.M., Ma, Y.: Quantitative analysis of properties and spatial relations of fuzzy image regions. IEEE Trans. Fuzzy Systems 1, 222–233 (1993)
66. Kulpa, Z.: Diagrammatic representation for a space of intervals. Machine Graphics and Vision 6, 5–24 (1997)
67. Kundu, S.: Preference relation on fuzzy utilities based on fuzzy leftness relation on interval. Fuzzy Sets and Systems 97, 183–191 (1998)
68. Kundu, S.: Min-transitivity of fuzzy leftness relationship and its application to decision making. Fuzzy Sets and Systems 86, 357–367 (1997)
69. Lee, S., Lee, K.H., Lee, D.: Ranking the sequences of fuzzy value. Information Sciences 160, 41–52 (2004)
70. Li, F., Lu, A., Cai, L.: Methods of multi-criteria fuzzy decision making based on vague sets. Journal of Huazhong University of Science and Technology 29, 1–3 (2001) (in Chinese)
71. Li, F., Rao, Y.: Weighted methods of multi-criteria fuzzy decision making based on vague sets. Computer Science 28, 60–65 (2001) (in Chinese)
72. Li, D.-F.: Multiattribute decision making models and methods using intuitionistic fuzzy sets. Journal of Computer and System Sciences 70, 73–85 (2005)
73. Liang, Q., Mendel, J.M.: Interval Type-2 Fuzzy Logic Systems: Theory and Design. IEEE Trans. on Fuzzy Systems 8, 535–550 (2000)
74. Lin, L., Yuan, X.-H., Xia, Z.-Q.: Multicriteria fuzzy decision-making methods based on intuitionistic fuzzy sets. Journal of Computer and System Sciences 73, 84–88 (2007)
75. Liu, H.-W., Wang, G.-J.: Multi-criteria decision-making methods based on intuitionistic fuzzy sets. European Journal of Operational Research 179, 220–233 (2007)
76. Liu, F., Mendel, J.M.: Aggregation Using the Fuzzy Weighted Average, as Computed by the KM Algorithms. IEEE Trans. on Fuzzy Systems 16, 1–12 (2008)
77. Liu, F., Mendel, J.M.: Encoding words into interval type-2 fuzzy sets using an interval approach. IEEE Trans. on Fuzzy Systems 16, 1503–1521 (2008)
78. Lowrance, J.D., Garvey, T.D., Strat, T.M.: A framework for evidential-reasoning systems. In: Proceedings of the 5 National Conference on Artificial Intelligence (AAAI 1986), Philadelphia, pp. 896–901 (1986)
79. Mannucci, M.A.: Quantum Fuzzy Sets: Blending Fuzzy Set Theory and Quantum Computation. Cornell University Library (2008) arXiv:cs/0604064v1
80. Markov, S.M.: A non-standard subtraction of intervals. Serdica 3, 359–370 (1977)
81. Melgarejo, M.: Implementing interval type-2 fuzzy processors. IEEE Computational Intelligence Magazine 2, 63–71 (2007)
82. Mendel, J.M.: Uncertainty, fuzzy logic, and signal processing. Signal Processing Journal 80, 913–933 (2000)
83. Mendel, J.M.: Uncertain Rule-Based Fuzzy Logic Systems: Introduction and New Directions. Prentice-Hall, Upper-Saddle River (2001)
84. Mendel, J.M., John, R.I.B.: Type-2 Fuzzy Sets Made Simple. IEEE Transactions on Fuzzy systems 10, 117–127 (2002)
85. Mendel, J.M.: Fuzzy Sets for Words: a New Beginning. In: Proc. IEEE FUZZ Conference, May 26-28, pp. 37–42. St. Louis, MO (2003)
86. Mendel, J.M.: Type-2 fuzzy sets and systems: an overview. IEEE Computational Intelligence Magazine 2, 20–29 (2007)

87. Mendel, J.M.: Advances in type-2 fuzzy sets and systems. Information Sciences 177, 84–110 (2007)
88. Montero, J., Gmez, D., Bustince, H.: On the relevance of some families of fuzzy sets. Fuzzy Sets and Systems 158, 2429–2442 (2007)
89. Moore, R.E.: Interval analysis. Prentice-Hall, Englewood Cliffs (1966)
90. Moore, R.E.: Automatic error analysis in digital computation. Technical Report Space Div. Report LMSD84821, Lockheed Missiles and Space Co. (1959)
91. Moore, R.E., Kearfott, R.B., Cloud, M.J.: Introduction to Interval Analysis. SIAM Press, Philadelphia (2009)
92. Murphy, C.K.: Combining belief functions when evidence coflicts. Decision Support Systems 29, 1–9 (2000)
93. Nakamura, K.: Preference relations on set of fuzzy utilities as a basis for decision making. Fuzzy Sets and Systems 20, 147–162 (1986)
94. Nguyen, H.T., Kreinovich, V., Longpre, L.: Dirty Pages of Logarithm Tables, Lifetime of the Universe, and (Subjective) Probabilities on Finite and Infinite Intervals. Reliable Computing 10, 83–106 (2004)
95. Nikolova, M., Nikolov, M., Cornelis, C., Deschrijver, G.: Survey of the research on intuitionistic fuzzy sets. Advanced Studies in Contemporary Mathematics 4, 127–157 (2002)
96. Parsons, S.: Some qualitative approaches to applying the Dempster-Shafer theory. Information and Decision Technologies 19, 321–337 (1994)
97. Pasi, G., Atanassov, K., Pinto, P.M., Yager, R., Atanassova, V.: Multi-person multi-criteria decision making: Intuitionistic fuzzy approach and generalized net model. In: Proc. of the 10th ISPE International Conference on Concurrent Engineering "Advanced Design, Production and Management Systems", Madeira, July 26-30, pp. 1073–1078 (2003)
98. Pasi, G., Yager, Y., Atanassov, K.: Intuitionistic fuzzy graph interpretations of multi-person multi-criteria decision making: Generalized net approach. In: Proceedings of 2004 Second International IEEE Conference Intelligent Systems, pp. 434–439 (2004)
99. Pedrycz, W.: Fuzzy Control and Fuzzy Systems. John Wiley and Sons, New York (1993)
100. Piegat, A.: Fuzzy Modeling and Control. Physica Verlag, Heidelberg (2001)
101. Ratschek, H., Rokne, J.: Computer methods for the Range of Functions. Ellis Horwood, Chichester (1984)
102. Rommelfanger, H.: Fuzzy Decision - Support System. Springer, Heidelberg (1994)
103. Rutkowski, L.: New Soft Computing Techniques for System Modelling, Pattern Classification and Image Proceesing. Springer, Heidelberg (2004)
104. Rutkowski, L.: Flexible Neuro-Fuzzy Systems. Kluwer Academic Publisher, Dordrecht (2004)
105. Sendov, B.: Segment arithmetic and segment limit. C.R. Acad. Bulgare Sci. 30, 995–998 (1977)
106. Sengupta, A., Pal, T.K.: On comparing interval numbers. European Journal of Operational Research 127, 28–43 (2000)
107. Sentz, K., Ferson, S.: Combination of evidence in Dempster-Shafer theory. Technical report, SANDIA National Laboratories (2002)
108. Sevastianov, P.: Numerical methods for interval and fuzzy number comparison based on the probabilistic approach and Dempster-Shafer theory. Information Sciences 177, 4645–4661 (2007)
109. Sevastianov, P., Róg, P.: Fuzzy modeling of manufacturing and logistic systems. Mathematics and Computers in Simulation 63, 569–585 (2003)

110. Sevastjanov, P., Róg, P., Karczewski, K.: A Probabilistic Method for Ordering Group of Intervals. Computer Science, Politechnika Częstochowska 2, 45–53 (2002)
111. Sevastjanov, P., Róg, P.: Fuzzy modeling of manufacturing and logistic systems. Mathematics and Computers in Simulation 63, 569–585 (2003)
112. Sevastjanov, P., Róg, P.: Two-objective method for crisp and fuzzy interval comparison in Optimization. Computers & Operations Research 33, 115–131 (2006)
113. Sevastjanov, P., Venberg, A.: Modeling and simulation of power units work under interval uncertainty. Energy (3), 66–70 (1998) (in Russian)
114. Sevastjanov, P., Venberg, A.: Optimization of technical and economic parameters of power units work under fuzzy uncertainty. Energy (1), 73–81 (2000) (in Russian)
115. Sevastjanov, P., Venberg, A., Róg, P.: A probabilistic approach to fuzzy and interval ordering. Task Quarterly, Special Issue "Artificial and Computational Intelligence" 7, 147–156 (2003)
116. Sewastianow, P., Róg, P., Venberg, A.: The Constructive Numerical Method of Interval Comparison. In: Wyrzykowski, R., Dongarra, J., Paprzycki, M., Waśniewski, J. (eds.) PPAM 2001. LNCS, vol. 2328, pp. 756–761. Springer, Heidelberg (2002)
117. Shafer, G.: A mathematical theory of evidence. Princeton University Press, Princeton (1976)
118. Sheen, J.N.: Fuzzy financial profitability analysis of demand side management alternatives from participant perspective. Information Sciences 169, 329–364 (2005)
119. Schubert, J.: Cluster-based specification techniques in Dempster-Shafer theory for an evidential intelligence analysis of multiple target tracks. Department of Numerical Analysis and Computer Science Royal Institute of Technology, S-100 44 Stockholm, Sweden (1994)
120. Smets, P.: The combination of evidence in the transferable belief model. IEEE Transactions on Pattern Analysis and Machine Intelligence 12, 447–458 (1990)
121. Smets, P.: Analyzing the combination of conflicting belief functions. Information Fusion 8, 387–412 (2007)
122. Sugeno, M.: Fuzzy measures and fuzzy integrals: a survey. In: Gupta, M.M., Saridis, D.N., Gaines, B.R. (eds.) Fuzzy Automata and Decision Process, pp. 89–102. North-Holland, Amsterdam (1977)
123. Szmidt, E., Kacprzyk, J.: Intuitionistic fuzzy sets in decision making. Notes *IFS* 2, 15–32 (1996)
124. Szmidt, E., Kacprzyk, J.: Remarks on some applications on intuitionistic fuzzy sets in decision making. Notes IFS 2, 22–31 (1996)
125. Szmidt, E., Kacprzyk, J.: Group decision making under intuitionistic fuzzy preference relations. In: Proceedings of Seventh International Conference (IMPU 1998), Paris, pp. 172–178 (1998)
126. Szmidt, E., Kacprzyk, J.: Applications of intuitionistic fuzzy sets in decision making. In: Proc. Eighth Cong. EUSFLAT 1998, Pampelona, pp. 150–158 (1998)
127. Szmidt, E., Kacprzyk, J.: A New Similarity Measure for Intuitionistic Fuzzy Sets:Straightforward Approaches network. In: Proc. of IEEE International Conference on Fuzzy Systems, pp. 1–6 (May 2008)
128. Szmidt, E., Kacprzyk, J.: A New Approach to Ranking Alternatives Expressed Via Intuitionistic Fuzzy Sets. In: Proc. of 2008 International Conference on Fuzzy Logic and Nuclear Safety, pp. 265–270 (2008)
129. Takeuti, G., Titani, S.: Intuitionistic fuzzy logic and intuitionistic fuzzy set theory. J. Symbolic Logic 49, 851–866 (1984)
130. Vasseur, P., Pegard, C., Mouad-dib, E., Delahoche, L.: Perceptual organization approach based on Dempster-Shafer theory. Pattern Recognition 32, 1449–1462 (1999)

131. Wadman, D., Schneider, M., Schnaider, E.: On the use of interval mathematics in fuzzy expert system. International Journal of Intelligent Systems 9, 241–259 (1994)
132. Walley, P.: Belief-function representations of statistical evidence. Annals of Statistics 10, 741–761 (1987)
133. Walster, G.W., Bierman, M.S.: Interval Arithmetic in Forte Developer Fortran. Technical Report., Sun Microsystems (2000)
134. Wang, X., Kerre, E.E.: Reasonable properties for the ordering of fuzzy quantities (I) (II). Fuzzy Sets and Systems 112, 387–405 (2001)
135. Wang, Y.M., Yang, J.B., Xu, D.L.: A preference aggregation method through the estimation of utility intervals. Computers & Operations Research 32, 2027–2049 (2005)
136. Warmus: Calculus of Approximations. Bull. Acad. Polon. Sci., Cl. III IV, 253–259 (1956)
137. Watanabe, T.: A generalized fuzzy set theory. IEEE Transactions on Systems, Man, and Cybernetics 8, 756–763 (1978)
138. Weber, S.: A general concept of fuzzy connectives, negations and implications based on t-norms and t-co-norms. Fuzzy Sets and Systems 11, 115–134 (1983)
139. Wilson, P.N.: Some theoretical aspects of the Dempster-Shafer theory. PhD Thesis, Oxford Polytechnic (1992)
140. Xu, Z., Yager, R.R.: Some geometric aggregation operators based on intuitionistic fuzzy sets. International Journal of General Systems 35, 417–433 (2006)
141. Xu, Z.: Models for multiple attribute decision making with intuitionistic fuzzy information. International Journal of Uncertainty, Fuzziness and Knowledge-Based Systems 15, 285–297 (2007)
142. Xu, Z.: Intuitionistic preference relations and their application in group decision making. Information Sciences 177, 2363–2379 (2007)
143. Xu, Z.: Intuitionistic fuzzy aggregation operators. IEEE Transactions on Fuzzy Systems 15, 1179–1187 (2007)
144. Xu, Z., Yager, R.R.: Dynamic intuitionistic fuzzy multi-attribute decision making. International Journal of Approximate Reasoning 48, 246–262 (2008)
145. Yager, R.: On general class of fuzzy connectives. Fuzzy Sets and Systems 4, 235–242 (1980)
146. Yager, R.R.: Fuzzy subsets of type II in decisions. Journal of Cybernetics 10, 137–159 (1980)
147. Yager, R.R.: A procedure for ordering fuzzy subsets of the unit interval. Information Sciences 24, 143–161 (1981)
148. Yager, R.R.: On the Dempster-Shafer framework and new combitanion rules. Information Sciences 41, 93–138 (1987)
149. Yager, R.R.: On ranking fuzzy numbers using valuations. International Journal of Intelligent Systems 14, 1249–1268 (1999)
150. Yager, R.R.: Modeling uncertainty using partial information. Information Sciences 121, 271–294 (1999)
151. Yager, R.R., Detyniecki, M.: Ranking fuzzy numbers using α-weighted valuations. International Journal of Uncertainty, Fuzziness and Knowledge-based Systems 8, 573–591 (2000)
152. Yager, R.R., Detyniecki, M., Bouchon-Meunier, B.: A context−dependent method for ordering fuzzy numbers using probabilities. Information Sciences 138, 237–255 (2001)
153. Yager, R.R., Kacprzyk, J., Fedrizzi, M.: Advances in Dempster-Shafer Theory of Evidence. Wiley, New York (1994)
154. Zadeh, L.A.: Fuzzy Sets. Information and Control 8, 338–353 (1965)

155. Zadeh, L.A.: Similarity relations and fuzzy orderings. Information Science 3, 177–200 (1971)
156. Zadeh, L.A.: The concept of a linguistic variable and its application to approximate reasoning. Information Science 8 (Part I), 199–249, 8 (Part II), 301–357, 9 (Part III), 43–80 (1975)
157. Zadeh, L.A.: Review of Books: A Mathematical Theory of Evidence. The AI Magazine 5, 81–83 (1984)
158. Zimmermann, H.-J.: Fuzzy Set Theory and Its Applications. Kluwer Academic Publishers, London (1991)
159. Zimmermann, H.J., Zysno, P.: Latest connectives in human decision making. Fuzzy Sets Systems 4, 37–51 (1980)

Chapter 4
MCDM with Applications in Economics and Finance

In this chapter, the problems typical for multiple criteria decision making (*MCDM*) are analyzed and new solutions of them are proposed as well. The problem of appropriate common scale for representation of objective and subjective criteria is solved using the simple subsethood measure based on the α-cut representation of fuzzy values. To develop an appropriate method for aggregation of aggregating modes, we use the synthesis of the tools of type 2 and level 2 fuzzy sets. As the result, the final assessments of compared alternatives are presented in the form of fuzzy valued membership function defined on the support composed of considered alternatives. To compare obtained fuzzy assessments we use the probabilistic approach to fuzzy values comparison. In is shown that investment evaluation problem is frequently a hierarchical one and a new method for solving such problems, different from commonly used fuzzy analytic hierarchy process (*AHP*) method, is proposed. The developed methods are used for the solution of the stock ranking problem based on the multiple criterion decision making and optimization in the fuzzy setting and for multiple criteria fuzzy evaluation and optimization in budgeting.

4.1 MCDM in the Fuzzy Setting

This section presents an analysis of the methodological problems of *MCDM* such as common representation of different types of local criteria, expert's opinions aggregation, aggregation of local criteria, aggregation of aggregating modes, fuzzy numbers comparison, hierarchical structure of local criteria set.

It is well known that the evaluation of important investment projects usually can not be successfully carried out using only financial parameters since the possible ecological, social and even political effects of project's implementation should be evaluated as well. The role of these effects rises along with the project's importance. Obviously, such effects as a rule can not be predicted with a high accuracy, moreover their estimations are usually based on the expert's opinions expressed in a verbal form. So the proper mathematical tools are needed to incorporate such ill defined estimations into the general evaluation of investment project. On the other hand,

L. Dymowa: Soft Computing in Economics and Finance, ISRL 6, pp. 107–186.
springerlink.com

the traditional approaches to the investment project evaluation are usually based on the budgeting, i.e., analysis of the discounted financial parameters of the considered projects such as Net Present Value (*NPV*), Internal Rate of Return (*IRR*) and so on.

It is easy to see that in this case also the estimation of the investment efficiency, as well as any forecasting, is rather an uncertain problem and the proper methods for operating in the uncertain setting should be used. Since the applicability of traditional probability methods is often restricted by the absence of objective probabilistic information about future events, during the last two decades the growing interest to the application of interval and fuzzy methods in budgeting has been observing (see [35, 55, 65, 135]).

On the other hand, when analyzing the investment project, we consider (sometimes implicitly) some local criteria based on the calculated financial parameters or quantitative evaluations of the project's implementation effects. Therefore, the project estimation is in essence a multiple criteria problem. As the examples of successful systematization, the local criteria sets proposed in [78, 135] may be considered. Even skin-deep analysis of these criteria systematization allows us to conclude that investment and project quality estimation is the complicated multiple criteria problem frequently with a certain hierarchical structure. There is a lot of multiple criteria methods proposed in the literature for solving economic and financial problems. Steuer [118] presented the widest review of this problem based on more than 250 literature indices. Nevertheless, we can cite the only few papers devoted to the multiple criteria financial project estimation (see [72, 87, 135]).

It is worth noting that in all these works the concepts of fuzzy sets theory were used. The method proposed in [87] is based on the representation of local criteria by membership functions and their aggregation using simple fuzzy summation. The ranks of local criteria and possible hierarchical structure of the problem were not taken into account. The hierarchical structure of the problem is considered in [136] and well known *AHP* method is used for its building, but only simple normalization of financial parameters (dividing them by their maximal values) is applied instead of natural local criteria. An interesting example of practical application of the multiple criteria hierarchical analysis is presented in [72]. The generalized *AHP* method was used for estimation of 103 mutually dependent investment projects proposed for the Tumen river region (China) industrial development.

We do not intend to make here the detailed review of these works, but as a result of the analysis we have done in the field of investment project estimation as well as in some other practical applications, we can say that generally the project's evaluation is a multiple criteria decision making hierarchical problem in the fuzzy setting. It is important that as it has been pointed out in [6] "The theory of *MCDM* is an open theoretical field and not a closed mathematical theory solving a specific class of problems". Nevertheless, there are some methodological problems which are common ones for almost all multiple criteria based approaches. They were considered and systemized in [5, 75, 99, 117, 118, 119]. There are different definitions of *MCDM* proposed in the literature, but regardless of what type of *MCDM* task is solving (choice, ranking, sorting, *ets*), two pivotal problems arise: how to evaluate alternatives and how to compare them? The last problem is especially important if the results of alternatives evaluations are presented by interval or fuzzy numbers. In this section, we focus on the class of *MCDM* problems which may be treated as the developing the methods helping the decision maker to choose the best alternative when several local criteria (sometimes antagonistic ones) effect to the decision. These problems usually are solved in a two phase process [25, 97, 128]: the *rating*, i.e., aggregation of the criteria values for comparing alternatives and *ranking* or *ordering* these alternatives. Last phase is not trivial in the fuzzy or interval setting. There are some problems we are faced in the rating phase especially when dealing with fuzzy local criteria and/or their fuzzy weights (ranks). They can be roughly clustered as follows:

(i) *Common representation of different types of local criteria.*
The local criteria may be constructed on the base of quantitative parameters such as financial ones, as well as using expert's subjective estimations (verbal assessments of project's scientific importance, technological level, etc). It is known that experts prefer to provide rather "fuzzy" advises on the linguistic level of presentation to avoid possible mistakes caused by the qualitative nature of predictions. However, human experience and intuition play an important role in the projects evaluation and cannot be ignored, although the specific uncertainty is their inherent property. This uncertainty is of subjective (fuzzy) nature and cannot be described in the usual probabilistic way. So a proper methodology is needed to take into account the uncertainty factors which will allow us to build a set of comparable local criteria based on directly measurable quantitative parameters as well as on linguistically formulated assessments. The mathematical tools of fuzzy sets theory developed for dealing with subjective kind of uncertainty [143] may be successfully used for this purpose. Thus, in the real-world problems we meet two group of local criteria [22, 105, 134]: objective criteria based on the numerical parameters and subjective ones based on the subjective expert's opinions. So the problem arises: how to find an appropriate common scale for representation of objective and subjective criteria?

(ii) *Expert's opinions aggregation.* Usually for the evaluation of important investment projects a number of experts in the relevant fields are involved into decision making process. Since in such cases we are dealing with the group *MCDM* [49, 69, 105, 134], we face with the problem of searching a compromise between expert's opinions available, especially if they are represented by different experts in linguistic form, e.g., as "low importance", " medium importance", " high importance", "large importance" [134].

(iii) *Aggregation of local criteria.* Real-world decision problems may involve a lot of local criteria to be analyzed simultaneously. Regrettably, the human ability to do this is strongly restricted by the known empirical law of psychology according to which a person can distinguish no more than 7 plus minus 2 classes or grades on some feature scale. If the number of grades is greater, the adjacent grades start to merge and cannot be clustered confidently (see [84] and [85]). To solve this problem, the relevant aggregation of local criteria taking into account their ranking can be used to create some generalized criteria. Therefore, the problem of choice of appropriate aggregation method is of perennial interest, because of its direct relevance to practical decision making [92, 138, 148]. The most popular aggregating mode is the weighted sum. It is used in many well known decision making models such as *AHP* [100], multi-attribute utility analysis [91] and so on, but often without any critical analysis. On the other hand, in some fields, e.g., in ecological modeling, the weighted sum is not used for aggregation [116]. The reason behind this is that in practice there are the cases when if any of local criteria is totally dissatisfied then considered alternative should be rejected from the consideration at all. Nevertheless, when dealing with a complex task characterized by a great number of local criteria, it seems reasonable to use all types of aggregations relevant to this task. If the results obtained using different aggregation modes are similar, this fact may be considered as a good confirmation of their optimality. In opposite case, an additional analysis of local criteria and their ranking should be advised.

(iv) *Aggregation of aggregating modes*. The natural consequence of problem (iii) is a growing interest in the methods for generalizing the aggregating operators (aggregation of aggregation modes) [98]. For this purpose, it is proposed to apply the possibility theory [39] as well as the weighted sum aggregation [141]. Also, Yager's *t*-norms are used in [47] and a hierarchical aggregation approach is developed in [40, 81]. Nowadays the most popular is the so-called γ-operator [148, 149]:

$$\eta = (\prod_i \mu_i)^{1-\gamma}(1-\prod_i(1-\mu_i))^{\gamma}, i = 1,2,\ldots,n, 0 \leq \gamma \leq 1, \tag{4.1}$$

where μ_i is a membership function corresponding to the local criterion. Since expression (4.1) is based only on the multiplicative aggregation, the more general approach was proposed in [86]:

$$\eta_{or} = \gamma \underset{i}{max}(\mu_i) + \frac{(1-\gamma)(\sum_i \mu_i)}{n}, \tag{4.2}$$

$$\eta_{and} = \gamma \underset{i}{min}(\mu_i) + \frac{(1-\gamma)(\sum_i \mu_i)}{n}. \tag{4.3}$$

These expressions were used in [104] to solve the multiple level decision making problem. As a key issue, the lack of strong rules for choosing the value of γ is mentioned. The work [29] is specifically devoted to this problem, but the method proposed by the authors demands too much additional information to be presented by a decision maker in the quantitative form. In practice, it is hard to get such information since it is not directly related to the decision maker's real problems. It is easy to see that local criteria in (4.3),(4.3) are considered to be not ranked, whereas their ranking seems to be a more important issue than choosing the value of γ. Finally, the generalizing modes (4.1)-(4.3) do not involve all possible approaches to the aggregation.

Of course, the set of above mentioned problems of *rating* in *MCDM* is not complete and exhausted. For example, in the realm of group *MCDM* the values of membership functions describing the local criteria may be the fuzzy values as well. So the problem of type-2 fuzzy representation of local criteria arises. To solve this problem the constructive method based on the so-called hyperfuzzy approach [41, 42] had been proposed recently. A nontrivial problem in the *ranking* stage of *MCDM* usually arises when as the result of previous rating phase the fuzzy (or interval) valued evaluations of alternatives are obtained. Generally, this problem can be formulated as follows.

(v) *Fuzzy numbers comparison*. It must be emphasized that the problem of intervals and fuzzy numbers comparison plays a pivotal role in the fuzzy *MCDM* and fuzzy optimization [18]. There exist numerous definitions of the ordering relation for fuzzy numbers (as well as crisp intervals) proposed in the literature. In most cases, the authors use some quantitative indices. The values of such indices present the degree to which one number (fuzzy or interval) is greater/smaller than the other number. In some cases, even several indices are used simultaneously. The widest review of this problem based on more than 35 literature indices was presented in [133] where the authors proposed a new interesting classification of methods for fuzzy numbers comparison. The separate group of methods is based on the so-called probabilistic approach to the intervals and fuzzy numbers comparison [18, 63, 66, 67, 89, 103, 112, 129, 142]. The attractiveness of this approach is caused by the possibility to build interval and fuzzy value relations using the minimum set of preliminary assumptions. In [109, 111], a new effective method for interval and fuzzy values comparison based on the probabilistic approach has been developed.

(vi) *Hierarchical structure of local criteria set*. The local criteria may compose a multilevel hierarchical structure when given set of local criteria consists of certain subgroups connected logically. Although the *AHP* method and its numerous fuzzy modification nowadays are commonly used to solve this problem, we can say that this problem is still open one.

The list of problems can be continued. For instance, some problems appear when fuzzy valued financial parameters are used as arguments of functions representing the local criteria [35] or when such functions are fuzzy as well [41, 42]. Nevertheless, the detailed analysis of these issues is out of scope of this book.

We can say that almost all components for building the efficient method that may be used for real-life project evaluations are already developed and described in the literature. What is needed is their critical analysis from the viewpoint of considered problem, and a proper synthesis of them into the integrated method for hierarchical multiple criteria evaluation of investment projects. The aim of this chapter is to present a synthetical approach to solve the above mentioned problems (i)-(vi) in context of investment evaluation. To make the presentation more transparent, it is illustrated throughout the chapter with the use of two examples. The first of them is well known tool steel material selection problem [134] which can be considered as the typical investment problem and relevant test charged by all difficulties concerned with the problems (i)-(v). The next example is a simplified investment project evaluation problem we have used to show that even when project's estimation is based on the budgeting, i.e., only financial parameters are taking into account, we are dealing with multiple criteria task in the fuzzy setting. The rest of the chapter is organized as follows. In Section 4.2, the tool steel material selection task is recalled and the expert's opinions aggregation problem (ii) concerned with the linguistic weights of local criteria is considered. Subsection 4.2.1 provides an exposition of the simple and transparent method for evaluation of fuzzy subsethood measure based on the α-cut representation of fuzzy numbers. This method has been developed for solving the above mentioned problems (i) and (ii). Subsection 4.2.2 is devoted to the common representation of different types of local criteria (i). Subsection 4.2.3 describes the probabilistic method for fuzzy numbers comparison (v). In Subsection 4.2.4, the problem of aggregation of local criteria (iii) is considered and a new approach to aggregation of aggregating modes (iv) based on the synthesis of type-2 and level-2 fuzzy sets is described. Proposed approach is illustrated with the use of the tool steel material selection problem as an example. In Section 4.3, we present the illustrative example of fuzzy multiple criteria investment project evaluation and an approach to the solution of hierarchical fuzzy *MCDM* problem which is different from the fuzzy *AHP* method. Section 4.4 is devoted to the fuzzy *MCDM* and optimization in the stock screening. In Section 4.5, the methods for multiple criteria fuzzy evaluation and optimization in budgeting are performed.

Finally, concluding section summarizes the chapter and discusses future research issues.

4.2 Tool Steel Material Selection Problem

This problem has been chosen as the test and illustrative example of investment problem since it is well known and had been discussed in the literature earlier [105, 134]. The most important is that it is charged with all problems (i)-(v) noted in Section 4.1.

Generally, the solution of this *MCDM* problem is organized as follows. The decision maker with a help of analyst considers expert's opinions and aggregates them to obtain a final conclusion.

Let us briefly recall the main assumptions and restrictions were made in the formulation of tool steel material selection problem in [134]. Suppose there are three experts involved in the decision process. The best among of five tool steel materials V_1, A_2, $\mathbb{D}_2$, γ_1, T_1 (classification of American Iron and Steel Institute, AISI) should be chosen. The local criteria are clustered into two groups: subjective criteria defined by experts on the base their experience and intuition (the properties of materials) and objective criteria based on the numerical parameters not dependent on the expert's opinions (the cost of material imposed by open market). In [134], the classification of criteria had been proposed as follows.

Subjective criteria:

- Non deforming properties for materials - local criterion C_1
- Safety in hardening for materials - local criterion C_2

Table 4.1 Linguistic values for criteria estimations

Linguistic terms	Corresponding fuzzy number
Worst (W)	(0,0,0.3)
Poor (P)	(0,0.3,0.5)
Fair (F)	(0.2,0.5,0.8)
Good (G)	(0.5,0.7,1)
Best (B)	(0.7,1,1)

- Toughness for materials - local criterion C_3
- Resistance to softening effect of heat for materials - local criterion C_4
- Wear resistance for materials - local criterion C_5
- Machinability for materials - local criterion C_6

Objective criterion:

- Cost -local criterion C_7

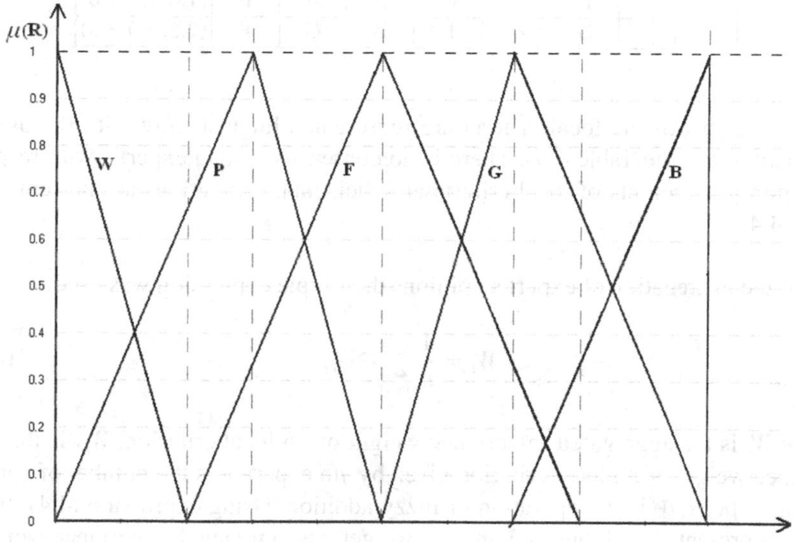

Fig. 4.1 Membership functions for linguistic values

The subjective criteria in [105, 134] were assessed linguistically (see Table 4.1 and Fig. 4.1). The conventional approach [33, 125] was used, which assumes that the meaning of each linguistic term (such as "good", "poor", when it is not directly connected with the concrete values) is given by a fuzzy subset defined in the [0, 1] interval. Of course, there may be different ways to represent the linguistic terms as the triangles or trapezoids in the [0, 1] interval, but usually the choice of such performance is the prerogative of analyst since experts as a rule are not familiar enough with fuzzy set theory. In contrast to subjective criteria, the cost of material is presented directly by the fuzzy number. The method for obtaining the fuzzy costs is not described in [105, 134], but here we suppose that such fuzzy numbers can be obtained from usual statistics of market prices. Resulting assessments are presented in Table 4.2. To make our result comparable with those obtained by other authors [134], we have used exactly the same initial data and their fuzzy representation as in [105, 134]. Obviously, the problem of common representation of different types of local criteria (i) takes a place.

Table 4.2 Linguistically and numerically represented local criteria

Steel quality	C_1	C_2	C_3	C_4	C_5	C_6	C_7
V_1	P	F	G	P	F	B	(1.5, 1.6, 1.7)
A_2	B	B	F	F	G	F	(1.8, 2.0, 2.2)
$\mathbb{D}_2$	B	B	F	F	G	P	(1.0, 2.0, 2.2)
γ_1	F	G	G	F	F	F	(1.0, 1.0, 1.0)
T_1	G	G	F	B	G	F	(2.5, 3.0, 3.5)

The weights of the local criteria are represented linguistically with the use of linguistic term (see Table 4.3). There is no consensus among expert's with respect to importance weights of local criteria and their final estimations are presented in Table 4.4.

For the aggregation of expert's opinions the simple expression was used:

$$W_i = \frac{1}{n} \sum_{j=0}^{n} \oplus W_{ij}, \tag{4.4}$$

where W_i is the aggregated importance weight of ith local criterion, W_{ij} is the importance weight of ith local criterion given by jth expert, n is the number of participating experts, $\oplus$ is the operation of fuzzy addition. Using expression (4.4), from the data presented in Tables 4.3 and 4.4 we get the aggregated importance weights shown in Table 4.5.

The fuzzy numbers from Table 4.5 may be used as the local criteria weights to calculate the aggregated assessments of compared tool steel materials. Nevertheless, in practice, often the consensus of experts in respect to the obtained weights is needed. To achieve such a consensus the special procedures, e.g., the Delphi method [48] were developed. They are based on the correction by the experts their individual opinions taking into account the results of aggregation. In our case, the problem is that the initial expert's assessments of weights are represented in the linguistic form. Usually experts, e.g., supplies engineers, are not enough familiar with the

Table 4.3 Linguistic terms and corresponding triangular fuzzy values

Linguistic terms	Corresponding fuzzy number
Very low (VL)	(0,0,0.3)
Low (L)	(0,0.3,0.5)
Medium (M)	(0.2,0.5,0.8)
High (H)	(0.5,0.7,1)
Very high (VH)	(0.7,1,1)

Table 4.4 Linguistic assessment of importance weight made by three decision makers

Criteria	Opinions		
	E_1	E_2	E_3
C_1	H	H	VH
C_2	M	H	M
C_3	VH	VH	H
C_4	H	H	M
C_5	M	M	M
C_6	H	H	VH
C_7	VH	VH	VH

Table 4.5 Aggregated importance weights

Criteria	Opinions in fuzzy form
C_1	$W_1 = (0.567, 0.800, 1.000)$
C_2	$W_2 = (0.300, 0.567, 0.867)$
C_3	$W_3 = (0.633, 0.800, 1.000)$
C_4	$W_4 = (0.400, 0.633, 0.933)$
C_5	$W_5 = (0.200, 0.500, 0.800)$
C_6	$W_6 = (0.567, 0.800, 1.000)$
C_7	$W_7 = (0.700, 1.000, 1.000)$

fuzzy sets theory and know nothing about the numerical representation of linguistic terms. Hence, if the problem of consensus arises, an analyst should represent the aggregated weights in the linguistic form to be understood by experts. Moreover, often the experts insist on using such aggregated linguistic weights (consensus) in further analysis, since initially only the linguistic assessments of weights have been presented. It is easy to see that the weighs in Table 4.5 are not coincide with the fuzzy numbers representing linguistic terms used initially by experts (see Table 4.3). So the problem of reasonable linguistic interpretation of aggregated importance weights arises. The natural way for its approximate solution is to estimate the degrees to which each of obtained W_i coincides with the corresponding fuzzy numbers from Table 4.3 and to choose the linguistic term with the most degree of coincidence. In other words, the subsethood measure should be estimated. The simple and transparent method for doing this is presented in the following subsection.

4.2.1 Subsethood Measure for Linguistic Representation of Fuzzy Numbers

As proposed by Kosko [62], the fuzzy subsethood $S(A \subset B)$ measures the degree to which a fuzzy subset A is a subset of a fuzzy subset B and is given by the expression

$$S(A \subset B) = \frac{\sum_{u \in U} min(\mu_A(u), \mu_B(u))}{\sum_{u \in U} \mu_A(u)}, \tag{4.5}$$

where U is the universe of discuss common for fuzzy subsets A and B. Although expression (4.5) is widely used in applications [9], some of its drawbacks should be noted which prevent from using it for our purposes. Expression (4.5) is formulated for discrete supports of fuzzy subsets A and B, whereas we deal with continuous ones. Besides, in the asymptotic case when the support of A is tending to zero (fuzzy number A is reducing to a real value) the problem of reasonable interpretation of (4.5) arises. That is why, in this subsection the simple and transparent approach free of above mentioned drawbacks is described. It is based on the α-cuts representation of fuzzy numbers [60]. So, if A is a fuzzy value then

$$A = \bigcup_{\alpha} \alpha A_{\alpha},$$

where αA_{α} is the fuzzy subset $(x \in U, \mu_A(x) \geq \alpha)$ and A_{α} is the support set of fuzzy subset αA_{α}, U is the universe of discourse. It was proved that if A and B are fuzzy numbers, then all the operations on them may be presented as operations on the set of crisp intervals corresponding to their α-cuts: $(A@B)_{\alpha} = A_{\alpha}@B_{\alpha}$, $@ \in \{+,, *, /\}$. In a similar way, the subsethood operation can be defined as the set of subsethood operations on corresponding α-cuts. So the definition of subsethood measure as a degree to which a crisp interval A_{α} is a subset of B_{α}, i.e., $S(A_{\alpha} \subset B_{\alpha})$, is needed. Since we deal with crisp intervals, an intuitively obvious measure of subsethood may be defined as follows:

$$S(A_{\alpha} \subset B_{\alpha}) = \frac{W(A_{\alpha} \cap B_{\alpha})}{W(A_{\alpha})}, \tag{4.6}$$

where $W(A_{\alpha})$ is the width of interval A_{α}, $W(A_{\alpha} \cap B_{\alpha})$ is the width of overlapping area of intervals A_{α} and B_{α}. Expression (4.6) has the reasonable asymptotic properties. For example, in the case of $W(A_{\alpha}) \to 0$ from (4.6) we have the intuitively obvious result

$$\lim_{W(A_{\alpha}) \to 0} S(A_{\alpha} \subset B_{\alpha}) = S(a \in B_{\alpha}) = \begin{cases} 1 \; if \; a \in B_{\alpha}, \\ 0 \; if \; a \notin B_{\alpha}, \end{cases}$$

where $a \in B_{\alpha}$ is a real value. To get an aggregated estimation of subsethood measure on the base of its α-cuts representation (4.6), we propose to use the following weighted sum:

$$S(A \subset B) = \frac{\sum_{\alpha}(\alpha S(A_\alpha \subset B_\alpha))}{\sum_{\alpha} \alpha} \tag{4.7}$$

The last expression indicates that the contribution of α- cut to the overall subsethood estimation is increasing along with the rise of its number. The proposed method is graphically illustrated in Fig. 4.2.

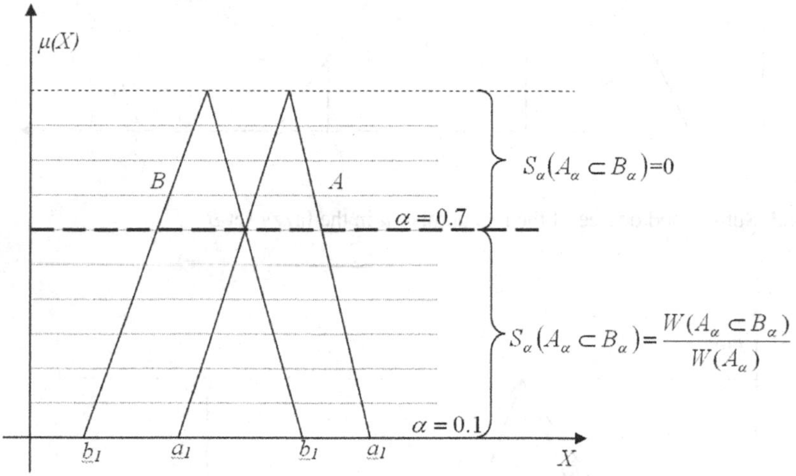

Fig. 4.2 Subsethood degree of fuzzy sets

In the asymptotic case, when a degree to which a real value a belongs to the interval B_α should be assessed, the expression (4.7) is reduced to

$$S(a \subset B) = \frac{\sum_{\alpha} \alpha S(a \in B_\alpha)}{\sum_{\alpha} \alpha}. \tag{4.8}$$

This case is shown in Fig. 4.3.

It is clear (even without calculations) that the following equalities are verified:

$$S(a^{***} \in B) > S(a^{**} \in B) > S(a^{*} \in B).$$

Of course, this intuitively obvious result well reflecting the inherent meaning of subsethood is numerically confirmed with the use of expression (4.8).

The other asymptotic case we meet when fuzzy subset A is completely enveloped by fuzzy subset B (see Fig.4.4). In this case, both the common sense and Exp.(4.8) provide the same results: $S(A \subset B) = 1$.

The described approach has been used to estimate the degrees to which each of aggregated importance weights W_i, obtained by averaging the expert's opinions

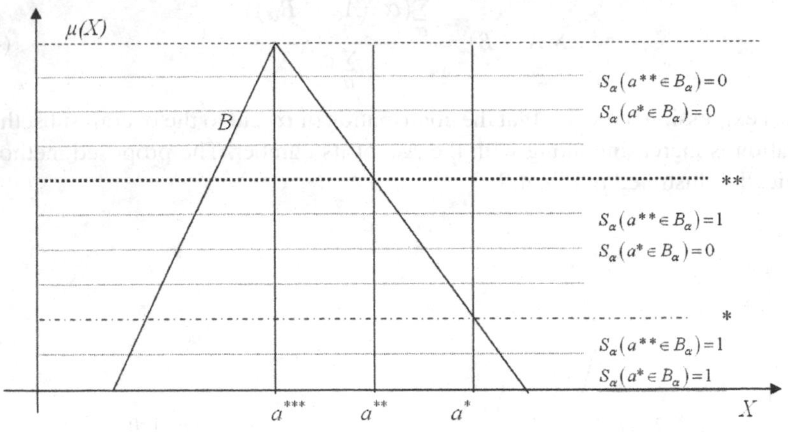

Fig. 4.3 Subsethood degree of the real number a in the fuzzy set B

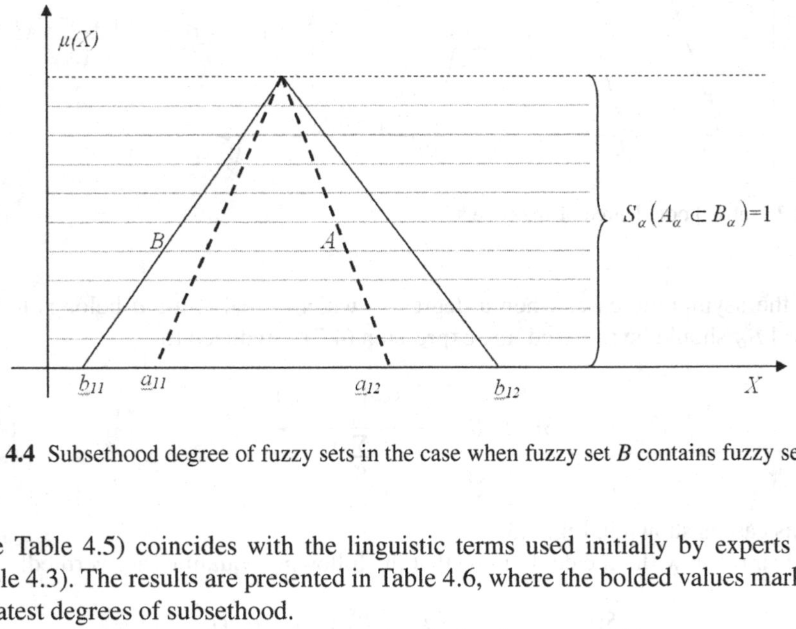

Fig. 4.4 Subsethood degree of fuzzy sets in the case when fuzzy set B contains fuzzy set A

(see Table 4.5) coincides with the linguistic terms used initially by experts (see Table 4.3). The results are presented in Table 4.6, where the bolded values mark the greatest degrees of subsethood.

So we can approximately represent the obtained aggregated importance weights by linguistic terms as shown in Table 4.7.

It is easy to see that the results obtained with the use of described approach (see screenshots in Fig. 4.5 and Fig. 4.6) are in a good conformity with our intuition.

Table 4.6 Degrees of confidence (subsethood degrees) of importance weight of the criteria with linguistic terms

Aggregated importance weight	Subsethood degree				
	VL (0.0,0.0,0.3)	*L* (0.0,0.3,0.5)	*M* (0.2,0.5,1.0)	*H* (0.5,0.7,0.8)	*VH* (0.7,1.0,1.0)
W_1	0%	0%	4.53%	**14.45%**	12.44%
W_2	0%	2.83%	**48.04%**	19.86%	0.92%
W_3	0%	0%	2.44%	**37.45%**	14.67%
W_4	0%	0.40%	24.53%	**35.69%**	2.82%
W_5	0.18%	8.98%	**100.00%**	8.98%	0.18%
W_6	0%	0%	4.53%	**14.45%**	12.44%
W_7	0%	0%	11.51%	11.51%	**100.00%**

Table 4.7 Final linguistic rating of ranks of local criteria

Criteria	Fuzzy number representation	Linguistic representation
C_1	$W_1 = (0.567, 0.800, 1.000)$	H
C_2	$W_2 = (0.300, 0.567, 0.867)$	M
C_3	$W_3 = (0.633, 0.800, 1.000)$	H
C_4	$W_4 = (0.400, 0.633, 0.933)$	H
C_5	$W_5 = (0.200, 0.500, 0.800)$	M
C_6	$W_6 = (0.567, 0.800, 1.000)$	H
C_7	$W_7 = (0.700, 1.000, 1.000)$	VH

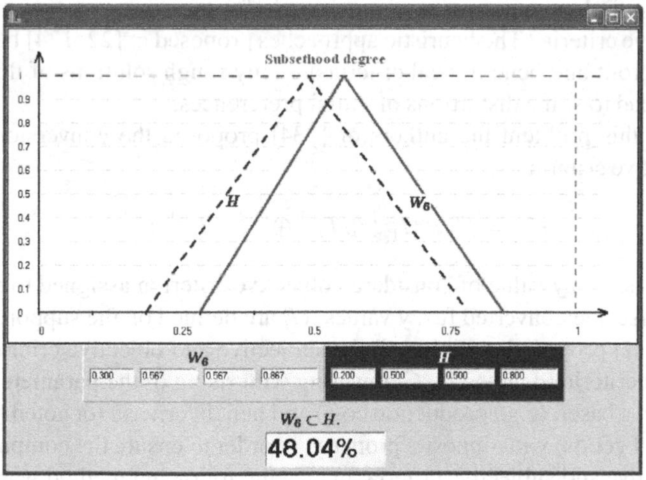

Fig. 4.5 Subsethood degree of $W_6 \subset H$

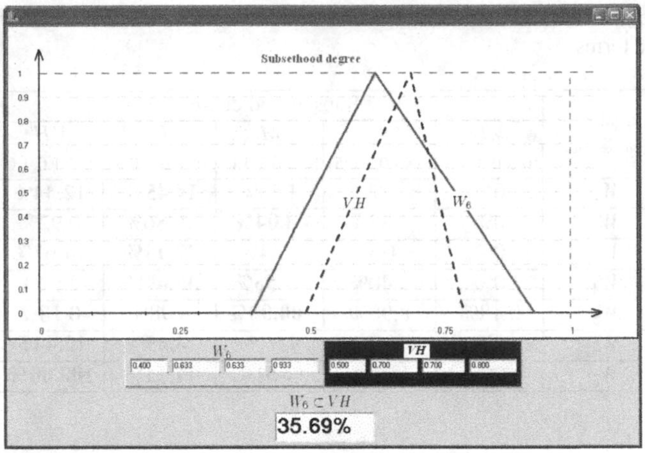

Fig. 4.6 Subsethood degree of $W_6 \subset VH$

4.2.2 *Common Representation of Different Types of Local Criteria*

When dealing with real-world problem, we can meet two group of local criteria: objective ones, expressed in the numerical form directly and subjective criteria represented by experts in the form of linguistic terms [134]. Linguistic terms are usually described by fuzzy numbers on the conventional common support [0,1], whereas objective criteria may have different supports. So the problem arises in such situations: how to find an appropriate common scale for the representation of objective and subjective criteria? The heuristic approaches proposed in [22, 134] based on the concept of profit and benefit local criteria are only rough solutions of this problem since they lead to some distortions of initial preferences.

To solve this problem the authors of [134] proposed the conversion of initial fuzzy objective criteria

$$RT_i = \{T_i \otimes [T_1^{-1} \oplus T_2^{-1} \oplus \ldots \oplus T_m^{-1}]\}^{-1}, \tag{4.9}$$

where T_i is the fuzzy value of considered objective criterion assigned to ith alternative. Of course, the converted fuzzy values RT_i are defined on the support $[0,1]$. The authors of [22] proposed to divide all the subjective and objective criteria into two groups: cost criteria (denoted by C) lowering with rising of the parameter on which the criterion is based (e.g., production cost) and benefit criteria (denoted by B) (e.g., the quality of goods) with opposite property. In order to ensure the compatibility between objective and subjective criteria, in [22] the following method was proposed.

Let $\tilde{x_{ij}}$ be the fuzzy rating of alternative A_i $(i = 1,\ldots,m)$ with respect to criterion $C_j(j = 1,\ldots,n)$. Suppose $\tilde{x_{ij}}$ are represented by triangular fuzzy numbers, i.e., $\tilde{x_{ij}} = (a_{ij}, b_{ij}, c_{ij})$. The following normalizations had been proposed in [22]: for the benefit criteria

$$\tilde{r_{ij}} = \left\{ \frac{a_{ij}}{c_j^*}, \frac{b_{ij}}{c_j^*}, \frac{c_{ij}}{c_j^*} \right\}, c_j^* = \max_i c_{ij}, \tag{4.10}$$

for cost criteria

$$\tilde{r_{ij}} = \left\{ \frac{\bar{a}_j}{c_{ij}}, \frac{\bar{a}_j}{b_{ij}}, \frac{\bar{a}_j}{a_{ij}} \right\}, \bar{a}_j = \min_i a_{ij} \tag{4.11}$$

The normalizations (4.10), (4.11) as well as the normalization (4.9) preserve the property that supports of normalized fuzzy numbers belong to the interval $[0,1]$. On the other hand, such normalizations lead to some distortions of initial preferences that were explicitly or implicitly taken into account on the stage of local criteria formalization. To clarify, consider an example.

Suppose there are two alternatives A_1, A_2, (the goods to be produced) and two local criteria for their assessment: investment cost (C_1) and expansion possibility (C_2) (see [22]).

Table 4.8 Fuzzy assessments of alternatives with respect to local criteria

	C_1	C_2
A_1	(6.0,7.0,8.0)	(6.3, 8.0, 9.0)
A_2	(3.6,4.0,4.4)	(9.0, 10.0, 10.0)

Suppose that evaluations of considered alternatives A_1, A_2 with respect to local criteria are represented by triangular fuzzy numbers shown in Table 4.8. According to [22], C_1 is the "cost" criterion and C_2 is the "benefit" criterion. That is why, the criterion C_1 was normalized using expression (4.11), whereas C_2 was normalized with the use of (4.10). The results of normalization are shown in Table 4.9.

Table 4.9 Fuzzy assessments of alternatives after normalization

	C_1	C_2
A_1	(0.45, 0.51, 0.60)	(0.63, 0.80, 0.90)
A_2	(0.81, 0.90, 1.00)	(0.90, 1.00, 1.00)

It could be noted that normalization can not transform the initial fuzzy evaluations of compared alternatives (see Table 4.8) into local criteria. Factually, in Table 4.9 we can see only triangular fuzzy numbers which can not be treated as the criteria since the values of cost criterion must decrease with rising of the parameter on which this criterion is based and the benefit criterion has the opposite property. Conversion (4.9) provides the similar results. Nevertheless, the data from Table 4.8 can be used to build the correct cost and benefit criteria.

For benefit criteria instead of (4.10) the following function may be used:

$$\mu_B(r_j) = \begin{cases} 0, & \text{if } r_j \leq r_{jmin} \text{ ,} \\ \frac{(r_j - r_{jmin})}{(r_{jmax} - r_{jmin})} & \text{if } r_{jmin} < r_j < r_{jmax}, \\ 1 & \text{if } r_j \geq r_{jmax}, \end{cases} \tag{4.12}$$

and for the cost criteria:

$$\mu_C(r_j) = \begin{cases} 1, & \text{if } r_j \leq r_{jmin} \text{ ,} \\ 1 - \frac{(r_j - r_{jmin})}{(r_{jmax} - r_{jmin})} & \text{if } r_{jmin} < r_j < r_{jmax}, \\ 0 & \text{if } r_j \geq r_{jmax}. \end{cases} \tag{4.13}$$

In (4.12) and (4.13), we have denoted $r_{jmin} = \min_i a_{ij}$ and $r_{jmax} = \max_i c_{ij}$.

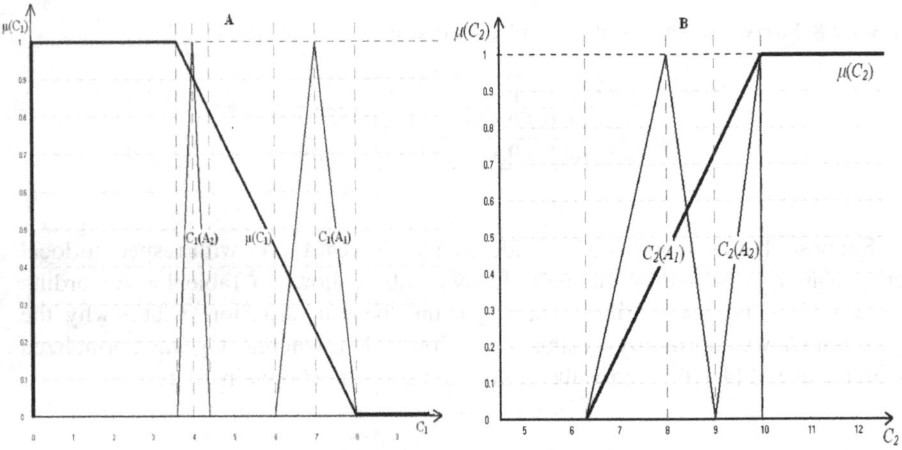

Fig. 4.7 The cost (*A*) and benefit (*B*) criteria based on the data from Table 4.8

In Fig. 4.7, the local criteria based on the data presented in Table 4.8 with the use of (4.12) and (4.13) are shown. It is easy to see that triangular fuzzy costs and benefits are only fuzzy arguments of membership functions representing corresponding local criteria.

Let us turn to the tool steel material selection problem. It is clear that linguistic terms in Table 4.2 represent the final linguistic assessments of criteria and the transformations of them are not required. On the other hand, the local criterion based on the cost of tool steel materials is needed. We have built it using the expression (4.13). As the result, the membership function representing the local cost criterion has been obtained. The next step is the calculation of fuzzy value of the cost criterion as the function of fuzzy argument for all compared tool steel materials. To do this, the well known procedure [60] of evaluation of function of fuzzy argument illustrated in Fig. 4.8 has been used.

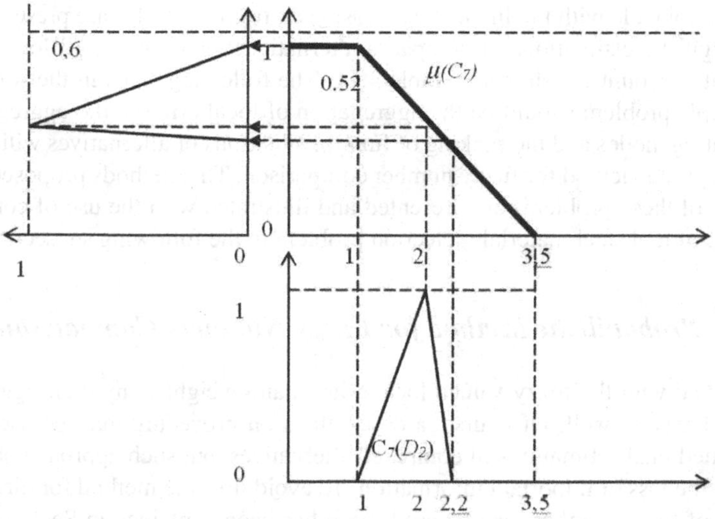

Fig. 4.8 Calculation of fuzzy value of the cost criterion C_7 for steel D_2

Table 4.10 Linguistic representation of fuzzy values of cost criterion for compared tool steel materials

Steel quality	Fuzzy cost	Fuzzy value of cost criterion	Subsethood degree	Linguistic approximation of cost criterion's value
V_1	(1.50, 1.60, 1.70)	(0.72, 0.76, 0.80)	63.67%	G
A_2	(1.80, 2.00, 2.20)	(0.52, 0.60, 0.68)	43.65%	F
D_2	(1.00, 2.00, 2.20)	(0.52, 0.80, 1.00)	43.97%	G
γ_1	(1.00, 1.00, 1.00)	(1.00, 1.00, 1.00)	100.00%	B
T_1	(2.50, 3.00, 3.50)	(0.00, 0.20, 0.40)	39.53%	P

Table 4.11 Fuzzy values of local criteria in the linguistic form

Steel quality	C_1	C_2	C_3	C_4	C_5	C_6	C_7
V_1	P	F	G	P	F	B	G
A_2	B	B	F	F	G	F	F
D_2	B	B	F	F	G	P	G
γ_1	F	G	G	F	F	F	B
T_1	G	G	F	B	G	F	P

Obviously, the resulting values of cost criterion are fuzzy values and the problem of common representation of linguistically and numerically defined local criteria arises. For its solution we have used the procedure of calculation of subsethood measure for the linguistic representation of fuzzy values described in Subsection 4.2.1.

In Table 4.10, the calculated maximal degrees to which the fuzzy values of cost criterion coincide with the linguistic assessments from Table 4.1 are presented. The final linguistic estimations of compared alternatives regarding to all local criteria taken into account are shown in Table 4.11. The following steps in the solution of our sample problem should be the aggregation of local criteria, the aggregation of aggregating modes and the ranking of fuzzy evaluations of alternatives with the use of appropriate method for fuzzy number comparison. The methods proposed for the solution of these problems are presented and illustrated with the use of considered example of tool steel materials selection problem in the following subsections.

4.2.3 Probabilistic Method for Fuzzy Numbers Comparison

As we deal with the fuzzy valued local criteria and weights, any their aggregation will be fuzzy as well. Of course, a defuzzification procedure may be used to get real valued final estimations of compared alternatives, but such approach obviously leads to the loss of important information. To avoid this, the method for direct comparison of fuzzy numbers can be used. As it has been explained in Section 4.1, we prefer to use the probabilistic approach to fuzzy numbers comparison. It was already successfully used in some applications [107, 110].

Let us recall the basics of this approach and developed method. We present firstly the probabilistic crisp interval relations and further extend them to the fuzzy numbers comparison. There are only two nontrivial situation of intervals setting deserve to be considered: the overlapping and inclusion cases (see Fig. 4.9).

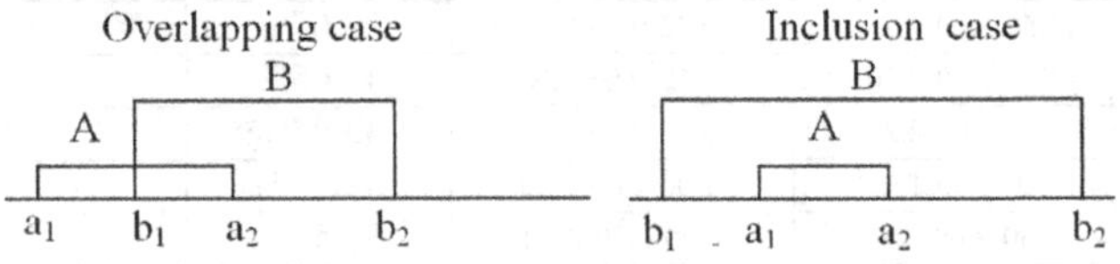

Fig. 4.9 Examples of interval relations

Let $A = [a_1, a_2]$ and $B = [b_1, b_2]$ be independent intervals and $a \in [a_1, a_2], b \in [b_1, b_2]$ be random values distributed on these intervals. As we are dealing with crisp intervals, the natural assumption is that the random values a and b are distributed uniformly. There are some subintervals, which play an important role in our analysis. For example, in overlapping case (see Fig. 4.9), the falling of random $a \in [a_1, a_2], b \in [b_1, b_2]$ into subintervals $[a_1, b_1], [b_1, a_2], [a_2, b_2]$ may be treated as a set of independent random events. Let us define the events $H_k : a \in A_i, b \in B_j$, where A_i and B_j are subintervals formed by the boundaries of compared intervals A and B such that $A = \bigcup A_i, B = \bigcup B_j$. It easy to see that events H_k form the complete set of events describing all the cases of falling random values a and b into the various subintervals A_i and B_j, respectively. Let $P(H_k)$ be the probability of event

H_k and $P(B > A/H_k)$ be the conditional probability of $B > A$. Hence, the composite probability can be presented as

$$P(B > A) = \sum_{k=1}^{n} P(H_k)P(B > A/H_k). \tag{4.14}$$

As we are dealing with uniform distributions of random values a and b in the given subintervals, the probabilities $P(H_k)$ can be easily obtained geometrically. These basic assumptions make it possible to infer the complete set of probabilistic interval relations involving separated equality and inequality relations and comparisons of real numbers with intervals and fuzzy numbers [111]. The complete set of expressions for interval relations is shown in Table 4.12, the cases without

Table 4.12 Probabilistic interval relations

$P(B>A)$	$P(B<A)$	$P(B=A)$
1. $b_1 > a_1 \wedge b_1 < a_2 \wedge b_1 = b_2$		
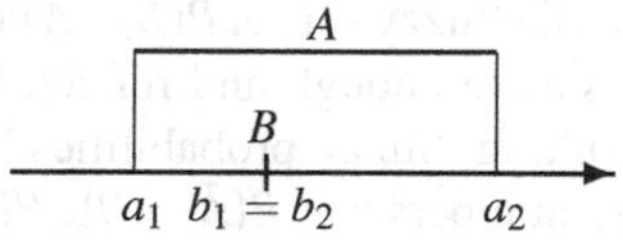		
$\frac{b_1 - a_1}{a_2 - a_1}$	$\frac{a_2 - b_1}{a_2 - a_1}$	0
2. $b_1 \geq a_1 \wedge b_2 \leq a_2$		
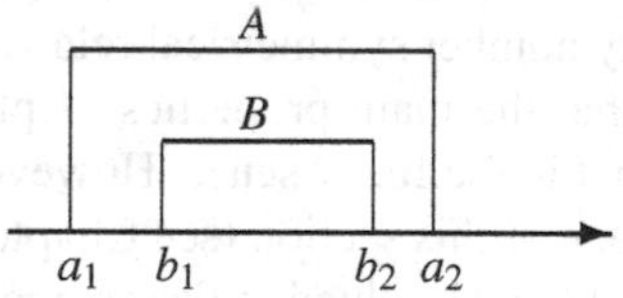		
$\frac{b_1 - a_1}{a_2 - a_1}$	$\frac{a_2 - b_2}{a_2 - a_1}$	$\frac{b_2 - b_1}{a_2 - a_1}$
3. $a_1 \geq b_1 \wedge a_2 \geq b_2 \wedge a_1 \leq b_2$		
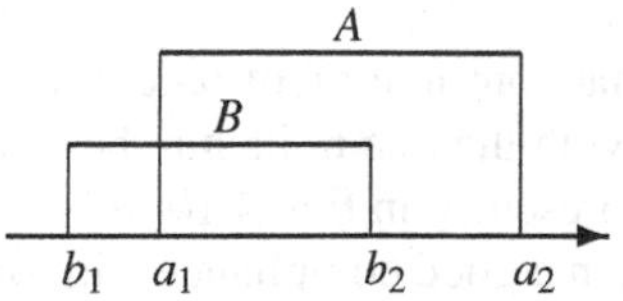		
0	$1 - \frac{(b_2 - a_1)^2}{(a_2 - a_1)(b_2 - b_1)}$	$\frac{(b_2 - a_1)^2}{(a_2 - a_1)(b_2 - b_1)}$

overlapping and inclusion are omitted. In Table 4.12, only half of cases that may be realized when considering interval overlapping and including are presented since the other cases, e.q., $b_2 > a_2$ for overlapping and so on, can be easily obtained by changing letter a through b and otherwise in the expressions for the probabilities.

It easy to see that in all cases $P(A < B) + P(A = B) + P(A > B) = 1$.

Let $\tilde{A}$ and $\tilde{B}$ be fuzzy values on X with corresponding membership functions $\mu_A(x), \mu_B(x) : X \to [0,1]$. We can represent $\tilde{A}$ and $\tilde{B}$ by the sets of α-cuts: $\tilde{A} = \bigcup_\alpha A_\alpha$, $\tilde{B} = \bigcup_\alpha B_\alpha$, where $A_\alpha = \{x \in X : \mu_A(x) \geq \alpha\}, B_\alpha = \{x \in X : \mu_B(x) \geq \alpha\}$ are crisp intervals. Then all fuzzy value relations $\tilde{A}\ rel\ \tilde{B}$, $rel = \{<,=,>\}$, may be presented by the sets of α-cut relations $\tilde{A}\ rel\ \tilde{B} = \bigcup_\alpha A_\alpha\ rel\ B_\alpha$. Since A_α and B_α are crisp intervals, the probability $P_\alpha(B_\alpha > A_\alpha)$ for each pair A_α and B_α can be calculated in the way presented in Table 4.12. The set of the probabilities $P_\alpha(\alpha \in (0,1])$ may be treated as the support of the fuzzy subset

$$P(\tilde{B} > \tilde{A}) = \{\alpha / P_\alpha(B_\alpha > A_\alpha)\},$$

where the values of α may be considered as grades of membership to the fuzzy number $P(\tilde{B} > \tilde{A})$. In this way, the fuzzy subset $P(\tilde{B} = \tilde{A})$ may also be easily defined.

The obtained results are simple enough and reflect, in some sense, the nature of fuzzy arithmetic. The resulting "fuzzy probabilities" can be used directly. For instance, let $\tilde{A}$, $\tilde{B}$, $\tilde{C}$ be fuzzy numbers and $P(\tilde{A} > \tilde{B})$, $P(\tilde{A} > \tilde{C})$ be fuzzy numbers expressing the probabilities of $A > \tilde{B}$ and $\tilde{A} > \tilde{C}$, respectively. Hence, the probability $P(P(\tilde{A} > \tilde{B}) > P(\tilde{A} > \tilde{C}))$ has a sense of probability's comparison and is expressed in the form of fuzzy number as well. Such fuzzy calculations may be useful at intermediate stages of analysis since they preserve the fuzzy information available. Indeed, it can be shown that in any case $P(\tilde{B} > \tilde{A}) + P(\tilde{B} = \tilde{A}) + P(\tilde{B} < \tilde{A}) =$"near 1", where "near 1" is a fuzzy number symmetrical relative to 1.

It is worth noting here that the main properties of probability are remained in the introduced operations, but in the fuzzy sense. However, a detailed discussion of these issues is out of the scope of this section (see Chapter 3 for more details).

Nevertheless, in practice, the real-valued indices sometimes are needed for fuzzy numbers comparison. For this purpose, some characteristic numbers of fuzzy sets could be used. But it seems more natural to use the following weighted sum:

$$\overline{P}(\tilde{B} > \tilde{A}) = \sum_\alpha \alpha P_\alpha(B_\alpha > A_\alpha) / \sum_\alpha \alpha. \tag{4.15}$$

This expression indicates that contribution of α- cut to the overall probability estimation is increasing along with the rise in its number. Some typical cases of fuzzy numbers comparison are represented in Fig. 4.10. It is easy to see that the resulting quantitative estimations are in a good compliance with our intuition.

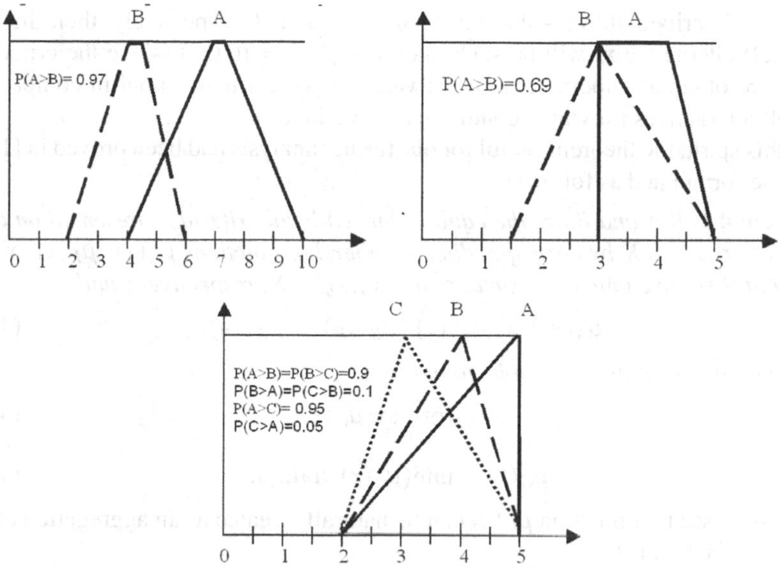

Fig. 4.10 Typical cases of fuzzy numbers comparison

4.2.4 Aggregation of Local Criteria and Aggregating Modes

There are many different methods for aggregation of local criteria proposed in the literature, but there is no method proved to be the best in all practical cases. Moreover, in [148] it is stated that the choice of aggregation scheme is a context dependent problem. Nevertheless, we contribute here in its consideration from some other point of view.

Firstly, it is possible to represent the membership functions of local criteria as some artificial functions of alternatives. Such a transformation can be carried out formally after calculation of the values of membership functions for all alternatives. More strictly, let A and B be the local criteria and μ_A, mu_B be their membership functions. Then for each $x \in X$, where X is a set of alternatives, the artificial functions $\mu_A(x), \mu_B(x)$ can be formally introduced. Of course, x is not a variable in the common sense. Factually, it is only a label (or number) assigned to the corresponding alternative. Hence, we can say that if for some $x_1 \in X$ we have $\mu_A(x_1) = \mu_B(x_1)$, then the alternative x_1 satisfies the local criteria A and B in the equal extent, and if for some $x_2 \in X$, we have $\mu_A(x_2) > \mu_B(x_2)$, then alternative x_2 satisfies the local criterion A in a greater degree than the criterion B. In this way, the initially multidimensional problem can be formally transformed into the one-dimensional one with the alternative number (label) as the variable.

To make our consideration more transferable, consider firstly the case of only two local criteria A and B which are equally important for a decision maker and therefore have equal ranks, i.e., $\alpha_A = \alpha_B = 1$. So if X is a set of the alternatives and $\mu_A(x)$, $\mu_B(x)$, $x \in X$ are membership functions representing formally – as it has been described above – the local criteria A and B, respectively, then the best (optimal) alternative x_o will be such that: a) $\mu_A(x_o) = \mu_B(x_o)$ (since the criteria A and B are of equal importance), b) the value of $\mu_A(x_o)$ is maximal in comparison with all alternatives for which condition a) is verified.

In this spirit, the theorem useful for our further analysis had been proved in [114]. It can be formulated as follows:

Theorem 4.1. *If A and B are the equally ranked local criteria represented on a set of the alternatives X by corresponding membership functions $\mu_A(x)$, $\mu_B(x)$, $x \in X$ such that they have unique maximal points $x_A, x_B \in X$, respectively, and*

$$\mu_A(x_A) > \mu_B(x_A),\ \mu_B(x_B) > \mu_A(x_B), \tag{4.16}$$

then optimal alternative x_o can be found as

$$x_o = arg \max_{x \in X}(\mu_C(x)), \tag{4.17}$$

$$\mu_C(x) = \min(\mu_A(x), \mu_B(x)). \tag{4.18}$$

It is easy to see that function $\mu_C(x)$ can be naturally treated as an aggregation of the local criteria A and B.

The results which can be achieved using some other popular aggregation methods are shown in Fig. 4.11.

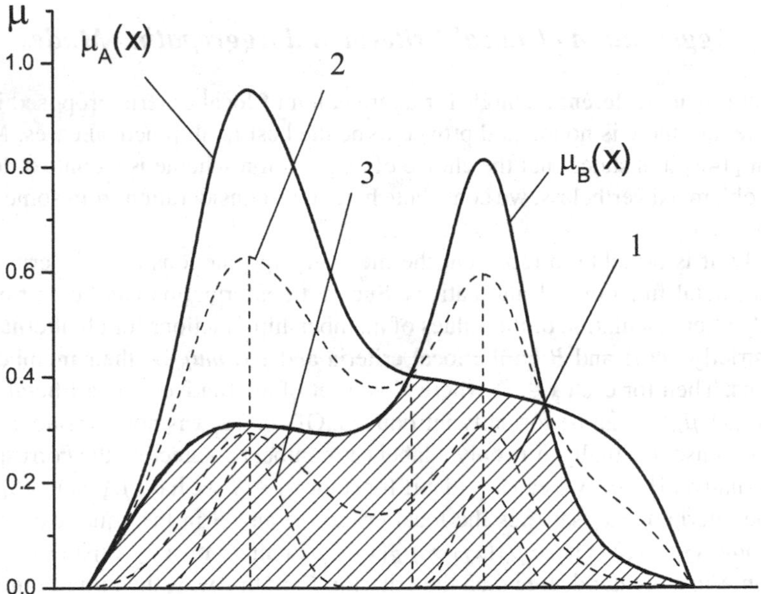

Fig. 4.11 Methods for equally ranked local criteria aggregation: 1: $\mu_C(x) = \mu_A(x) \cdot \mu_B(x)$; 2: $\mu_C(x) = 0.5\mu_A(x) + 0.5\mu_B(x)$; 3: $\mu_C(x) = \max(0, \mu_A(x) + \mu_B(x) - 1)$; x_1 is the optimal alternative for 1, 2 and 3 types of aggregation; x_2 is the optimal alternative for the aggregation $\mu_C(x) = min(\mu_A(x), \mu_B(x))$.

Obviously, only the *min*–type aggregation (4.17) derives the optimal alternative x_2 (see Fig.4.11) fulfilling the natural restriction a). All other considered aggregation methods provide the optimal alternative x_1, which is in Pareto region, but far from the actual optimum (see Fig. 4.11). In practice, we can meet more complicated situations than considered above. For example, the membership functions can have several points of extreme. This problem can be solved by clustering a subset of the alternatives into some Pareto regions as it is shown in Fig. 4.12. Obviously, within such regions (I, II, III in Fig. 4.12) all the conditions of Theorem 4.1 are verified.

Assume that the local criteria A and B are of different importance for a decision maker, i.e., for their ranks we have $\alpha_A \neq \alpha_B$. Since the additive and multiplicative aggregations $\mu_C(x) = \alpha_A\mu_A(x) + \alpha_B\mu_B(x)$ and $\mu_C(x) = (\alpha_A\mu_A(x)) \cdot (\alpha_B\mu_B(x))$ in the case of $\alpha_A \approx \alpha_B$ result in inappropriate decisions (see Fig. 4.11), we looked for more correct aggregation rules.

In [147], the following aggregation has been proposed:

$$\mu_C^1(x) = \min(\alpha_A\mu_A(x), \alpha_B\mu_B(x)). \qquad (4.19)$$

It is easy to see that in the asymptotic case $\alpha_A = \alpha_B = 1$, the expression (4.19) reduces to the optimal *min*–type aggregation (4.18) and the weighting in (4.19) appears to be logically justified. Nevertheless, in practice such aggregation can produce completely absurd results. For example, let $\alpha_A = 0.8, \alpha_B = 0.2$, i.e., the local criterion A is more important than B. As it is shown in Fig. 4.13, x_0 is the optimal alternative for the case of equally ranked criteria A and B, i.e., $x_0 = arg\max_{x\in X}(\min(\mu_A(x), \mu_B(x)))$, and x_0^1 is the optimal alternative for the case of the weighted criteria, i.e., $x_0^1 = arg\max_{x\in X}(\min(\alpha_A\mu_A(x), \alpha_B\mu_B(x)))$. It is easy

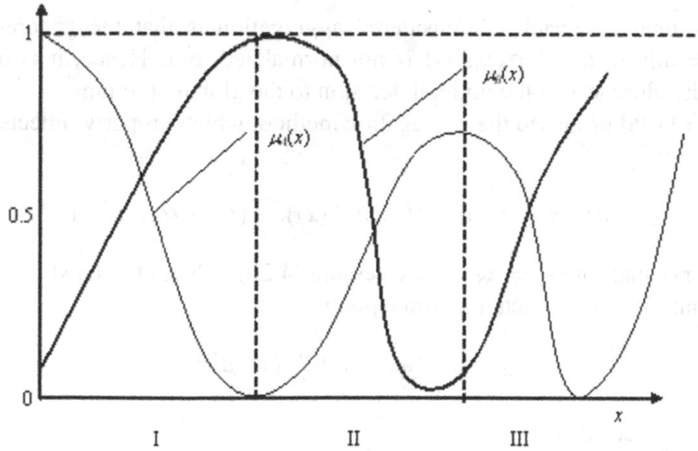

Fig. 4.12 Multiple- extreme membership functions

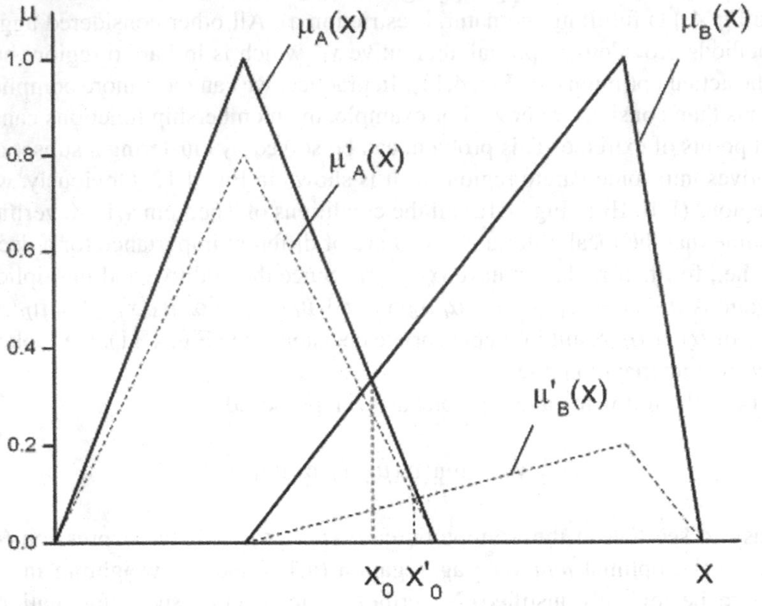

Fig. 4.13 Min–type aggregation of ranked local criteria: $\mu\prime_A(x) = 0.8\mu_A(x)$, $\mu\prime_B(x) = 0.2\mu_B(x)$

to see that the alternative x_0^1 satisfies the criterion B in a greater extent than the criterion A. This is in contradiction with initial assumption for A to be more important than B.

Additional drawback of considered aggregation is that the general criterion $\mu_C^1(x) = \min(\alpha_A \mu_A(x), \alpha_B \mu_B(x))$ is not normalized to 1. Hence, it is difficult to assess the closeness of the optimal decision to the global optimum.

Yager [138] proposed the aggregation method, which properly reflects the sense of ranking:

$$\mu\prime_C(x) = \min(\mu_A^{\alpha_A}(x), \mu_B^{\alpha_B}(x)), \quad (\alpha_A + \alpha_B)/2 = 1. \tag{4.20}$$

Observe that for $\alpha_A = \alpha_B = 1$ expression (4.20) reduces to (4.18). It is shown in [138] that if $\alpha_A > \alpha_B$ then in optimal point

$$x_0^1 = \arg\max_{x \in X} \min\left(\mu_A^{\alpha_A}(x), \mu_B^{\alpha_B}(x)\right)$$

the natural inequality

$$\mu_A(x_0^1) > \mu_B(x_0^1)$$

always takes place. In the general form, the expression (4.20) can be rewritten as follows:

$$\mu\prime_C(x) = \mu_1^{\alpha_1}(x) \wedge \mu_2^{\alpha_2}(x) \wedge \ldots \wedge \mu_n^{\alpha_n}(x), \quad (4.21)$$
$$\alpha_1, \alpha_2, \ldots, \alpha_n > 0, \quad \frac{1}{n}\sum_{i=1}^{n} \alpha_i = 1,$$

where $\wedge$ is the *min*-operator, n is the number of the local criteria.

Therefore, it can be asserted that the aggregation (4.20) is the best one, but it is true only when the conditions of Theorem 4.1 are verified. In practice, this is not always the case. Moreover, the *min*-type aggregation sometimes does not comply with intuitive concepts of decision makers about optimality [39] since sometimes it does not properly represent the contributions of local criteria to the overall estimation.

The detailed analysis of the advantages and drawbacks of aggregating modes can be found in [114], where with a help of two proved theorems and on the base on the author's experience it was shown that, in general, the most reliable aggregation approach lies in the use of Yager's *min*-type operator (4.20). The multiplicative mode appears to be somewhat less reliable and, finally, the additive (weighted sum) method may be considered as unreliable and insensitive when choosing an alternative in Pareto-region (see also Fig. 4.15 below).

On the other hand, as all known aggregation modes have their own advantages and drawbacks it seems impossible to choose the best one especially when dealing with a complicated hierarchical problem. Therefore, when dealing with a complex task characterized by a great number of local criteria, it seems reasonable to use all types of aggregation relevant to the considered problem.

Since the different final results may be obtained on the base of different aggregating modes, the problems arises: how to aggregate such results?
For this purpose a new method for aggregation of aggregating modes was developed in [36, 108].

For further analysis we choose the aggregation modes that are usually used as the atomic ones for building more complex aggregating operations:

$$D_1 = min(\mu_1(C_1)^{W_1}, \mu_2(C_2)^{W_2}, \ldots, \mu_n(C_n)^{W_n}), \quad (4.22)$$
$$D_2 = \mu_1(C_1)^{W_1} \otimes \mu_2(C_2)^{W_2}, \ldots, \otimes \mu_n(C_n)^{W_n}, \quad (4.23)$$
$$D_3 = W_1 \otimes \mu_1(C_1) \oplus W_2 \otimes \mu_2(C_2), \ldots, \oplus W_n \otimes \mu_n(C_n), \quad (4.24)$$

where $\oplus$ and $\otimes$ are the operations of fuzzy addition and multiplication, respectively.

Let us turn to our example of tool steel material selection problem. Since the values of all local criteria and their weights are presented as the linguistic terms (see Table 4.7 and 4.11) with the corresponding numerical representation (see Tables 4.1 and 4.2), the final aggregated assessments of compared tool steel materials obtained using the aggregating modes (4.22)–(4.24) should be fuzzy numbers as well. All arithmetical operations on fuzzy numbers needed to do this are well defined in [60], and the method for fuzzy numbers comparison has been described in the previous subsection. The results of calculations are presented in Table 4.13.

Table 4.13 Results of aggregations with the use of modes (4.22)–(4.24)

Steel quality	D_1	D_2	D_3
V_1	(0.000,0.382,0.675)	(0.020,0.106,0.186)	(0.000,0.191,0.577)
A_2	(0.200,0.500,0.855)	(0.077,0.186,0.257)	(0.140,0.500,0.855)
$\mathbb{D}_2$	(0.000,0.382,0.675)	(0.107,0.214,0.286)	(0.350,0.700,1.000)
γ_1	(0.200,0.574,0.881)	(0.086,0.200,0.257)	(0.140,0.574,0.881)
T_1	(0.000,0.300,0.616)	(0.040,0.123,0.214)	(0.000,0.226,0.616)

Obviously, the different final fuzzy estimations for the compared tool steel materials were obtained with the use of different aggregating modes. Usually when the results obtained using different aggregation modes are similar, this fact may be considered as a good confirmation of their optimality. In opposite case, an additional analysis of local criteria and their ranking should be advised. It is easy to see that such approach seems to be based on inexact reasoning, which implicitly takes into account all aggregating modes relevant to the considering task.

To build the method for doing this more rigorously on the base of aggregation of aggregating modes, the use of synthesis of type–2 and level–2 fuzzy sets defined on the support composed of compared alternatives has been proposed in [36, 108]. It is worth noting that in the framework of this approach we avoid the use of *min*, *sum* and *multiplication* operations for aggregation of aggregating modes themselves, since the use of them leads inevitably to the unlimited sequence of aggregation problems.

Let us recall briefly the basic definitions of type–2 and level–2 fuzzy sets.

As we deal with a restricted number of compared alternatives and aggregating modes, it seems reasonable to consider only the discrete representation of type–2 and level–2 fuzzy sets.

Type–2 fuzzy sets were introduced by L. Zadeh in [146] as the framework for mathematical formalization of linguistic terms. In essence, these sets are the extension of usual fuzzy sets (type–1) to the case when the membership function of fuzzy subset is performed by another fuzzy subset.

More strictly, let A be the fuzzy set of type–2 on the support subset X. Then, for any $x \in X$ the membership function $\mu_A(x)$ of A is the fuzzy set with the membership function $f_x(y)$. As the result, for the discrete set we get

$$\mu_A(x) = \left\{ \frac{f_x(y_i)}{y_i} \right\}, i = 1, \ldots, n. \tag{4.25}$$

Further elaboration of the theory of fuzzy sets of type–2 was presented in [59, 79, 140], where the main mathematical operations on such sets were defined. In [140], it was proved that using Zadeh's fuzzy extension principle, it is possible to build the fuzzy sets of types–3,4 and so on.

Originally, level–2 fuzzy sets were introduced by Zadeh [144] and were more elaborately studied in [46, 127].

As proposed by Zadeh [145], the level– 2 fuzzy set is a fuzzy set defined on the support, elements of which are ordinary fuzzy sets. So if the fuzzy subset A is defined on a discrete set of x_i, $i = 1,\ldots,N$, with the membership function $\mu_A(x_i)$ and x_i are represented by ordinary fuzzy sets defined on discrete universe set of z_j, $j = 1,\ldots,M$, with the corresponding membership function $h_i(z_j)$ then A is a level–2 fuzzy subset defined by the following expressions:

$$A = \left\{ \frac{\mu_A(x_i)}{x_i} \right\}, x_i = \left\{ \frac{h_i(z_j)}{z_j} \right\}. \tag{4.26}$$

$$A = \left\{ \frac{max\,[\mu_A(x_i)h_i(z_j)]}{z_j} \right\}, i = 1,\ldots,N, j = 1,\ldots,M. \tag{4.27}$$

It follows from expression (4.27) that the final degree of membership of z_j in A may be presented as:

$$\mu_A(z_j) = \max_i\,[\mu_A(x_i)h_i(z_j)],\ j = 1,\ldots,M. \tag{4.28}$$

Suppose there is a set of alternatives z_j, $j = 1,\ldots,M$, and N types of aggregating modes D_i, $i = 1,\ldots,N$. Since usually in practice it is possible to estimate the relative reliability of aggregating modes at least on a verbal level, it seems natural to introduce the membership function $\mu(D_i)$, $i = 1,\ldots,N$, representing expert's opinions about closeness of considered aggregating operator D_i to the some perfect type of aggregation, which can be treated as the best one or even "ideal" method of aggregation. Then such an "ideal" method D_{ideal} can be represented by its membership function defined on the set of compared aggregation modes as follows:

$$D_{ideal} = \left\{ \frac{\mu(D_i)}{D_i} \right\}, i = 1,\ldots,N. \tag{4.29}$$

As for all alternatives z_j, $j = 1,\ldots,M$, their estimations $D_i(z_j)$ with the use of aggregation modes D_i, $i = 1,\ldots,N$ can be calculated, each D_i can be formally defined on the set of comparing alternatives z_j . As the result, each D_i can be represented by the fuzzy subset

$$D_i = \left\{ \frac{D_i(z_j)}{z_j} \right\}, j = 1,\ldots,M, \tag{4.30}$$

where $D_i(z_j)$ is treated as the degree to which alternative z_j belongs to the set of "good" ones estimated with use of aggregating mode D_i. Substituting (4.30) into (4.29) with the use of definitions (4.26)-(4.28) we get

$$D_{ideal} = \left\{ \frac{\mu_{ideal}(D_i)}{D_i} \right\}, i = 1,\ldots,N, \tag{4.31}$$

where

$$\mu_{ideal}(z_j) = \max_i\,[\mu(D_i)D_i(z_j)], \tag{4.32}$$

Finally, the best alternative can be found as:

$$z_{best} = arg\max_{j} \mu_{ideal}(z_j) \tag{4.33}$$

In the case when $\mu(D_i)$ or/and $D_i(z_j)$ are fuzzy values, the expression (4.29) represents an object which can be treated simultaneously as level–2 and type–2 fuzzy subset.

To illustrate, let us continue the consideration of our example. At first, we have to calculate the values $\mu(D_i)$, $i = 1,2,3$. As it has been stated above, the *min*-type aggregation D_1 is more reliable than the multiplicative aggregation D_2 and both are noticeably more reliable than additive aggregation D_3. Such linguistic assessments may be represented in the numerical form using the linguistic reciprocal pair comparison matrix [100] which in our case can be presented as in Table 4.14.

Table 4.14 Pair comparison of aggregating modes

Aggregation modes	D_1	D_2	D_3
D_1	1	3	9
D_2	$\frac{1}{3}$	1	9
D_3	$\frac{1}{9}$	$\frac{1}{9}$	1

The number 3 in this Table indicates that *min*-type aggregation D_1 is more reliable than multiplicative aggregation D_2 an so on. Using the method proposed in [32], from this matrix we get

$$\mu(D_1) = 0.7, \mu(D_2) = 0.25, \mu(D_3) = 0.05.$$

Thus, the "ideal" method of aggregation in our case can be presented as follows:

$$D_{ideal} = \left\{ \frac{\mu(D_1)}{D_1}, \frac{\mu(D_2)}{D_2}, \frac{\mu(D_3)}{D_3} \right\}. \tag{4.34}$$

In our case, the compared alternatives are tool steel materials V_1, A_2, $\mathbb{D}_2$, γ_1, T_1. Hence, expressions (4.30) can be rewritten as follows:

$$D_i = \left\{ \frac{D_i(V_1)}{V_1}, \frac{D_i(A_2)}{A_2}, \frac{D_i(\mathbb{D}_2)}{\mathbb{D}_2}, \frac{D_i(\gamma_1)}{\gamma_1}, \frac{D_i(T_1)}{T_1} \right\}, i = 1,2,3. \tag{4.35}$$

The fuzzy values of $D_i(steel\ quality)$, i=1,2,3, *steel quality* $\in \{V_1, A_2, \mathbb{D}_2, \gamma_1, T_1\}$ are presented in Table 4.13. Finally, expression (4.31) in our example takes the form:

$$D_{ideal} = \left\{ \frac{\mu_{ideal}(V_1)}{V_1}, \frac{\mu_{ideal}(A_2)}{A_2}, \frac{\mu_{ideal}(\mathbb{D}_2)}{\mathbb{D}_2}, \frac{\mu_{ideal}(\gamma_1)}{\gamma_1}, \frac{\mu_{ideal}(T_1)}{T_1} \right\}, \tag{4.36}$$

where $\mu_{ideal}(steel\ quality) = \max_i(\mu(D_i) \cdot D_i(steel\ quality))$, i=1,2,3 and *steel quality* $\in \{V_1, A_2, \mathbb{D}_2, \gamma_1, T_1\}$.

After calculations we get

$$\mu_{opt}(V_1) = max((0.000, 0.267, 0.472), (0.005, 0.026, 0.046), (0.000, 0.009, 0.028)), \quad (4.37)$$

$$\mu_{opt}(A_2) = max((0.140, 0.350, 0.598), (0.019, 0.046, 0.064), (0.007, 0.025, 0.420)), \quad (4.38)$$

$$\mu_{opt}(\mathbb{D}_2) = max((0.000, 0.267, 0.472), (0.026, 0.053, 0.071), (0.017, 0.035, 0.050)), \quad (4.39)$$

$$\mu_{opt}(\gamma_1) = max((0.140, 0.401, 0.616), (0.021, 0.050, 0.064), (0.007, 0.028, 0.044)), \quad (4.40)$$

$$\mu_{opt}(T_1) = max((0.000, 0.210, 0.431), (0.008, 0.030, 0.053), (0.000, 0.011, 0.030)). \quad (4.41)$$

To find the maximal triangular fuzzy numbers in these expressions, the probabilistic approach described in Subsection 4.2.3 has been used. The resulting

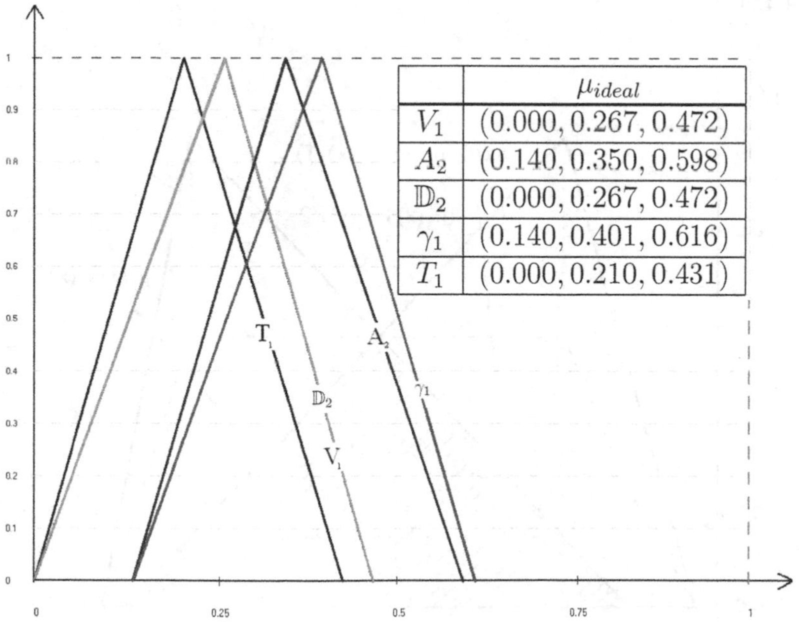

Fig. 4.14 Resulting evaluations of compared tool steel materials

Table 4.15 Final probability relations

$P(\gamma_1 > A_2)$	77.31%
$P(A_2 > V_1)$=$P(A_2 > \mathbb{D}_2)$	92.67%
$P(V_1 > T_1)$	80.28%

evaluations of compared tool steel materials are shown in Fig. 4.14. Comparing them (see Table 4.15) we conclude that the best one is the steel γ_1. The final ranking order is $\gamma_1, A_2, V_1, \mathbb{D}_2, T_1$, but $\mathbb{D}_2$ and V_1 have the same ranking.

This result is quite opposite to that obtained in [105, 134] where the steel γ_1 is only on the fourth place in ranking: A_2, $\mathbb{D}_2$, T_1, γ_1, V_1. This difference is easy to explain. In the framework of our approach, the final ranking is based on aggregation of aggregating modes producing own and different ranking (see Table 4.15). Therefore, in general, our final ranking should differ from those obtained using particular aggregating modes. Besides, we have assigned the lowest weight to the addition type of aggregation used in [105, 134]. Hence, the contribution of this aggregation to the final ranking is minimal.

The reasons for such weighting are partially pointed out at the beginning of this subsection. What is worth noting in addition is that addition-type aggregation and the so-called bounded difference aggregation $\max(0, \mu_A(x) + \mu_B(X) - 1)$ might be used only with a great prudence, since even in the case of only two local criteria such aggregations frequently cannot reveal any preferences in Pareto-region (see Fig. 4.15).

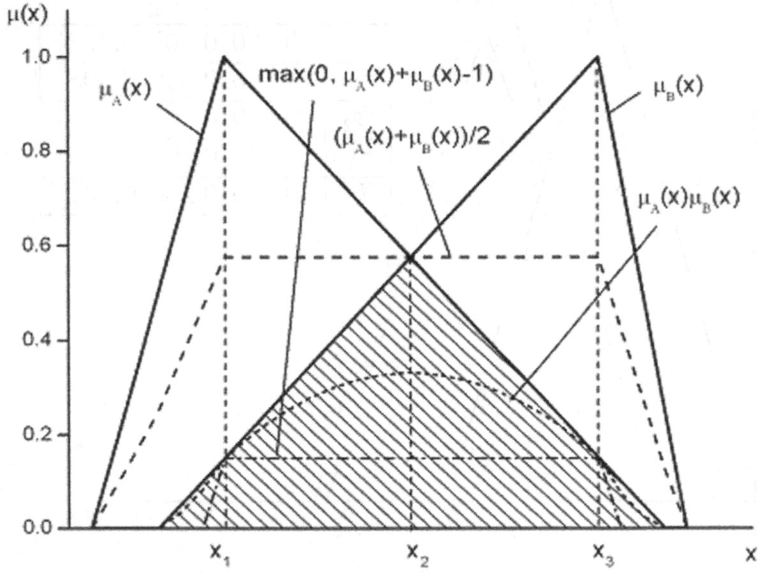

Fig. 4.15 Different types of aggregating operator

We think that just this negative feature of weighted sum type of aggregation was the underlying cause of identical ranking of tool steel materials obtained in [105, 134] in spite of the fact that in [105] contrary to the work [134] instead of fuzzy values their real valued representations were used (for the same initial data).

4.3 Multiple Criteria Investment Project Evaluation in the Fuzzy Setting

In this section, we present the simplified example of investment project evaluation using mainly financial parameters to stress the multiple criteria nature of this problem and the need for the use of fuzzy set theory tools even when we are dealing with the traditional budgeting.

4.3.1 Local Criteria Building

Let us consider such important quantitative financial parameters as Internal Rate of Return (*IRR*) and Net Present Value (*NPV*). Since *IRR* is measured in percentages, whereas *NPV* accounted in currency units, it seems impossible to compare them while estimating the project. However, it can be done using the functions presenting the local criteria based on these financial parameters. It is worth noting that introducing of such function does not lead to the loss of the original information, quite the contrary. Really, if x is some parameter of analyzed system representing its quality to certain extent, then some scale of preference - presented on numerical or verbal level - for this parameter inevitably exists at least in the decision maker's mind. Indeed, such preference type information is the key of decision-making and local criteria are the mathematical tools for its formalization. Often they can be built using the following simple and natural procedure.

For example, when considering *IRR*, it is easy to see that there always exists some lower bound for permissible values of *IRR*, usually being equal to the bank rate r. Further, there is some interval $r \leq IRR \leq IRR_m$, where project's quality rises gradually along with increasing of *IRR*. Finally, it is expected that if $IRR \geq IRR_m$, the project's efficiency with respect to *IRR* is so high that it is difficult to make a reasonable choice among such excellent projects.

To transform this description into mathematical form, the membership function, which is the pivotal concept of fuzzy sets theory, may be used. The membership function $\mu(IRR)$ representing the local criterion based on *IRR* for some analyzed example is shown in Fig. 4.16.

The values of membership functions change from zero (for the worst values of quality parameters) to maximum value equal to 1 in the area of best values of analyzed quality parameters. Hence, in the context of considered problem the values of membership functions may be treated as degrees of quality parameter's preference. The linear form of membership function is not a dogma. However, in practice what we usually know is only that some value is more preferable than other without any certain quantitative estimation of this preference. In other words, in most cases we

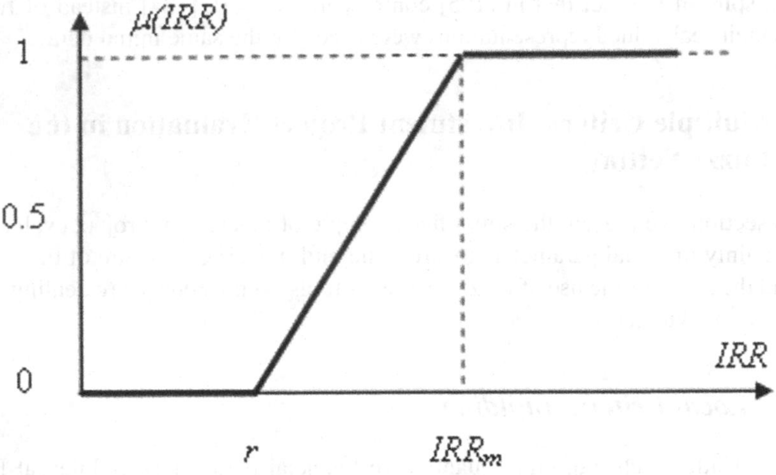

Fig. 4.16 Membership function representing the local criterion based on *IRR*

deal with the so-called "linear ordering" and for such situations the linear form of criterion function is proved to be the best one [139]. Since the membership function must be convex and normalized to 1, the only few forms of such function, presented in Fig. 4.17, may usually be used. Of course, if the probability distribution of quality parameter is known, the corresponding membership function may have a more complex form.

Consider a qualitative parameter presented in a verbal form, e.g., such as "ecological impact on region". It can be described by set of statements that linguistically represent the degrees to which an analyzed project may affect the ecology of region: "not significant", "slightly significant", "noticeably significant" and so on. As it was shown in [84], no more then 9 such linguistic degrees may be used in practice. The linguistic variables may be translated into mathematical form by presenting them in the form of triangular of trapezoidal fuzzy numbers. This approach is in accordance with the spirit of fuzzy sets theory and undoubtedly is very fruitful in a great number of applications, especially in fuzzy logic. Nevertheless, when dealing with the decision-making problems, we usually do not have enough reliable information to build such fuzzy numbers. Frequently, it is hard to choose the base for such triangular or trapezoidal fuzzy numbers, due to the absence of any evident reasons to prefer some base as the best one. In fact, the set of linguistic terms such as "not significant", "slightly significant", "noticeably significant" and so on in practice often represents only some labels signed to the levels of the decision maker's preference scale.

So the membership function shown in Fig. 4.18 seems to be a quite sufficient level of abstraction for the formalization of local criteria in the majority of real-life situations (of course, in some cases the use of more complex descriptions, e.g., on

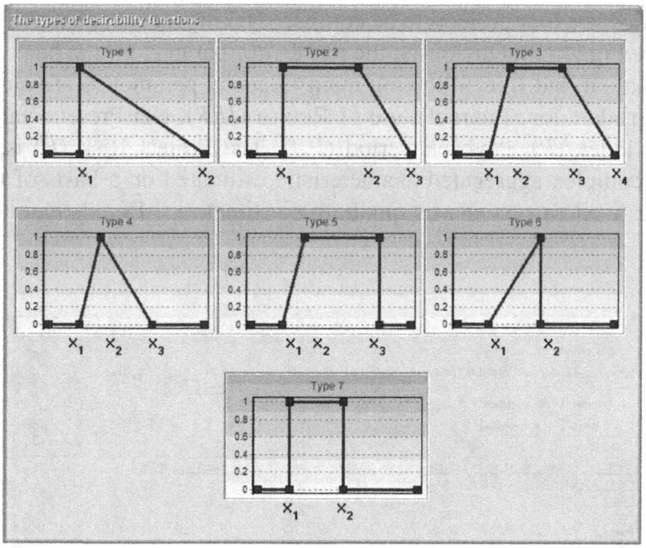

Fig. 4.17 Typical forms of membership functions

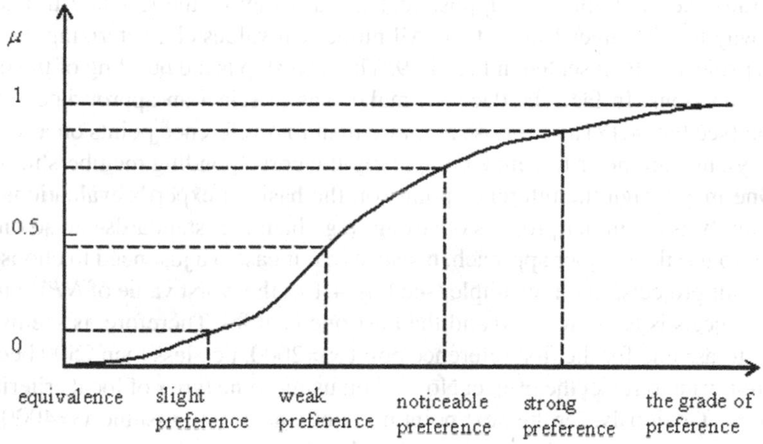

Fig. 4.18 Membership functions of qualitative local criterion

the base of type–2 fussy sets, may be preferable [41, 42]. Finally, we can see that all the qualitative and quantitative local criteria can be naturally presented within the universal scale of membership function.

To make our further analysis more transparent, consider simple example which is not the description of any real projects estimation situation, but reflects good the main advantages of the proposed approach.

Assume we have four projects to be compared taking into account five quality parameters (see Fig. 4.19). Obviously, the number of parameters can be much greater in a real situation, but four following main financial parameters are almost obligatory for consideration: Internal Rate of Return (*IRR*), Net Present Value (*NPV*), Profitability Index (*P*), Payback period (*PB*). The Project Risk (*R*) is, generally speaking, a complex aggregated characteristic estimated on a basis of the quantitative data as well as on expert's qualitative estimations. Details can be found in [37].

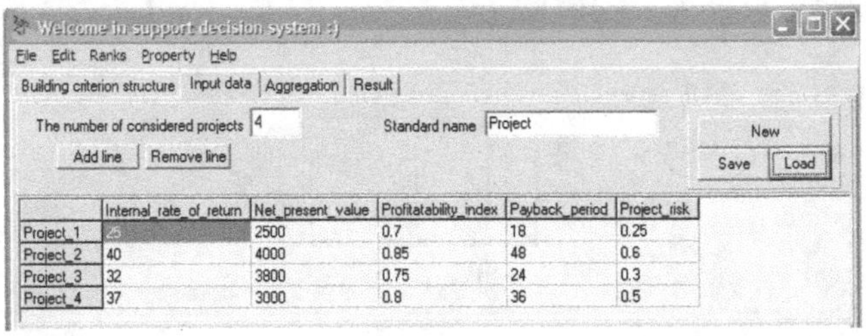

Fig. 4.19 Parameters of the analyzed projects

For the sake of simplicity, suppose that in our example the risk is estimated in such a way that it ranges from 0 to 1. All numerical values characterizing the considered projects are presented in Fig. 4.19. The next step is the building of the membership functions. In fact, the decision maker has to select an appropriate type of function (see Fig. 4.17) and choose no more than four reference points on a scale of the analyzing parameter to define completely the corresponding membership function. One may assign the reference points on the basis of expert's evaluations, statistical analysis of similar projects or strong (e.g., banking) standards and so on. We shall try to use the simpler approach, because in our case we just need to choose the best of four projects. In our example (see Fig. 4.19), the worst value of *NPV* among all the projects is equal to 2500 and the best one is 4000. Therefore, is seems reasonable to assume for the first reference point x_1=2000, i.e., less then 2500 because we do not want to reject the project No 1 when using some types of local criteria aggregations. Obviously, for the next point it is quite natural to assume x_2=4000, and to obtain a complete description of membership function we introduce an auxiliary point x_3=6000 (see Fig. 4.20).

The other membership functions for the considered example are built in a similar way.

4.3.2 *Ranking the Local Criteria*

Since in the real-world decision problems local criteria are usually expected to provide different contributions to the final aggregated estimation of alternatives, an appropriate method for the local criteria ranking is needed. It should be emphasized

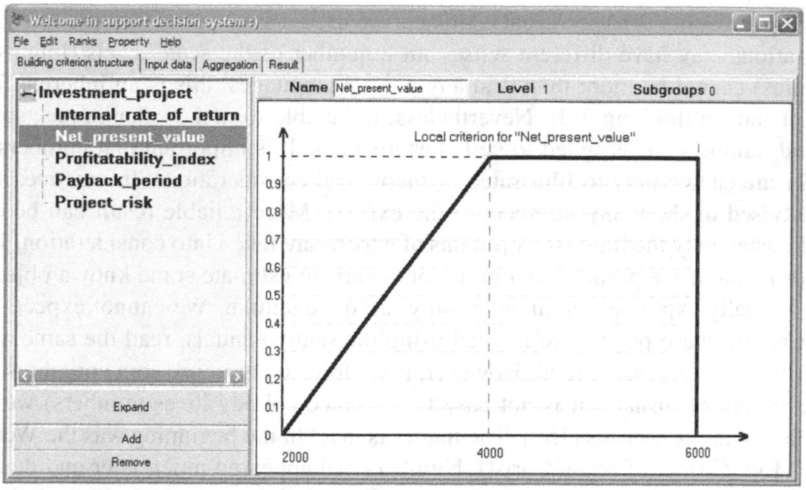

Fig. 4.20 Building the membership functions

that the simple qualitative ordering of local criteria is not enough for practical purposes: usually some quantitative indices representing the local criteria contributions to the overall alternative estimation are necessary. The experience shows that quantitative ranking of criteria is more difficult task for a decision maker than building the membership function. Though it is usually hard for decision makers to rank a set of local criteria as a whole, they usually can confidently state preference, at least verbally, when comparing only a pair of criteria. Therefore, the proper criteria ranking technique should use this pair comparison in a verbal form. Such technique is based on the so-called matrix of linguistic pair comparisons, proposed by T.Saaty [100]. The procedure of building this matrix for our example is illustrated in Fig. 4.21.

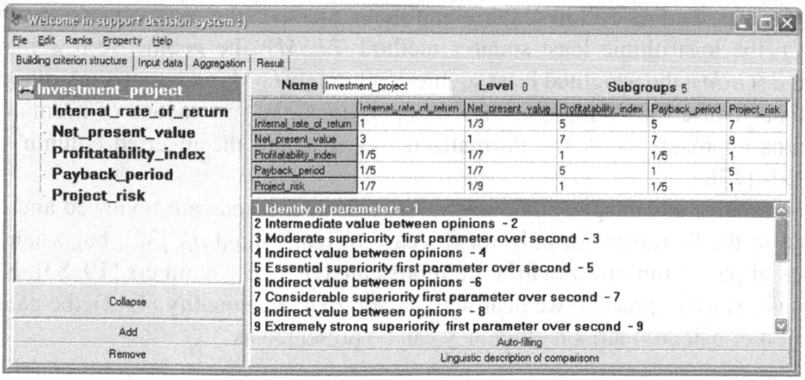

Fig. 4.21 Matrix of pair comparison

Of course, only 9 basic verbal estimates are used. Linguistic scales used in such estimations may have different sense, but a number of the scale levels (linguistic granules) cannot be more than 9 in any natural languages: this is an inherent feature of human thinking [13]. Nevertheless, to be able to make calculations, some natural numbers are assigned to verbal estimations. It is important that numbers in Fig.21 are shown only to illustrate our theoretical considerations. In practice, it is not advised to show any numbers to the experts. More reliable result can be obtained when only the linguistic opinions of experts are taken into consideration. The matter is that if we propose to a group of experts to estimate some known objects, their verbally expressed opinions usually are quite similar. We cannot expect any other result: these people are learned using the same manuals, read the same articles, worked in the same field. However, if we force them to use some numbers for the estimations (usually it is not easy to do, since nobody loves numbers) we do not obtain any consensus [150]. The matter is that "In the beginning was the Word" (The Holy Gospel of Jesus Christ). Numbers had appeared much later and during last few millenniums, which are a minute from historical viewpoint, people have not learned to use numbers properly yet. So far we are thinking using words, not numbers, and even trying to teach our computers this trick.

More strictly, let C_i, $i = 1$ to n, be local criteria to be ranked. Then the pair comparison matrix A can be created such that any entry $a_{ij} \in A$ represents the relative preference of criterion C_i when it is compared with the criterion C_j. The pair comparison matrix is reciprocal, meaning that $a_{ij} = 1/a_{ji}$ and $a_{ii} = 1$. If α_i and α_j represent values of ranks, then in the perfect case with a consistent matrix we have $a_{ij} = \frac{\alpha_i}{\alpha_j}$, $a_{ij} = a_{ik}a_{kj}$ and the ranks α_i , $i = 1$ to n, can be calculated easily. Unfortunately, in real-world situations, we usually have only approximate estimates of a_{ij} and actual values of a_{ij} may be unknown. Then the question arises: how to find α_i, $i = 1$ to n, such that

$$a_{ij} \approx \frac{\alpha_i}{\alpha_j}. \tag{4.42}$$

Various approaches have been proposed to obtain these ranks. In [19], they were roughly classified as follows: the eigenvector method; the least squares method (*LSM*); the logarithmic least squares method (*LLSM*); the geometric row means method (*GRM*); the weighted least squares method (*WLSM*) and a category of methods that involve only simple arithmetic operations: the row means of normalized columns approach [100], the normalized row sum and the inverted column sum methods [77].

The relative advantages and drawbacks of these methods are reviewed and discussed in the literature not only for the case of real valued a_{ij} [32], but when the entries of pair comparison matrix are represented by fuzzy numbers [19, 83]. Nevertheless, since in practice we deal with an approximate equality (4.42), the natural criterion of pair comparison efficiency can be presented as

$$S = \sum_{i=1}^{n} \sum_{j-1}^{n} (a_{ij} - \frac{\alpha_i}{\alpha_j})^2, \tag{4.43}$$

where α_i are the ranks obtained using considered method. For this reason, it is quite natural to use the method based on the minimization of S, i.e.,

$$S=\sum_{i=1}^{n}\sum_{j-1}^{n}(a_{ij}-\frac{\alpha_i}{\alpha_j})^2 \rightarrow \min,\ s.t. \sum_{i=1}^{n}\alpha_i = 1 \tag{4.44}$$

(sometimes the restriction $\sum_{i=1}^{n}\alpha_i = n$ is used). Using rich experimental data, it was shown in [130] that this method (*LSM*) is the best one as it provides the least final values of S. On the other hand, in some cases the simplest row means of normalized columns method [100]

$$\alpha_i = \frac{\sqrt[n]{\prod_{j=1}^{n} a_{ij}}}{\sum_{i=1}^{n}\sqrt[n]{\prod_{j=1}^{n} a_{ij}}} \tag{4.45}$$

may produce results good from practical viewpoint [130]. Of course, the criteria weights in some special cases may be interpreted differently. Some not exhaustive classification of such possible interpretations and the overview of corresponding ranking methods are presented in [30]. However, in our case the method based on the expressions (4.44) seems to be completely corresponding to the nature of the problem and sufficiently justified mathematically.

4.3.3 Numerical Evaluation of the Comparing Investment Projects

As in Subsection 4.2.4, we shall use three most popular aggregation modes: $D_1 = \min(\mu_1(C_1)^{\alpha_1}, \mu_2(C_2)^{\alpha_2}, ..., \mu_n(C_n)^{\alpha_n})$, $D_2 = \prod_{i=1}^{n}\mu_i(C_i)^{\alpha_i}$, $D_3 = \sum_{i=1}^{n}\alpha_i\mu_i(C_i)$.

The results of multiple criteria evaluation of compared projects using the data and methods described in previous subsections are presented in Fig. 4.22.

Observe that the *min*-type and multiplicative aggregations give us similar resulting evaluations, which are far from those obtained using weighted sum type of aggregations. It is important that the proposed approach allows us to estimate directly the contribution of each local criterion to the generalized project assessment (see Fig. 4.23).

As the aggregation modes D_1, D_2, D_3 provide different numerical evaluations of compared four projects, a natural approach to this problem is the aggregation of these modes to get the compromise final evaluation. For this purpose we use the method presented in Subsection 4.2.4. This method is based of the synthesis of type–2 and level–2 fuzzy sets defined on the support composed of compared alternatives, which in the considered case are the analyzed investment projects. The central in this approach is a concept of "ideal" method of aggregation presented in the form of mathematical object which can be treated as the level–2 fuzzy set of type–2:

$$D_{ideal} = \left\{ \frac{\mu(D_1)}{D_1}, \frac{\mu(D_2)}{D_2}, \frac{\mu(D_3)}{D_3} \right\},$$

where the values of $\mu(D_i)$ represent expert's opinions about closeness of considering aggregating operators to the some perfect type of aggregation which can be treated as the best or "ideal" method for aggregation. As it has been shown in Subsection 4.2.4, the *min*-type aggregation D_1 is more reliable one than the multiplicative aggregation D_2 and both are noticeably more reliable than the additive aggregation D_3. Therefore, we have used the matrix of pair comparison of these aggregation modes presented in Table 4.14. As in Subsection 4.2.4, using this matrix we get: $\mu(D_1) = 0.7$, $\mu(D_2) = 0.25$, $\mu(D_3) = 0.05$.

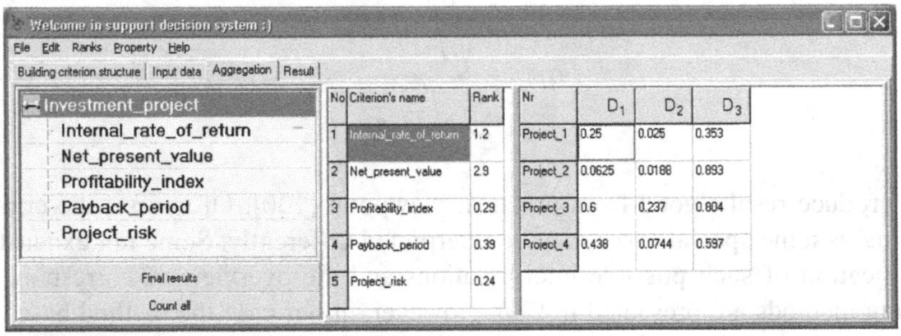

Fig. 4.22 Resulting evaluations of the projects presented in Fig. 4.19 (min-type, multiplicative and additive aggregations are denoted by D_1,D_2,D_3, respectively)

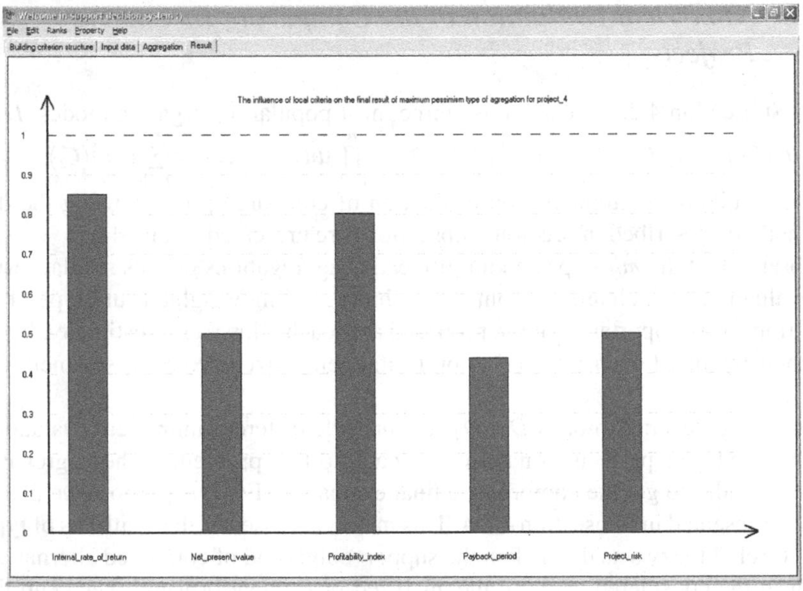

Fig. 4.23 The contributions of local criterion to the final project's evaluation

In the considered case, Exp.(4.30) takes the form

$$D_i = \left\{ \frac{D_i(Project_1)}{Project_1}, \frac{D_i(Project_2)}{Project_2}, \frac{D_i(Project_3)}{Project_3}, \frac{D_i(Project_4)}{Project_4} \right\}, i = 1,2,3.$$

Hence, D_{ideal} can be presented as

$$D_{ideal} = \left\{ \frac{\mu_{ideal}(Project_1)}{Project_1}, \frac{\mu_{ideal}(Project_2)}{Project_2}, \frac{\mu_{ideal}(Project_3)}{Project_3}, \frac{\mu_{ideal}(Project_4)}{Project_4} \right\},$$

where $\mu_{ideal}(Project_j) = \max_i(\mu(D_i) \cdot D_i(Project_j)), i = 1,2,3.$
Finally, after all calculations using the results from Fig. 4.22 we obtain

$$D_{ideal} = \left\{ \frac{0.18}{Project_1}, \frac{0.05}{Project_2}, \frac{0.42}{Project_3}, \frac{0.31}{Project_4} \right\}.$$

Thus, the third project is the best one. Factually, we can infer the same result analyzing the data presented in Fig. 4.22, but it would be hard to do this for a greater number of projects or/and aggregating operators under consideration.

4.3.4 Hierarchical Structure of Local Criteria

As it has been noticed above, many of real-life problems of investment evaluation are not only multiple criteria, but also multi-level (hierarchical) ones. The method presented in previous subsections allows us to build in a natural way the branched hierarchical structures as in Fig. 4.24.

It can be seen that each criterion of upper k level is built on a basis of local criteria of underlying $(k-1)$ levels using one of the aggregation methods or their generalization as proposed in Subsection 4.2.4. The general expression for the calculation of criteria on intermediate levels of hierarchy is as follows:

$$D_{k,i_{n-1},i_{n-2},\dots,i_k} = f_{k,i_{n-1},i_{n-2},\dots,i_k}(D_{k-1,i_{n-2},\dots,i_k,1}, \alpha_{k-1,i_{n-2},\dots,i_k,1}, \dots, \\ D_{k-1,i_{n-1},i_{n-2},\dots,i_k,m_{k,i_{n-1},i_{n-2},\dots,i_k}}, \alpha_{k-1,i_{n-1},i_{n-2},\dots,i_k,m_{k,i_{n-1},i_{n-2},\dots,i_k}})$$

where $f_{k,i_{n-1},i_{n-2},\dots,i_k}$ is an operator of criteria aggregation, $m_{k,i_{n-1},i_{n-2},\dots,i_k}$ is the number of of local criteria on $(k-1)$ level, aggregated to intermediate local criterion $D_{k,i_{n-1},i_{n-2},\dots,i_k}$. It is worth noting that the values of $D_{n-1,i_{n-1}}$ are always in the interval $[0,1]$ and may be interpreted as some intermediate local criteria assessments. On the lowest level of hierarchy the initial membership functions representing the local criteria based on origin parameters of project's quality are used, i.e.,

$$D_{1,i_{n-1},i_{n-2},\dots,i_1} = f_{1,i_{n-1},i_{n-2},\dots,i_1}(\mu_{0,i_{n-1},i_{n-2},\dots,i_1,1}, \alpha_{0,i_{n-1},i_{n-2}\dots,i_1,1}, \dots, \\ \mu_{0,i_{n-1},i_{n-2},\dots,i_1,m_{1,i_{n-1},i_{n-2},\dots,i_1}}, \alpha_{0,i_{n-1},i_{n-2},\dots,i_1,m_{1,i_{n-1},i_{n-2},\dots,i_1}})$$

where $m_{1,i_{n-1},i_{n-2},\dots,i_1}$ is the number of initial local criteria on the lowest level.

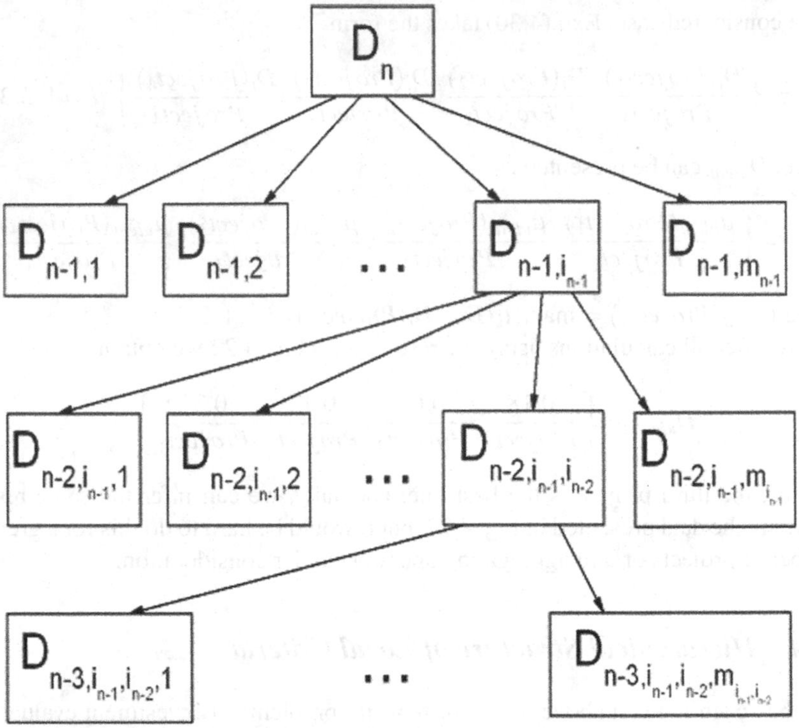

Fig. 4.24 Hierarchical structure of local criteria

The method for building the hierarchical local criteria systems is based on the multiple criteria approach described in previous subsections and generalizes it. The method is implemented in the form of user-friendly software (see Fig. 4.25).

It is important that presented method not only gives us some generalized quantitative evaluation of a project as a whole, but makes it possible to assess the contribution of each local criterion to this final evaluation.

4.4 Fuzzy MCDM and Optimization in the Stock Screening

This section is devoted to the application of the methods presented above in sections 4.1-4.3. A new method for the stock ranking based on the multiple criterion decision making and optimization is proposed. Two general criteria are used in the analysis. The first of them is based on the financial indices and may be treated as the criterion of firm's "health" or its financial performance. The second one is the two-criteria performance of firm based on the stock prices. It represents the firm's market success. The method rests on the selection of the stocks with a great correlation of the firm's financial performance and its market success. The local criteria are built in the form of membership functions of corresponding fuzzy subsets. Two

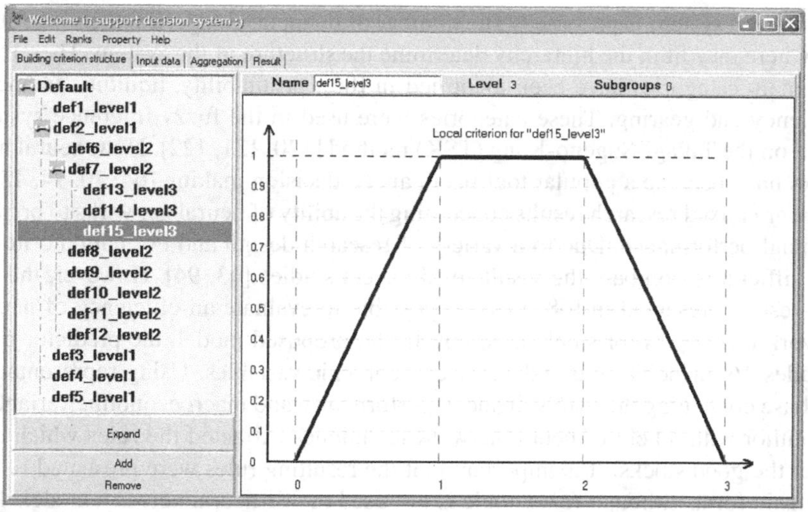

Fig. 4.25 Building the hierarchical local criteria system

different strategies for stock ranking and three most popular methods for local criteria aggregation are compared. The method is addressed to those who prefer to select for a portfolio only the firms which demonstrate the closeness of their overall financial performance in the past year and success in the stock exchange in the following year.

Observed onrush of global stock market provides the new possibilities for the successful investment. On the other hand, the pressing problem -the consequence of market growth-is the choosing of appropriate stocks or stock selection. Since the financial performance as well as the stock returns of almost all firms, which are worth mentioning, is now presented in Internet, the stock screening programs become popular. They make it easier and quicker to tailor a portfolio to fit the desired style and preferences of the investor [3]. The user can customize a model to include up to 20-40 different indicators simultaneously and can scan on relative values, e.g., the stocks with the highest relative strength, or on fundamental variables, like earnings per share momentum. In many cases, such approaches based on the single selecting criterion provide good results [44], but in general, the stock selection problem seems to be a multiple criteria one. In practice, an investor deals with a set of local criteria based explicitly or implicitly on the different financial rations and stock return. Since the preferences of the investor concerned such criteria often are presented in the verbal form, the problem is usually charged by the uncertainty of subjective type. Therefore, now the researches are focused on the synthesis of fuzzy logic methods and multiple criteria approaches in the fuzzy setting. In [80], the problem is considered in context of corporate acquisition process. In the proposed fuzzy system, its hierarchical structure is confirmed by the user along with the financial categories and ratios to populate it. Each financial category or ratio is given a certain degree of influence over the decision-making process. The two level hierarchical system is proposed. The number of financial categories, their priorities and where they fit in the hierarchy determine the structure of the system. Therefore, four main categories have been identified in [80]: profitability, liquidity, financial efficiency and gearing. These categories were used in the fuzzy inference system based on the Takagi-Sugeno-Kang (TSK) method [120, 121, 122]. Now, neural networks have become a popular tool for financial decision making [64, 70, 74, 131]. There are mixed research results concerning the ability of neural networks to predict financial performance. Due to a variety of research design and evaluation criteria, it is difficult to compare the results of different studies [43, 94]. However, the recent results presented in [68] make it possible to evaluate an efficiency of neural network techniques for stock selection. In the proposed model, the predictor base includes 16 financial ratio and 11 macroeconomic variables. Using representative database containing the firm's financial performance and macroeconomic variables, the author with a help of neural network techniques extracted the rules which can select the good stocks. It is important that the resulting rules were presented in the linguistic form, although fuzzy logic is not used in this neural network model. The direct fuzzy multiple criteria method for stock selection was used in [123]. This paper proposes a new method for group decision making in the fuzzy environment. Some modification of Chen's method [21] is proposed and it is shown how this approach can be used for stocks selection on *ISE* (Istanbul Stock Exchange). The rating of each stock and the weight of each criterion are presented in the form of linguistic terms represented by triangular fuzzy numbers. Six well known financial rations (market value of firm/earnings before amortization, interest and taxes ratio, return on equity (*ROE*), dept/equity ratio, current ratio market value/net sales, price earnings ratio (P/E)) were used as the base for the multiple criteria performance of analyzed firms. It is easy to see that in the framework of this approach there is no intention to confront the parameters characterizing in general the firm's financial "health" with those performing its success in the stock exchange.

Summarizing, we can say that there is no common point of view on the problem of stock selection. Moreover, in our opinion, the searching for the "best" method of stock selection is rather senseless endeavor since in any case the method should reflect the decision makers' preferences, him/her acceptation of risk and so on.

Nevertheless, there are common points in the analyzed papers. It is easy to see that the stock selection problem is a multiple criteria one. There are many approaches to the solution of this problem, but in practice, the best results provides the synthesis of different methods. Bollinger and Pictet [12] used the synthesis of known *ELECTREIII*, *SURMESURE* and *AGATHA* methods for multiple criteria decision analysis and land-filling technologies for waste incineration residues in Switzerland. Gomes et al. [45] proposed a multicriteria decision support system based on multiattribute utility theory and preference modeling. Chen et al. [24] proposed a case-based distance model to solve screening problems in multiple criteria decision aid.

In this section, we factually use the felicitous definition proposed in [24]: "Screening is a process of multiple-criteria decision aid (*MCDA*) in which a large set of alternatives is reduced to a smaller set that most likely contains the best choice".

Beynon et al. [8] proposed a synthesis of *AHP* and the Dempster-Shafer theory (*DST*). A major advantage of incorporating *DST* into the *AHP* lies in the reduction of the complexity of the multiple criteria problem.

As the local criteria and methods for their mathematical formalization and aggregation depend on the subjective decision maker's preferences, the use of fuzzy sets theory tools seems quite natural. Therefore, the fuzzy extensions of different classical *MCDM* approaches are now frequently used for the solution of real- world problems. The good examples of application of the fuzzy *AHP* in different fields are presented in [17, 50]. The use of the specificity of fuzzy sets theory methods makes it possible to enhance the performance of multiple criteria analysis. Chiou et al. [26] proposed to use the so-called non-additive fuzzy integral, to cope with the *MCDM* problem particularly while there is dependence among local criteria.

Two deferent sources of information can be used for stock selection: financial rations and stock prices. The final results of selection can be obtained with the use of teaching the decision making system based on the comparison of firm's financial performance with its success in the stock exchange. The teaching of decision making system can be carried out with the use of optimization methods. The incorporation of optimization methods into *MCDM* is widely used in different applications. The good results of such synthesis recently have been obtained in multiobjective decision making process for supplier selection and order allocation [34], optimal blending for beneficiation of coal [16], aggregate production planning [115], fuzzy multi-objective transportation problems [1], fuzzy R&D portfolio selection [132], and in the land-use planning in agricultural system [10].

Here we propose a new approach to stock selection and screening taking into account above mentioned circumstances. The method we have developed is not the "best" one. As any other method, it can be accepted as appropriate one by some investors and completely rejected by other ones. We address our method to those who prefer to choose for a portfolio only such firms which demonstrate the closeness of their overall financial performance in the past year and the success in the stock exchange in the following year. Of course, our approach can reject many of possible market championships, but we point out that proposed method is addressed to rather cautious investors.

4.4.1 Multiple Criteria Performance of Firms

To evaluate the firm's financial performance, we have used the financial rations from database comprising the data of 162 firms (subsector of the biotechnology of US economy). The next 12 financial rations were chosen for analysis:

- EPSTTM2TTM (Earnings per share (EPS) Percent Change, TTM vs. Prior TTM (%))
- RTTM2TTM (Revenue Percent Change, TTM vs. Prior TTM (%))
- ChPTMgn (Change in Pretax Margin Positive)
- ChROI (Change in Return on Investment)
- ChSO (Change In Shares Outstanding)

- EPSQ2Q (EPS Percent Change, Most Recent Quarter vs. Prior Quarter
- DCS (Depreciation of Capital Spending)
- MKTCAP (Market Capitalization)
- PTMgn (Pretax Margin)
- ROI (Return On Investment)
- EPSQ2TTM (EPS Percent Change, Most Recent Quarter vs. Last Twelve Months)
- RQ2TTM (Revenue Percent Change, Most Recent Quarter vs. Last Twelve Months)

The detailed descriptions of these rations can be found in [51, 52]. It is worth noting here that proposed set of financial rations is not the "best" and exhaustive one. When choosing the rations, we have used mainly the well known textbooks such as [95] and other literature sources [38, 61, 96, 124]. The other criterion of choice was the availability of the data in Internet. This criterion is essentially important as we deal with the stocks screening. The results of correlation analysis we have made for one year show that the strong correlation between some rations takes place.

Therefore, using the results of correlation analysis and some additional information from [38, 61, 95, 96, 124] concerned with an appropriateness of correlating rations the next 7 most uncorrelated financial rations were selected for the subsequent analysis: EPSTTM2TTM, ChSO, EPSQ2Q, DCS, MKTCAP, EPSQ2TTM, RQ2TTM.

4.4.2 General Criterion of Firm's "Health" Based on Financial Rations

To build the local criteria based on financial rations r they were divided into two groups: the rations enhancing the firm's performance along with rising of their values and the rations depressing the performance with rising of their values. For all rations we have found the lowest and highest values in the available database considering all analyzing firms. These values were used as the base for building the local criteria. For the first group of rations, the so-called profit type local criteria [21] were built in the form of membership functions $\mu(r)$ (in terms of fuzzy sets theory) rising from 0 to 1 in intervals between the lowest and highest values of the corresponding financial rations we have found in the database. Similarly the cost type local criteria were built. Examples are presented in Fig. 4.26.

It is worth noting that sometimes, e.g., when analyzing the earnings per share (EPS) ratio, the linear form of membership function does not well reflect the financial performing of firm and trapezoidal shape of function seems to be a more appropriate one. On the other hand, such situations may be considered rather as the exceptional cases, although there are no strong restrictions on the form of membership functions is the framework of the proposed approach.

When all introduced seven local criteria $\mu_i(r_i)$ of firm's financial performance based on the selected financial rations r_i, i=1 to 7, are calculated, the problem arises: how to aggregate them into the general criterion of financial performance taking into

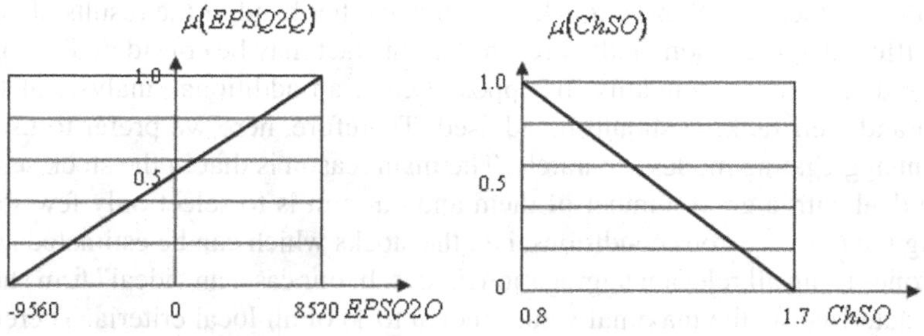

Fig. 4.26 Profit type criterion μ(EPSQ2Q) based on financial ratio EPSQ2Q and cost type criterion μ(ChSO) based on financial ratio ChSO

account they ranks (weights), α_i, i=1 to 7 ? To make the results of our analysis more reliable, three most popular aggregating modes have been used:

$$D_1 = \frac{1}{7}\sum_{i=1}^{7} \alpha_i \mu_i, \quad D_2 = \prod_{i=1}^{7} \mu_i^{\alpha_i}, \quad D_3 = \min\left(\mu_1^{\alpha_1}, \mu_1^{\alpha_1}, \ldots, \mu_7^{\alpha_7}\right), \qquad (4.46)$$

where $\sum_{i=1}^{7} \alpha_i = 7$.

Since there exists a pluralism in choosing of appropriate aggregating mode, some comments are needed. Generally, the choice of aggregating mode is a context dependent problem [148]. Nevertheless, now the most popular method of aggregation is the weighted sum (D_1 in our case). It is used in many well known decision making models such as *AHP* [100], multi-attribute utility analysis [91] and so on, but often without any critical analysis. On the other hand, in some fields, e.g., in ecological modeling, the weighted sum is not used for aggregation [116]. The reason behind this is that in practice there are cases when if any of local criteria is totally dissatisfied then analyzed alternative should be rejected from consideration at all. The detailed analysis of the advantages and drawbacks of aggregating modes can be found in [36, 108]. It is shown in these papers that, in general, the most reliable aggregation approach is the use of Yager's [138] *min*-type operator D_3. The multiplicative mode D_2 appears to be somewhat less reliable and, finally, the additive (weighted sum D_1) method may be considered as unreliable and insensitive when choosing an alternative in Pareto-region (see also Subsection 4.2.4). On the other hand, as all known aggregation modes have their own advantages and drawbacks, it seems impossible to choose the best one especially when dealing with complicated hierarchical problem. Therefore, when dealing with a complex task characterized by a great number of local criteria, it seems reasonable to use all relevant to the considered problem types of aggregations. Since the different final results may be obtained on the base of different aggregating modes, the problems arises: how to aggregate such results? For this purpose, recently a new method for aggregation of aggregating modes [36, 108] based on the synthesis of type–2 and level–2 fuzzy sets was proposed (see also Subsection 4.2.4). On the other hand, if the results obtained using different aggregation modes are similar, this fact may be considered as a good confirmation of their optimality. In opposite case, an additional analysis of local criteria and their ranking should be advised. Therefore, here we prefer to use the different aggregating modes separately. The main reason is that in the stock screening we deal with a great number of them and our aim is to select only few stocks meeting the most rigorous conditions, i.e., the stocks which can be estimated as the "best" ones using all relevant aggregation modes. In our case, an "ideal" firm should be characterized by the maximal values (equal to 1) of all local criteria. Therefore, for such "ideal" firm, the values of general criteria obtained using the aggregation modes D_1, D_2, D_3 should be equal to 1. Obviously, the performance of firms which we can qualify as the "good" ones should be close to performance of "ideal" firm (probably not existing). Hence, the values of the general criteria D_1, D_2, D_3 for the "good" firms should be close to 1.

To calculate the general criteria of firm's "health" (4.46) based on the financial rations, the values of local criteria ranks α_i, i=1 to 7, are needed. The are two approaches to estimate α_i we shall compare: the first one is based on the expert's opinions and the so-called matrix of linguistic pair comparisons, the second one uses the teaching procedure formulated as the multiple criteria optimization task.

4.4.3 Two-Criteria Performance of Firm Based on Stocks Prices History

In addition to the general criteria of the firm's "health" described in previous subsection, we introduce here the general criterion of firm's performance based on its stocks prices history during the year after the date when we have estimated the firm's "health" taking into account only financial rations.

It worth noting that such investment horizon (one year) is chosen only for the simplicity. There are no any restrictions on the investment horizon in the framework of the proposed approach. Moreover, the analysis could be carried out using the firm's history in the last 3-5 years. Theoretically, such approach based on more historical data should provide more reliable results. On the other hand, we can see the decreasing of the durations between great market crises (financial crisis of 1997-1998, dot.com crisis of 2000, the great crisis of 2008 and the local crises of less importance). Therefore, the choice of relatively short investment horizon (one year) and analyzing the firm's "heals" only in the last year seems as an acceptable compromise, especially taking into account that the aim of this section is to present only a new approach to the stock screening. Thereupon, all the results of stock screening we have obtained analyzing the subsector of biotechnology of US economy should be treated only as an illustration.

As it is impossible to make only proper decisions in the stock market, the total result of trading should range from maximal losses ML (if all the decisions are worst ones) to the maximal possible returns MR that can be received if an investor makes only the best decisions during a year. We have used the values ML and MR to build the local criteria characterizing our intentions to reduce the risk and maximize incomes, respectively. To estimate ML and MR the month bars (high and low month stock prices) were used with the assumption that investor makes a decision monthly. For the simplicity, suppose that the starting investment is of $100 000 and each month an investor should buy or sell using the actual fund he/she possesses as the result of all previous financial transaction based on the initial investment of $100 000. As usually there may be restrictions on the short sells for some stocks, we except here the possibility of short sells at all. The month bars for the ticker(firm) BBY and illustration of the best and worst decisions are presented in Fig. 4.27.

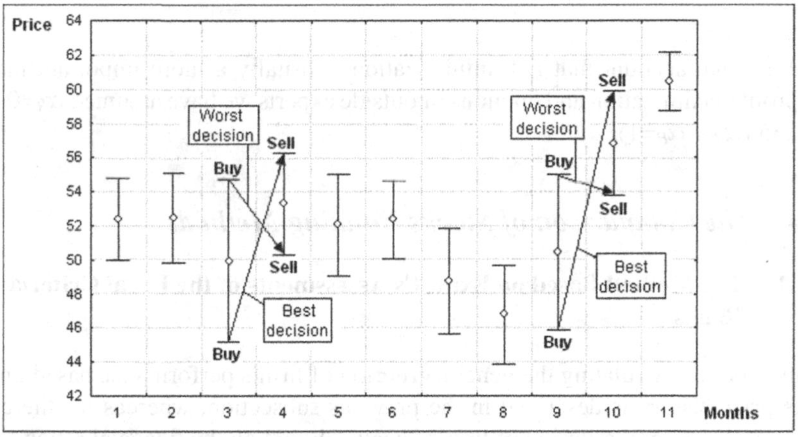

Fig. 4.27 Month bars of the ticker BBY (2004)

For this ticker, the interval of possible results of trading is $[ML, MR]$ = [-65918, 291353]. Such intervals have been calculated for all considered stocks (subsector of the biotechnology of US economy). The left bounds of these intervals represent the financial risk as a consequence of poor decisions, whereas the right bounds are maximal possible returns. Hence, they can be used for building the local criteria of risk minimization μ_R and profit maximization μ_P. For this purpose, we calculate the values $ML_{max} = max_i\{\text{abs}(ML_i)\}$ and $MR_{max} = max_i\{\text{abs}(MR_i)\}$, $i = 1$ to m, where m is the number of tickers in the considered subsector. The corresponding local criteria are presented in Fig. 4.28.

The general criterion of firm's performance D_P based on its stocks prices history we have built as the aggregation of local criteria μ_R and μ_P. As in the previous subsection, three types of aggregation have been considered for each ith stock:

$$D_{1Pi} = \alpha_R \mu_R(ML_i) + \alpha_P \mu_P(MR_i), \quad D_{2Pi} = \mu_R(ML_i)^{\alpha_R} \cdot \mu_P(MR_i)^{\alpha_P},$$
$$D_{3Pi} = \min\left(\mu_R(ML_i)^{\alpha_R}, \mu_P(MR_i)^{\alpha_P}\right). \tag{4.47}$$

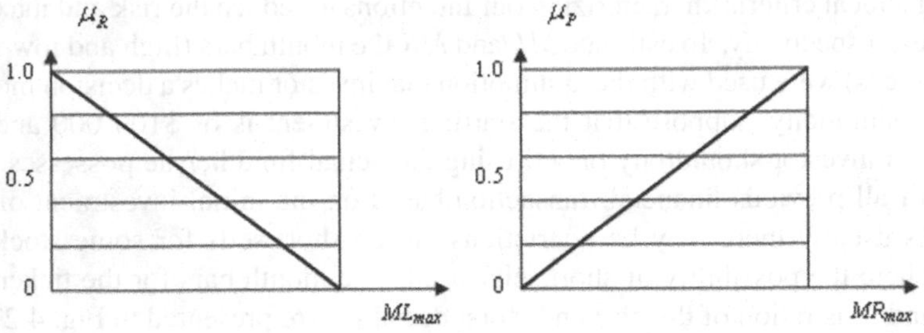

Fig. 4.28 Local criteria of risk minimization μ_R and profit maximization μ_P

Taking into account that risk minimization is usually a more important intent than profit maximization and opinions of outside experts we have assumed α_R=0.65, α_P=0.35 ($\alpha_R + \alpha_P$=1).

4.4.4 *The Comparison of Stocks Ranking Methods*

4.4.4.1 The Method Based on Expert's Assessments of the Local Criteria Ranks

The method for calculating the general criterion of firm's performance based on its stocks price history is described in the previous subsection, whereas for the estimation of the general criterion of firm's "health" based on the financial rations, the values of local criteria ranks are needed. These ranks may be obtained with the use of expert's opinions about the local criteria relative importance. Experience shows that quantitative ranking of the criteria is a more difficult task for a decision maker than building membership functions. Though it is hard for decision makers to rank the set of local criteria as a whole, they usually can confidently determine the preference, at least verbally, when comparing only a pair of criteria. Therefore, a proper criteria ranking technique should use this pair comparison in the verbal form. Such technique is based on the so-called matrix of linguistic pair comparisons which is the basic mathematical tool of sound *AHP* method developed by Saaty [100]. As this technique is described in details in the textbooks [101], here we only illustrate it briefly.

Let X,Y,Z be comparing local criteria. If X is rather more important than Y then to this relation the number 3 is assigned and the value $\frac{1}{3}$ is assigned to the opposite statement "Y is rather more important than X". If Y is definitely more important than Z then to this relation the number 5 is assigned and the value $\frac{1}{5}$ is assigned to the opposite statement. Indeed, the number 7 is assigned to the linguistic assessment "X is strongly more important than Z" and the value $\frac{1}{7}$ is assigned to the opposite statement.

Usually only 9 basic verbal estimates are in use. Linguistic scales used in such estimations may have different sense, but a number of the scale levels (linguistic granules) cannot be more than 9 in any natural languages: this is an inherent feature of human thinking [84]. There were different methods for obtaining the ranks from matrix of pair comparisons proposed in the literature. For our purposes, we prefer to use the method presented in [32]. The reasons behind this choice are based on the recent literature review [36]. After analysis of the literature [38, 61, 95, 96, 124] and opinions of outside experts the matrix of pair comparisons of local criteria based on the financial rations has been built (see Table 4.16).

Table 4.16 Reciprocal matrix of pair comparisons of local criteria

	EPS TTM2TTM	ChSO	EPS Q2Q	DCS	MKT CAP	EPS Q2TTM	RQ2 TTM
EPSTTM2TTM	1	9	1/2	2	8	1/3	3
ChSO	1/9	1	1/7	1/5	1/2	1/8	1/4
EPSQ2Q	2	7	1	3	6	1/2	4
DCS	1/2	5	1/3	1	4	1/4	2
MKTCAP	1/8	2	1/6	1/4	1	1/7	1/3
EPSQ2TTM	3	8	2	4	7	1	5
RQ2TTM	1/3	4	1/4	1/2	3	1/5	1

Finally, using this matrix and the method proposed in [32] (see also Subsection 4.2.4) the following ranks had been calculated: α_1 = 1.081 (EPSTTM2TTM), α_2 = 0.205 (ChSO), α_3 = 1.666 (EPSQ2Q), α_4 = 0.67 (DCS), α_5 = 0.243 (MKTCAP), α_6 = 2.663 (EPSQ2TTM), α_7 = 0.477 (RQ2TTM). The ranks were normalized as follows:

$$\frac{1}{7}\sum_{i=1}^{7} \alpha_i = 1.$$

Using the obtained ranks and values of local criteria based on the financial rations and stocks price history, the values of generalized criteria D_1, D_2, D_3 and D_{1P}, D_{2P}, D_{3P} described in previous subsection were calculated for each of 162 considered firms (subsector of the biotechnology of US economy). The results are presented in Fig. 4.29-Fig. 4.31.

It is seen that rather there is no considerable correlation between D_i, D_{iP}, i=1 to 3. We have confirmed this (null) hypothesis using standard statistical methods. For all pairs D_i, D_{iP}, i=1 to 3, the Pearson product-moment correlation coefficient is equal to 0 with significance level of $\alpha = 0.05$. Since the Pearson's approach concerns with the linear relationship between variables, in addition the Spearman correlation coefficient has been calculated. Its value is close to 0 too. Generally, this result is not surprising as the firm's market success is not wholly determined by its financial performance because of strong influence of "external" factors (public relations, macroeconomic situation, rumors and so on).

Therefore, our "cautious" approach to the stocks selection is based on the assumption that the "good" firms should demonstrate the closeness of their overall financial performance in the past year and success on the stock exchange in the following year. In such a way, it is possible to reject from consideration all "unsafe" firms, i.e., such ones that their market success is based rather on the public relations, subjective opinions of market experts, rumors and other sometimes unreliable information. Hence, taking into account that $0 \leq D_1, D_2, D_3, D_{1P}, D_{2P}, D_{3P} \leq 1$, the "good" firms are those that gathered around the diagonal lines in Fig. 4.29-Fig. 4.31.

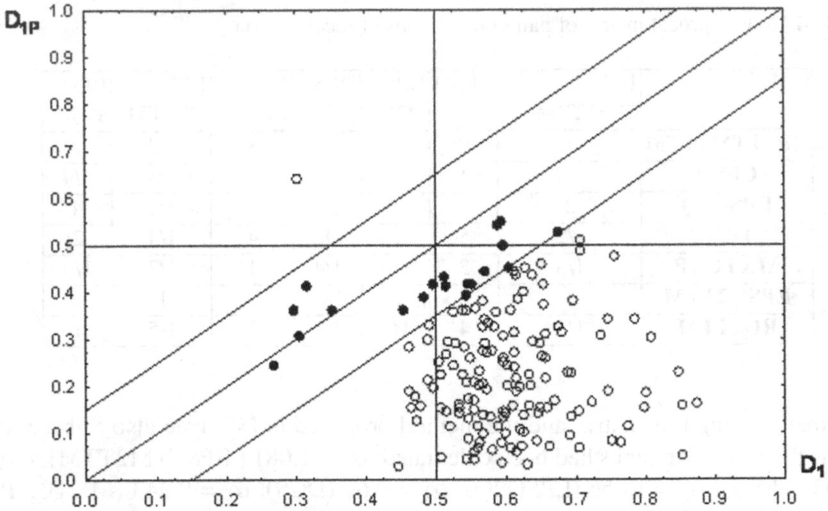

Fig. 4.29 "Good" firms (black points) selected using weighted sum type of aggregation and pair comparison matrix for local criteria ranking

The channels in these figures are chosen (using the corresponding numerical algorithm) in such a way that each of them contains only 20 "good" in above sense firms (black points). Therefore, only 20 firms with minimal difference between their generalized criteria based on the financial rations and stocks price history were selected using all considered methods for local criteria aggregation. It is seen that the widths of the channels of "good" firms are small enough to conclude that the selected firms are really "good" ones in our case. Of course, we choose here the 20 "good" firms in the channel only to illustrate our method. In practice, the choice of the number of "good" firms depends on the investment policy and many other objective and subjective factors. Obviously, the greater the values of D, D_P, the better is the firm contained in the channel. On the other hand, we have three analyzing channels and the most of firms can be placed only in one of channels shown in Fig. 4.29-Fig. 4.31. Therefore, to make the results of our analysis more reliable, we are looking for the firms containing simultaneously at lest in two channels. The results of analysis are presented in Table 4.17. It is worth noting that in our case there are no firms placed simultaneously in three considered channels.

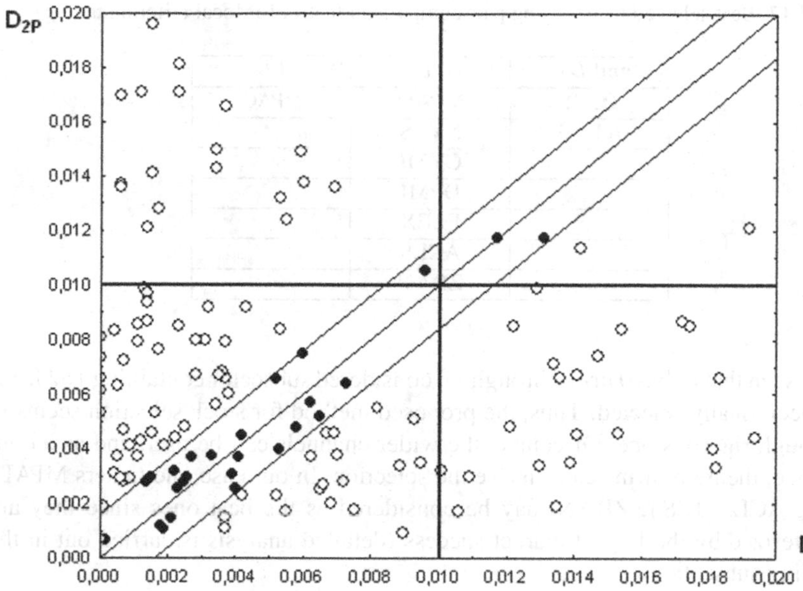

Fig. 4.30 "Good" firms (black points) selected using multiplicative type of aggregation and pair comparison matrix for local criteria ranking

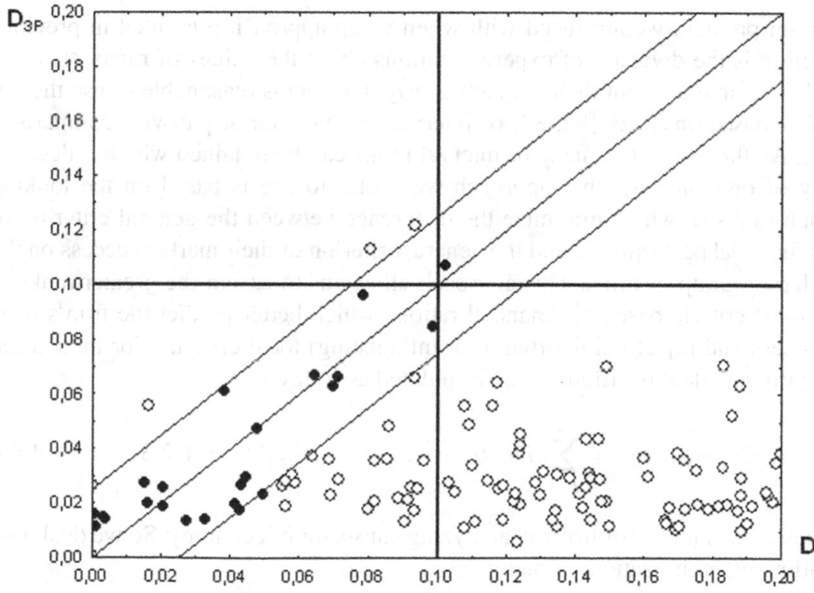

Fig. 4.31 "Good" firms (black points) selected using Yager's aggregation and pair comparison matrix for local criteria ranking

Table 4.17 Best selected firms (using pair comparison matrix for local criteria ranking)

D_1 **and** D_2	D_1 and D_3	D_2 and D_3
ICST	MPAD	DPAC
KLAS	NVLS	
	CYMI	
	DPMI	
	FLEX	
	ACLS	
	ZRAN	

It is seen that only 10 firms throughout considered subsector containing 162 firms have been finally selected. Thus, the proposed method for stock selection seems to be enough rigorous one. Of course, the wider channels can be used and as a consequence, the more firms can survive the selection. In our case, the tickers MPAD, DPAC, ACLS, ICST, ZRAN may be considered as the best ones since they are characterized by the largest market success (detailed analysis is carried out in the following subsection).

4.4.5 *Stock Ranking with the Use of Multiple Criteria Optimization*

The main problem we are faced with when using approach presented in previous subsection is the diversity of experts opinions about the values of ranks α_i, $i = 1$ to 7. Taking into account their subjective origin, it seems reasonable to use the approaches based on fuzzy [82, 83] or interval presentation of pairwise comparison matrix. As the result, the fuzzy or interval ranks can be obtained which reflect the variety of opinions. Another approach we prefer to use is based on the looking for such ranks α_i which minimize the difference between the general criterion of firm's financial performance and the general criterion of their market success on the considering group of firms. This approach allows us to assign the greater ranks to those local criteria based on financial rations which better predict the firm's market success and reject unimportant (not influencing) local criteria. For the chosen aggregating mode this difference is formulated as follows:

$$S_i(\alpha_1,\alpha_2,...,\alpha_7) = \sum_{j=1}^{m} (D_{ij}(\alpha_1,\alpha_2,...,\alpha_7) - D_{iPj})^2, i = 1.2.3, \quad (4.48)$$

where m is the number of firms in analyzing subsector of economy. So we deal with the following optimization problem:

$$(\alpha_1,\alpha_2,...,\alpha_7)_{i,opt} = \arg\min(S_i(\alpha_1,\alpha_2,...,\alpha_7), i = 1.2.3, \quad (4.49)$$

$$\frac{1}{7}\sum_{i=1}^{7}\alpha_i = 1. \tag{4.50}$$

The well known direct random search method [126] has been modified to develop the numerical algorithm for the solution of this optimization problem. The special procedure has been used for random choosing the vector $(\alpha_1, \alpha_2, ..., \alpha_7)$ in such a way that the constraint (4.50) is fulfilled on each step of algorithm. Of course, any other modern optimization method, e.g., genetic algorithm can be used as well. Nevertheless, it is shown in [2] that "when the optimizing function is nonlinear, non-differentiable and non-smooth, direct search methods are the methods of choice". The results obtained using different aggregating modes are presented in Table 4.18.

Table 4.18 Optimized ranks α_i, $i = 1$ to 7, of the local criteria based on financial ratios

Financial ratio/aggregating mode	D_1	D_2	D_3
EPSTTM2TTM	0	0	0
ChSO	0,65	3,39	0
EPSQ2Q	0	0,83	4,25
DCS	0,91	0	0
MKTCAP	4,3	0,11	0,21
EPSQ2TTM	0	0,76	0
RQ2TTM	1,11	1,92	2,54
S_{opt}	2,56	0,02	0,14

It is seen that the use of weighted sum aggregation D_1 rejects the local criteria based on financial ratios EPSTTM2TTM, EPSQ2Q and EPSQ2TTM as insignificant ones; multiplicative aggregation D_2 eliminates the local criteria based on EPSTTM2TTM and DCS; Yager's type aggregation D_3 excludes the local criteria based on EPSTTM2TTM, ChSO, DCS and EPSQ2TTM. These results show that the local criterion based on EPSTTM2TTM should be definitely excluded from further analysis as it has been rejected with the use of all compared aggregating modes. It is interesting that this criterion has been estimated as an important one using the ranking based on the pair comparison matrix. The values of the difference $S(\alpha_1, \alpha_2, ..., \alpha_7)$ for the optimal $(\alpha_1, \alpha_2, ..., \alpha_7)$, i.e., S_{opt}, presented in the lower row of Table 4.18, marginally show that the more reliable results of optimization are those obtained with use of multiplicative D_2 aggregation. This was the reason to reject additionally the local criterion based on DCS (zero ranks with the use of D_2 and D_3), but reserve the local criterion based on EPSQ2TTM.

Using the obtained optimal ranks of local criteria, the generalized criteria D_1, D_2, D_3 and D_{1P}, D_{2P}, D_{3P} have been calculated for the firms of analyzing subsector. To select the "good" firms, the approach based on building the channels containing at least 20 "best" firms, described in previous subsection has been used. The results are presented in Fig. 4.32-Fig. 4.34.

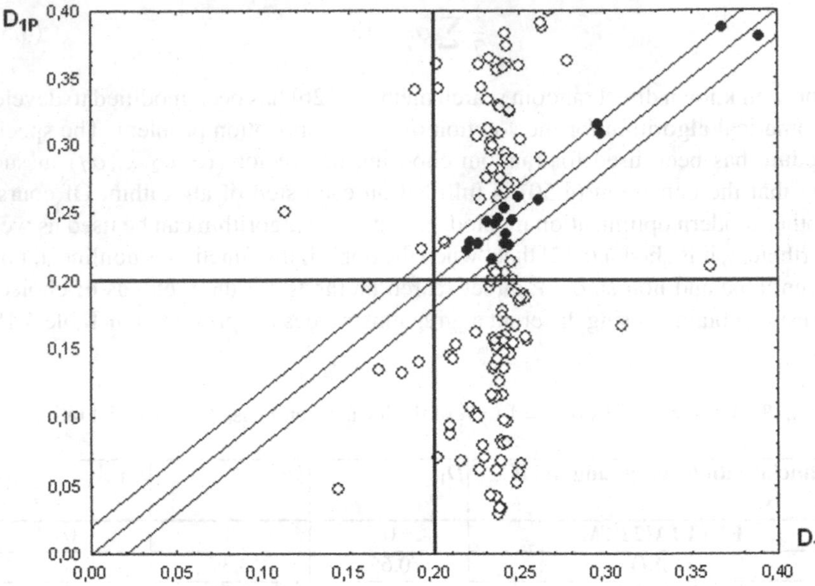

Fig. 4.32 "Good" firms (black points) selected using weighted sum type of aggregation and optimized ranks of local criteria

In is easy to see that obtained distributions of firms around the channels containing the "good" firms are substantially different from those presented in Fig. 4.29-Fig. 4.31, especially for multiplicative and Yager's aggregations. Moreover, the set of the "best" firms (see Table 4.19) we have selected as those contained simultaneously at lest in two channels is absolutely different from the set of the "best" firms selected using the pair comparison matrix for local criteria ranking (Table 4.17). This difference can be explained as follows. When ranking the local criteria based on financial ratios, an expert treats these criteria rather as those characterizing internal financial, production, innovation "health" of firm, not its market potential. Quite different treatment of local criteria is implicitly used in the framework of proposed method for stock ranking based on the multiple criteria optimization. Factually, optimized ranks of local criteria are the degrees to which the local criteria of firm's "health" can predict its success in the stock market. Another merit of this approach is that there is no need for the subjective expert's opinions in the ranking procedure. It is important that most of the "best" firms selected using optimized ranks of local criteria are in the channels of "good" firms obtained on the base of multiplicative D_2 and Yager's D_3 aggregating modes which are more reliable and rigorous ones than weighted sum aggregation D_1. So we can say that the results obtained with use of optimized ranks are more reliable. In Table 4.20, the extremal benefits and losses of the "best" firms selected using comparing approaches are presented in the form of intervals $[ML, MR]$ (see Subsection 4.4.3).

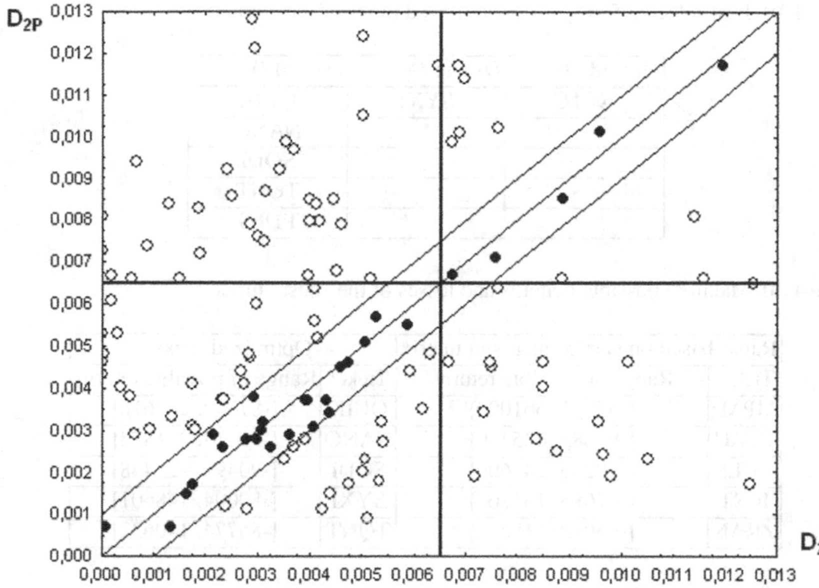

Fig. 4.33 "Good" firms (black points) selected using multiplicative type of aggregation and optimized ranks of local criteria

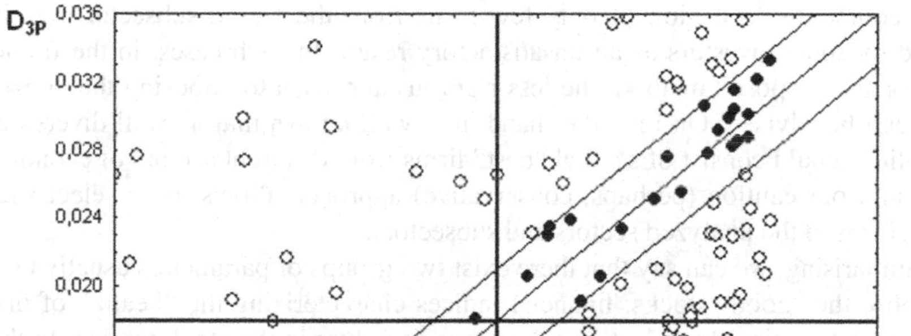

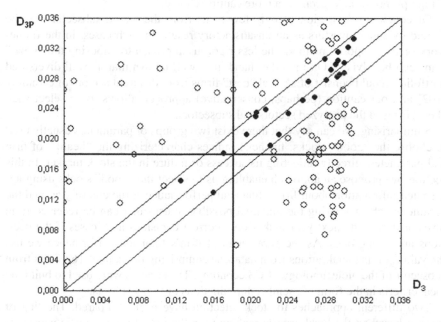

Fig. 4.34 "Good" firms (black points) selected using Yagers' aggregation and optimized ranks of local criteria

Table 4.19 Best selected firms (using optimized ranks of local criteria)

D_1 **and** D_2	D_1 and D_3	D_2 and D_3
KLIC	SYXI	QUIK
		NANO
		SODI
		TQNT
		PDFS

Table 4.20 Maximal possible benefits and losses of the "best" firms

Ranks based on pair comparison matrix		Optimized ranks	
Ticker	Range of possible returns	Ticker	Range of possible returns
MPAD	[-55271, 661000]	QUIK	[-87039, 243618]
DPAC	[-97683, 425429]	NANO	[-90414, 231828]
ACLS	[-70286, 241600]	SODI	[-90494, 223448]
ICST	[-62685, 131867]	SYXI	[-80004, 148601]
ZRAN	[-65938, 115862]	TQNT	[-85774, 140632]

It is seen that using the approach based on the pair comparison matrix in ranking procedure, the firms with substantially greatest possible returns can be selected. Nevertheless, taking into account the above consideration, the "best" firms selected using optimized ranks seem as a more caution choice.

Of course, the selection of only few firms from the whole subsector may be treated by some investors as an unsatisfactory result. In such cases, in the framework or the proposed method, the less rigorous approach to choosing the "good" firms can be advised. On the other hand, it is well known that a "well diversified portfolio" should consist of several "best" firms from different sectors of economy [102] and our caution (perhaps, conservative) approach allows us to select such "best" firms in the analyzed sectors and subsectors.

Summarising, we can say that there exist two groups of parameters usually used to choose the "good" stocks: financial indices characterizing the "health" of firm and parameters directly reflecting its success or failure in the stock market. In this section, we propose an approach enabling us to select the "good" stocks using aggregated information about firm's financial performance at the end of year and the parameters characterizing the maximal possible returns that can be received if an investor during the next year makes only correct decisions and losses if the decisions are wrong ones. As the parameters of firm's financial performance we use the values of financial rations from database comprising the data of 162 firms from subsector of the biotechnology of US economy. These ratios were used to build the local criteria in the form of membership functions of fuzzy subsets.

Two different approaches to stock selection have been compared. The first of them is based on the local criteria ranking with the use of the linguistic pair comparison matrix, the second one allows us to find the optimal ranks of local criteria minimizing the difference between the general criterion of firm's financial performance and the general criterion of their market success on the considered group of firms. Three most popular aggregating modes- weighted sum, multiplicative and *min* operator- were used for building the general criteria of firm's financial performance and market success. It is shown that a proper choice of method for the local criteria aggregation plays a key role in the success of stock selection irrespective of used approach to the local criteria ranking.

The proposed method makes it possible to select a group of "good" stocks with a great coincidence of their financial performance and market success and to reject simultaneously the analysis of all "unsafe" firms, i.e., such ones that their market success is based rather on the public relations, subjective expert's opinions, rumors and other, sometimes, unreliable information.

4.5 Multiple Criteria Fuzzy Evaluation and Optimization in Budgeting

Capital budgeting is based on the analysis of some financial parameters of projects. It is clear that estimation of investment efficiency, as well as any forecasting, is rather an uncertain problem. In this section, the techniques for fuzzy evaluation of financial parameters and estimation of risk of an investment are presented. Another problem is that one usually should consider a set of local criteria based on financial parameters of investments. As its possible solution, a numerical method for optimization of future cash-flows based on the generalized project's quality criterion is proposed.

4.5.1 The Problem Formulation

Consider common non-fuzzy approaches to the solution of capital budgeting problem. There are is lot of financial parameters proposed in the literature [7, 14, 20, 73] for budgeting. The main are: Net Present Value (*NPV*), Internal Rate of Return (*IRR*), Payback Period (*PB*), Profitability Index (*PI*). These parameters are usually used for the project quality evaluation, but in practice they have different importance. It is earnestly shown in [11] that the most important parameters are *NPV* and *IRR*.

Therefore, further consideration will be based only on the analysis of *NPV* and *IRR*. The good review of other useful financial parameters can be found in [4]. Net Present Value is usually calculated as follows:

$$NPV = \sum_{t=t_n}^{T} \frac{P_t}{(1+d)^t} - \sum_{t=0}^{t_c} \frac{KV_t}{(1+d)^t}, \tag{4.51}$$

where d is the discount rate, t_n is the first year of production, t_c is the last year of investments, KV_t is the capital investment in the year t, P_t is the income in the year t, T is the duration of an investment project in years. Usually, the discount rate is taken equal to the average bank interest rate in the country of investment or other value corresponding to the profit rate of alternate capital investments.

An economic nature of *IRR* can be explained as follows. Suppose that as an alternative to analyzed project, the deposit under some bank interest distributed in time the same way as the analyzed investments is considered. All earned profits are also deposited with the same interest rate. If the discount rate is equal to *IRR*, the investment in the project will provide the same total income as in a case of the deposit. Thus, both alternatives are economically equivalent. If the actual bank discount rate is less then *IRR*, the investment into the project is more preferable. Therefore *IRR* is a threshold discount rate dividing effective and ineffective investment projects.

The value of *IRR* is the solution of the following nonlinear equation with respect to d:

$$\sum_{t=t_n}^{T} \frac{P_t}{(1+d)^t} - \sum_{t=0}^{t_c} \frac{KV_t}{(1+d)^t} = 0. \tag{4.52}$$

The estimation of *IRR* is frequently used as a first step of the financial analysis. Only projects with *IRR* not below of some accepted threshold value, e.g., 15–20%, can be chosen for further consideration.

There are two conjoint discussable points in the budgeting realm. The first one is the multiple roots of Eq. (4.52), i.e., the so-called multiple *IRR* problem. The second one is the negative *NPV* problem. The problem of multiple roots of Eq. (4.52) rises when the negative cash flows take place. In practice, appearance of some negative cash flow after initial investment is usually treated as a local "force majeur" or even a total project's failure. That is why, on the stage of planning, investors try to avoid situations when such negative cash flows are possible, except the cases when they are dealing with long-term projects consist of some phases. Let us see to the Fig. 4.35. This is a typical two-phase project: after initial investment the project

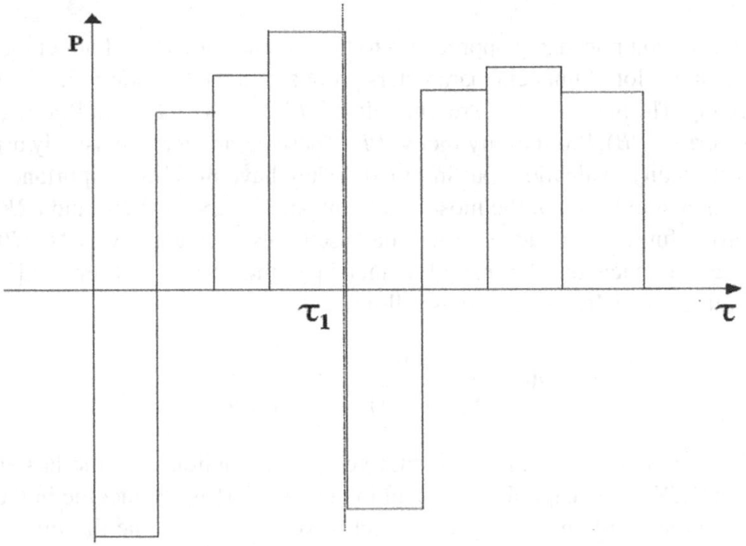

Fig. 4.35 Two stage investment project

provides considerable profits and at the time τ_1 a part of accumulated earnings and, perhaps, an additional banking credit are invested once again. Factually, an investor buys new production equipment and housings (in fact creating a new enterprise) and from his/her point of view a quite new project is started. It is easy to see that investor's creditors which are interested in repayment of a credit, always analyze the phases $\tau < \tau_1$ and $\tau > \tau_1$ separately. It is worth noting that what we describe is only the investment planning routine, not some theoretical considerations we can find in financial books. On the other hand, a separate assessment of different project's phases reflects better the economic sense of capital investment. Indeed, if we consider a two phase project as a whole, we often get the *IRR*s performed by two roots of (4.52) so different that it is impossible to make any decision. For example, we can obtain IRR_1 =4% and IRR_2 =120%. It is clear that average *IRR*= (4+120)/2=62% seems as rather fantastic estimation, whereas when considering the two phases of project separately we usually get quite acceptable values, e.g., for the first phase IRR_1 =20% and for the second phase IRR_2 =25%. So we can say that the problem of "multiple *IRR* values" exists only in some financial textbooks, not in the practice of capital investment. Therefore, only the case when Eq. (4.52) has a single root will be analyzed in the current section. Similarly, the problem of negative *NPV* seems to be rather an artificial one. Obviously, any investment project with negative *NPV* should be rejected at the planning stage. On the other hand, all possible undesirable events leading to the financial losses or even to the failure of the projects should be taken into account. In the framework of probabilistic approach, e.g., when using the Monte-Carlo method, there may be local results of calculations with negative *NPV* and the problem of their interpretation in terms of risk management or in other contexts arises.

The different situation we meet when future cash flows are presented by fuzzy numbers. It is clear that the full body of uncertainty is involved in such a description. So if the decision maker find some negative part in predicted cash flow he/she consider such a case as a source of risk and try to improve the project to avoid this risk. As the result, in the fuzzy budgeting the negative cash flows and especially *NPV*, seem to be rather exotics ones. Nevertheless, the probabilistic approach to interval and fuzzy numbers comparison we have described in Subsection 4.2.3, makes it possible to deal with such situation as well, i.e., to compare *NPV* comprising negative part with some real or fuzzy number representing acceptable risk associated with future *NPV* .

It should be noted that nowadays the traditional approach to the evaluation of *NPV*, *IRR* and other financial parameters is subjected to the quite deserved criticism, since the future incomes P_t, capital investments KV_t and rates d are rather uncertain parameters. Uncertainties which one meets in capital budgeting differ from those in a case of share prices forecasting and cannot be adequately described in terms of the probability theory. In the capital investment, one usually deals with a business-plan that takes a long time — as a rule, some years — for its realization. In such cases, the description of uncertainty within the framework of traditional probability methods usually is impossible due to the absence of objective information about probabilities of future events. Thus, what really is available in such cases are some expert's estimates. In real-world situations, investors or experts involved are able to predict confidently only intervals of possible values of P_t, KV_t and d and sometimes the most expected values inside these intervals. Therefore, during the last two decades the growing interest in applications of interval arithmetic [88] and fuzzy sets theory methods [143] in budgeting was observing.

After pioneer works by T.L.Ward [135] and J.U. Buckley [15], some other authors contributed to the development of the fuzzy capital budgeting theory [23, 27, 28, 31, 35, 54, 55, 56, 57, 58, 65, 71, 93, 113].

It is safe to say that almost all problems of the fuzzy *NPV* estimation are solved now, but an interesting and important problem of project risk assessment using fuzzy *NPV* gains higher priority.

An unsolved problem is a fuzzy estimation of the *IRR*. Ward [135] considered Eq. (4.52) and stated that such an expression cannot be applied to the fuzzy case because the left hand side of Eq. (4.52) is fuzzy, 0 is crisp and an equality is impossible. Hence, the Eq. (4.52) is senseless from fuzzy viewpoint.

In [65], a method for the fuzzy *IRR* estimation is proposed where the α-cut representation of fuzzy numbers [60] is used.

The method is based on the assumption (see [65, p. 380]) that the set of equations for *IRR* determination on each α-level may be presented (in our notation) as follows:

$$(CF_0^\alpha)_1 + \sum_{i=1}^{n} \frac{(CF_i^\alpha)_1}{(1+IRR_1^\alpha)^i} = 0, \quad (CF_0^\alpha)_2 + \sum_{i=1}^{n} \frac{(CF_i^\alpha)_2}{(1+IRR_2^\alpha)^i} = 0, \tag{4.53}$$

where $CF_i^\alpha = [(CF_i^\alpha)_1, (CF_i^\alpha)_2]$, $i = 0$ to n, are crisp interval representations of fuzzy cash flows on α-levels. Of course, from Eqs. (4.53) all crisp intervals $IRR^\alpha = [IRR_1^\alpha, IRR_2^\alpha]$ expressing the fuzzy valued *IRR* may be obtained. Regrettable, there is a little mistake in Eqs. (4.53). Taking into account the conventional interval arithmetic rules, the correct crisp interval representation of fuzzy Eq. (4.52) on α-cuts should be written as follows:

$$(CF_0^\alpha)_1 + \sum_{i=1}^{n} \frac{(CF_i^\alpha)_1}{(1+IRR_2^\alpha)^i} = 0, \quad (CF_0^\alpha)_2 + \sum_{i=1}^{n} \frac{(CF_i^\alpha)_2}{(1+IRR_1^\alpha)^i} = 0. \tag{4.54}$$

There is no way to get regular (non inverted) intervals IRR^α from Eq. (4.54), but some real valued estimates of *IRR* may be obtained (see Subsection 4.5.3 below). Another problem not presented in the literature is an optimization of cash flows.

4.5.2 Fuzzy NPV and Risk Evaluation

The technique is based on the fuzzy extension principle [143]. According to it, the values of uncertain parameters P_t, KV_t and d are substituted for corresponding fuzzy intervals. In practice, this means that an expert sets lower — P_{t1} (pessimistic value) and upper — P_{t4} (optimistic value) boundaries of the intervals and internal intervals of the most expected values [P_{t2}, P_{t3}] for analyzed parameters (see Fig. 4.36). The function $\mu(P_t)$ is usually interpreted as a membership function, i.e., a degree

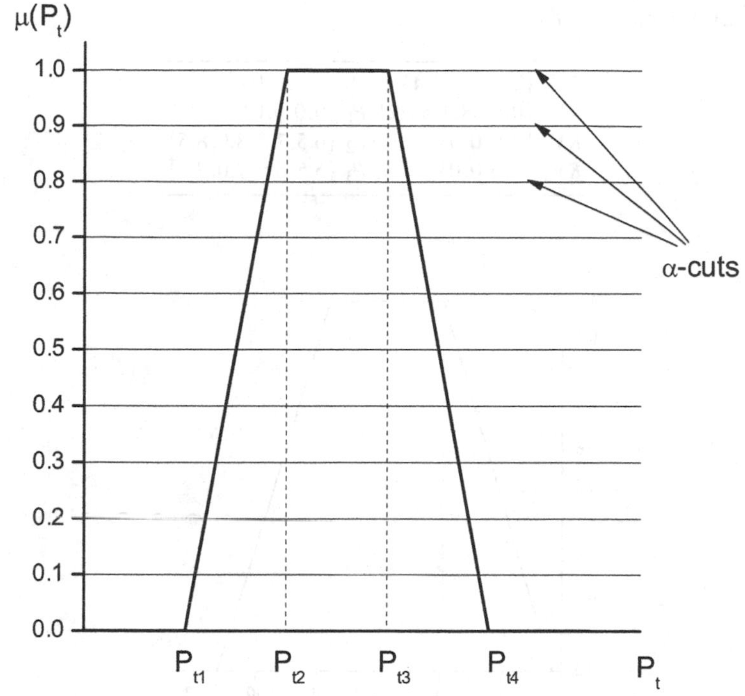

Fig. 4.36 Fuzzy interval of the uncertain parameter P_t and its membership function $\mu(P_t)$

to which values of a parameter belong to the interval (in this case $[P_{t1}, P_{t4}]$). A membership function changes continuously from 0 (area out of the interval) up to maximum value 1 in the area of the most possible values.

The linear character of the function is not obligatory, but such a mode is most frequently used and allows us to represent the fuzzy intervals in the convenient form of quadruple $P_t = \{P_{t1}, P_{t2}, P_{t3}, P_{t4}\}$. Then all necessary calculations are carried out using fuzzy arithmetic rules based on the α-cut representation of fuzzy numbers. These rules have been presented in Chapter 3.

To illustrate, consider the sample investment project, in which the building phase proceeds two years with investments KV_0 and KV_1, accordingly. The profits are expected only after the end of the building phase and will be obtained during two years (P_2 and P_3). It is suggested that the fuzzy interval for the discount d remains stable during the time of project realization. The sample trapezoidal initial fuzzy intervals are presented in Table 4.21.

It was assumed that $d = \{0.08, 0.13, 0.22, 0.35\}$. The resulting fuzzy NPV calculated using fuzzy extension of Eq. (4.51) is presented in Fig. 4.37.

The obtained fuzzy interval allows us to estimate the boundaries of possible values of predicted NPV, the interval of the most expected values, and also — that is very important — to evaluate a degree of financial risk of investment.

Table 4.21 Parameters of the sample project

KV_0 $\{2,2.8,3.5,4\}$	P_0 $\{0,0,0,0\}$
KV_1 $\{0,0.88,1.50,2\}$	P_1 $\{0,0,0,0\}$
KV_2 $\{0,0,0,0\}$	P_2 $\{6.5,7.5,8.0,8.5\}$
KV_3 $\{0,0,0,0\}$	P_3 $\{5.5,6.5,7.0,7.5\}$

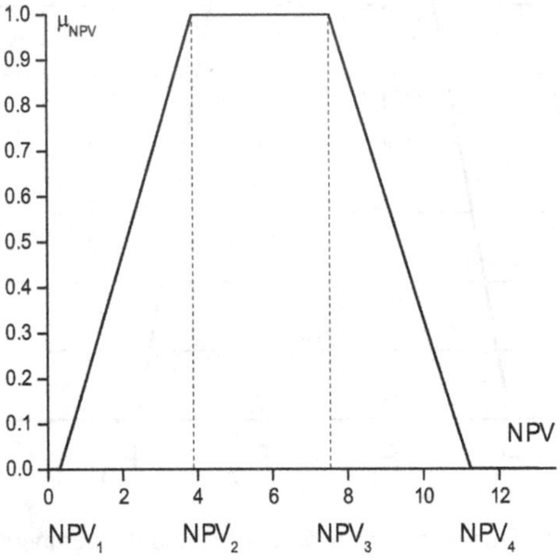

Fig. 4.37 Resulting fuzzy *NPV*

There may be different ways to define the measure of financial risk in the framework of fuzzy sets based methodology. Therefore we consider here only the three, in our opinion, most interesting and scientifically grounded approaches.

Let us consider the first approach.

To estimate the financial risk, the following inherent property of fuzzy sets has been taken into account. Let A be some fuzzy subset of X, being described by the membership function $\mu(A)$. Then complementary fuzzy subset $\bar{A}$ has the membership function $\mu(\bar{A}) = 1 - \mu(A)$. The principal difference between a fuzzy subset and a usual precise one is that the intersection of fuzzy subset A and $\bar{A}$ is not empty, that is $A \cap \bar{A} = B$, where B is also not an empty fuzzy subset. It is clear that the closer A to $\bar{A}$, the more power of set B and more A differs from ordinary sets.

Using this circumstance Yager [137] proposed the set of grades of nonfuzziness of fuzzy subsets

$$D_p(A,\bar{A}) = \frac{1}{n}\left|\sum_{i=1}^{n} |\mu_A(x_i) - \mu_{\bar{A}}(x_i)|^p\right|^{\frac{1}{p}}, p = 1,2,\ldots,\infty. \tag{4.55}$$

Hence, the grade of fuzziness may be defined as follows:

$$dd_p(A,\bar{A}) = 1 - D_p(A,\bar{A}). \qquad (4.56)$$

The definition (4.56) is in compliance with obvious requests to a grade of fuzziness. If A is a fuzzy subset on X, $\mu(A)$ is its membership function and dd is a corresponding grade of fuzziness, then following properties should be observed:

1. $dd(A) = 0$, if A is a crisp subset.
2. $dd(A)$ has a maximum value if $\mu(A) = 1/2$ for $x \in X$.
3. $dd(A) > dd(B)$ if $\mu(x) < \mu(y)$ $(x \in A, y \in B)$.

It is proved that introduced measure is similar to the Shannon entropy measure [137].

In the most useful case $(p = 1)$, expression (4.56) is transformed to

$$dd = 1 - \frac{1}{n}\sum_{i=1}^{n} |2\mu_A(x_i) - 1|. \qquad (4.57)$$

It is clear (see Eq. (4.57)) that the grade of fuzziness is rising from 0 when $\mu(A) = 1$ (crisp subset) up to 1 when $\mu(A) = 1/2$ (maximum degree of fuzziness).

With respect to considering problem, the grade of nonfuzziness of a fuzzy interval NPV can linguistically be interpreted as a risk of obtaining the crisp interval $[NPV_1, NPV_4]$. Really, the more precise, (more "rectangular") is the obtained fuzzy interval, the more is the degree of uncertainty and risk. At first glance, this assertion seems to be paradoxical. However, any precise (crisp) interval contains no additional information of the relative preference of values inside it. Therefore, it contains less useful information than any fuzzy interval being constructed on its basis. In this (fuzzy interval) case an additional information reducing uncertainty is derived from the membership function of considered fuzzy interval.

The second approach to the risk evaluation is based on the α-cut representation of fuzzy numbers and the measure of their fuzziness.

Let A be fuzzy value and A_r be rectangular fuzzy number defined on the support of A and represented by the characteristic function $\eta_A(x) = 1, x \in A; \eta_A(x) = 0, x \notin A$. Obviously, such rectangular value is not a fuzzy value at all, but it is the asymptotic limit (object) we obtain when the fuzziness of A tends to zero. Hence, it seems quite natural to define a measure of fuzziness of A as its distinction from A_r. To do this we define primarily the measure of non fuzziness as

$$MNF(A) = \int_0^1 ((A_{\alpha 2} - A_{\alpha 1})/(A_{02} - A_{01}))d\alpha,$$

where A_{02} and A_{01} are the right and left bounds of the support of A, respectively, $A_{\alpha 2}$ and $A_{\alpha 1}$ are the right and left bounds of crisp intervals on the corresponding α-cuts. Of course, last expression makes sense only for the fuzzy or interval values, i.e., only for the non zero width of support $A_{02} - A_{01}$. It is easy to see that if $A \to A_r$

then $MNF(A) \to 1$. Obviously, the measure of fuzziness can be defined as $MF(A) = 1 - MNF(A)$.

We can say that the rectangular value A_r defined on the support of A is a more uncertain object than A. Really, only what we know about A_r is that all $x \in A$ belong to A_r with equal degrees, whereas the membership function $(0 \leq \mu(x) \leq 1)$ characterizing the fuzzy value A, provides more information to the description and as a consequence, represents a more certain object. Therefore, we can treat the measure of non fuzziness MNF as the uncertainty measure. Hence, if some decision is made concerning fuzzy NPV, the uncertainty and, consequently, the risk of such decision can be calculated as $MNF(NPV)$.

Consider the third approach to the risk evaluation.

The authors of [90] proposed an approach that can be treated as a fuzzy analogue of the sound VAR method [76]. According to this approach the risk associated with the fuzzy NPV can presented as follows:

$$Risk = Prob(NPV < G),$$

where G is a fuzzy, interval or real valued effectiveness constrain [90]. In other words, G is the low bound on acceptable values of NPV. It is clear that the focus of this approach is the method for interval and fuzzy value comparison. In [90], such method based on the geometrical reasoning has been proposed which leads to the resulting formulas nearly the same as earlier were obtained in [142] with a help of probabilistic approach to fuzzy numbers comparison.

In [109, 111] (see also Chapter 3), the overview of existing methods for fuzzy numbers comparison based on probabilistic approach is presented. It is shown in [109, 111] that analyzed methods have a common drawback- the lack of separate equality relations- leading to the absurd results in the asymptotic cases and some others inconsistencies. The same can be said about of non- probabilistic method proposed in [90]. To solve the problem, in [109, 111] a new method based on the probabilistic approach has been developed, which generates the complete set of probabilistic interval and fuzzy numbers relations involving separated equality and inequality relations, comparisons of real numbers with interval or fuzzy values. This method has been presented above in Subsection 4.2.3.

Obviously, other approaches to the risk evaluation in the budgeting can be proposed and can be relevant in the specific situations. It is clear that they should lead to the different results of investment projects estimation or optimization tasks as they reflect the different decision maker's attitudes to the risk and its importance in the considered problem . Therefore, we think that for methodical purposes it is quite enough to consider only one of above approaches. So in further analysis the first model of risk, based on the Exp.(4.57) will be used. It is important that all considered approaches based on an evaluation of fuzzy NPV inevitably generate two criteria for the estimation of future profits: the fuzzy interval NPV and the degree of its uncertainty (degree of risk).

Therefore, the problem of evaluation of investment efficiency on a base of *NPV* becomes two-criteria and requires the special approach and technique. Such technique has been presented above in Section 4.3.

4.5.3 The Set of Crisp IRR Estimations Based on Fuzzy Cash Flows

In general, the problem of fuzzy *IRR* evaluation is based on the fuzzy solution of the Eq. (4.52) with respect to d.

It is proved that the solution of equations with fuzzy parameters (in this case, P_t, KV_t and d) is possible using the α-cuts representation of fuzzy parameters. For the evaluation of fuzzy *IRR*, the system of non-linear crisp-interval equations can be obtained:

$$\sum_{t=t_n}^{T} \frac{[P_t]\alpha}{(1+[d]\alpha)^t} - \sum_{t=0}^{t_c} \frac{[KV_t]\alpha}{(1+[d]\alpha)^t} = [0,0], \tag{4.58}$$

where $[Pt]_\alpha, [KV_t]_\alpha$ and $[d]_\alpha$ are crisp intervals on corresponding α-cuts.

Of course, it can be claimed that assumption that the degenerated zero interval $[0,0]$ should be placed in the right hand side of Eq. (4.58) does not ensure the obtaining of adequate outcomes since a non-degenerated interval expression is in the left hand side of Eq. (4.58). Nevertheless, this situation needs a more thorough consideration.

As a simplest example, consider a two-year project when all investments are finished in the first year and all revenues are obtained in the second year. Then from (4.58) on each α-cut we get the equation (index α is dropped for the simplicity):

$$\frac{[P_{11}, P_{12}]}{(1+[d_1,d_2]} - [KV_{01}, KV_{02}] = [0,0], \tag{4.59}$$

where P_{11}, P_{12} are the left and right bounds of interval income on the α-cut in the second year, KV_{01}, KV_{02} are the left and right bounds of interval investment on the α-cut in the first year, d_1, d_2 are the left and right bounds of interval *IRR* on the α-cut.

The formal solution of Eq. (4.59) with respect to d_1 and d_2 is trivial:

$$d_1 = \frac{P_{12}}{KV_{01}} - 1, \quad d_2 = \frac{P_{11}}{KV_{02}} - 1,$$

however it is senseless, as the right bound of the interval $[d_1, d_2]$ always appears to be less than the left one.

This absurd, at first glance, result is easy to explain from common methodological positions. Really, the rules of interval arithmetic are constructed in such a manner that any arithmetical operation on intervals results in an interval as well. These rules fully coincide with well known common viewpoint stating that any arithmetical operation with uncertainties should increase the total uncertainty and the entropy of a system. Therefore, placing the degenerated zero interval in the right hand side of (5.58) and (4.59) is equivalent to the request of reducing the uncertainty of the left sides down to zero. This is possible only in the case of inverse character of the interval $[d_1, d_2]$, and this in turn can be interpreted as a request to introduce negative entropy into a system.

Thus, the presence of the degenerated zero interval in the right hand sides of interval equations is incorrect. A more acceptable approach to solving this problem has been constructed with a help of following reasons. When analyzing expressions (4.59) it is easy to see that for any value d_1 the minimal width of the interval *NPV* is reached when $d_2 = d_1$. This is in accordance with a common viewpoint: the minimum uncertainty of an outcome (*NPV*) is reached when uncertainty of all system parameters is minimal. It is clear (see Fig. 4.38) that the most reasonable decision of "zero" problem is a request for the middle of the interval *NPV* to be placed on a zero point (request of symmetry of the interval in relation to zero). An obvious, at first glance, intention to minimize the length of interval *NPV* results in the deriving positive or negative intervals with minimum widths, but not containing zero point that does not correspond to the natural definition of zero containing interval. Besides, it can be easily proved that only the request of symmetry of zero containing interval ensures an asymptotically valid outcome when contracting bounds of all considered intervals to their centers. Thus, the problem is reduced to a search of exact (non-interval) values *d* that will provide a symmetry of zero containing resulting intervals *NPV* on each α-cut in the equations (4.58), i.e., would guarantee the fulfillment of $(NPV_1 + NPV_2) = 0$, for each $\alpha = 0, 0.1, 0.2 \ldots, 1$.

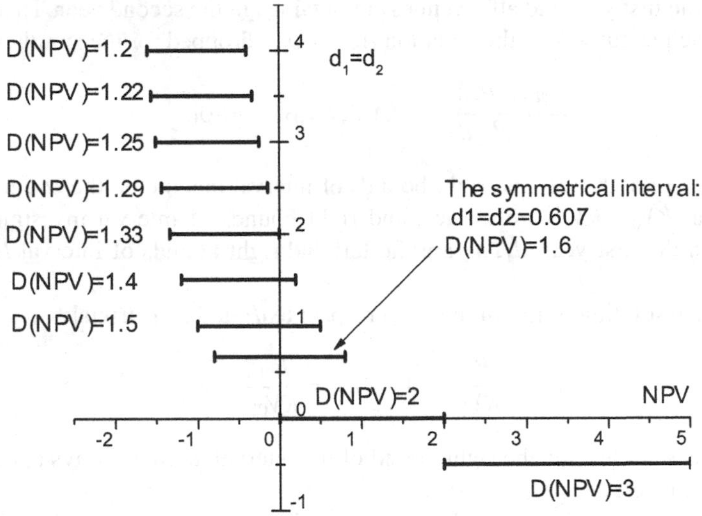

Fig. 4.38 Interval *NPV* for different real valued discounts *d* for the case when the investments in the first year is $KV_0 = [1, 2]$, the income in the second year is $P_1 = [2, 3]$, $D(NPV)$ is the width of interval *NPV*

Obviously, the problem is solved using numerical methods. To illustrate the previous theoretical considerations, compare two investment projects of 4 years duration. Fuzzy cash flows $K_t = P_t - KV_t$ are defined with a help of the four-reference point form described above (see Table 4.22). It is worth noting that the data of the first project are more certain.

Table 4.22 Results of IRR_α calculation

Project 1		Project 2	
Year	Cash flow	Year	Cash flow
1	$\{-6.95, -6,95, -7,05, -8.00\}$	1	$\{-6.00, -6.95, -7.50, -8.00\}$
2	$\{4.95, 4.95, 5.05, 6.00\}$	2	$\{4.00, 4.95, 5.50, 6.00\}$
3	$\{3.95, 3.95, 4.05, 5.00\}$	3	$\{3.00, 3.95, 4.50, 5.00\}$
4	$\{1.95, 1.95, 2.05, 3.00\}$	4	$\{1.00, 1.95, 2.50, 3.00\}$

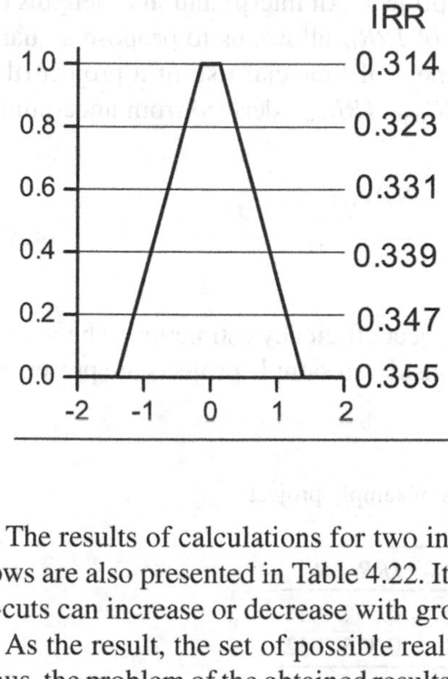

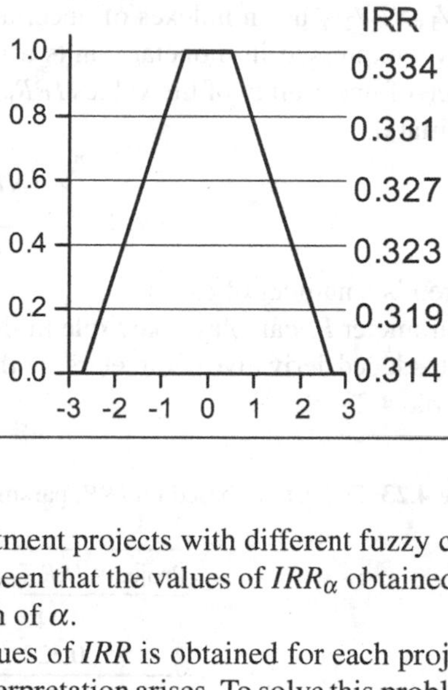

The results of calculations for two investment projects with different fuzzy cash flows are also presented in Table 4.22. It is seen that the values of IRR_α obtained on α-cuts can increase or decrease with growth of α.

As the result, the set of possible real values of IRR is obtained for each project. Thus, the problem of the obtained results interpretation arises. To solve this problem, it was proposed to reduce the sets of IRR_α obtained on each α-cut to a small set of parameters which can be easily interpreted.

The first elementary parameter — average value IRR_{m} — is certainly convenient, however it does not take into account that with growth of α, the reliability of an outcome increases as well, i.e., IRR_α, obtained on higher α-cuts are more reliable than those obtained on lower α-cuts according to the α-cut definition. On the other hand, the width of the crisp interval $[NPV_1, NPV_2]_\alpha$ corresponding to the IRR_α can be considered as a measure of uncertainty for the obtained crisp value IRR_α, since such width quantitatively characterizes the difference of the left side of Eq.(4.58) from the degenerated zero interval $[0,0]$. This allows us to introduce two weighted estimations of IRR on the set of IRR_α: least expected (least reliable) $IRR_{\min}$ and most expected (most reliable) $IRR_{\max}$:

$$IRR_{\min} = \frac{\sum_{i=0}^{n-1} IRR_i (NPV_{2i} - NPV_{1i})}{\sum_{i=0}^{n-1} (NPV_{2i} - NPV_{1i})}, \tag{4.60}$$

$$IRR_{\max} = \frac{\sum_{i=0}^{n-1} IRR_i \alpha_i}{\sum_{i=0}^{n-1} \alpha_i}, \tag{4.61}$$

where n is the number of α-cuts.

In the decision making practice, all three proposed parameters IRR_{m}, $IRR_{\min}$, $IRR_{\max}$ can be used when choosing the best project. An interpretation of lengths of $[NPV_1, NPV_2]_\alpha$ as an indexes of uncertainty of IRR_α allows us to propose a quantitative, expressed in monetary units assessment of financial risk of a project (the degree of uncertainty of the values IRR_{m}, $IRR_{\min}$, $IRR_{\max}$ derived from uncertainty of initial data):

$$R_{\mathrm{r}} = \frac{\sum_{i=0}^{n-1} (NPV_{2i} - NPV_{1i})}{n}, \tag{4.62}$$

where i is a number of α-cut.

Parameter R_{r} can play a key role in the project efficiency estimation. The values of introduced derivative parameters for the considered sample projects are presented in Table 4.23.

Table 4.23 Derivative (based on *IRR*) parameters of sample projects

Project#	$IRR_{\min}$	$IRR_{\max}$	IRR_{m}	R_{r}
1	0.34	0.327	0.335	1.56
2	0.322	0.329	0.325	3.52

It is seen that the projects have rather the close values of IRR_{m}, $IRR_{\min}$, $IRR_{\max}$. At the same time, the risk R_{r} for the second project is considerably higher than the risk of the first one. Hence, the first project is the best one. In addition, some other useful parameters have been proposed: IRR_{mr} — most reliable value of IRR_α — derived from the minimum interval $[NPV_1, NPV_2]_{mr}$ among all $[NPV_1, NPV_2]_\alpha$ and IRR_{lr} — the least reliable value of IRR_α — derived from the maximum interval $[NPV_1, NPV_2]_{\mathrm{lr}}$ among all $[NPV_1, NPV_2]_\alpha$. It is clear, that $[NPV_1, NPV_2]_{\mathrm{mr}}$ and $[NPV_1, NPV_2]_{\mathrm{lr}}$ are the risk estimations for the considering IRR_{mr} and IRR_{lr}. It should be noted (see Table 4.23) that the difference between values of IRR_{mr} for the projects is rather small, but the difference in risk estimations is considerable.

4.5.4 A Method for Numerical Solution of the Project Optimization Problem

Here we propose an approach to the solution of optimization problem which is based on the consideration of all initial fuzzy P_t and KV_t as the constraints on controlled input data, as well as on assumption that d_t is a random parameter describing external, in relation to a considered project, uncertainty. The fact that some preferences for the interval of possible values of d may be expressed by a certain membership function $\mu_d(d)$ is also taken into account. Thus, while describing the discount factor one deals with uncertainties of both random and fuzzy nature.

The problem is solved in two steps. At first, according to the fuzzy extension principle all parameters P_t , KV_t and d_t in Eq. (7.45) are substituted for corresponding fuzzy numbers. As a result the fuzzy NPV is obtained. On the next step, the obtained fuzzy NPV is considered as a restriction on a profit when building the local criterion for NPV maximization. For the mathematical description of local criteria, the so-called desirability functions are used. In essence, they can be described as a special interpretation of usual membership functions. Briefly, the desirability function rises from 0 (in area of unacceptable values of its argument) up to 1 (in area of the most preferable values). Thus, a construction of desirability function for NPV is rather obvious: the desirability function $\mu_{NPV}(NPV)$ can be considered only on the interval of possible values restricted by the interval $[NPV_1, NPV_4]$. Hence, the more value of the NPV, the more degree of desirability (see Fig. 4.39).

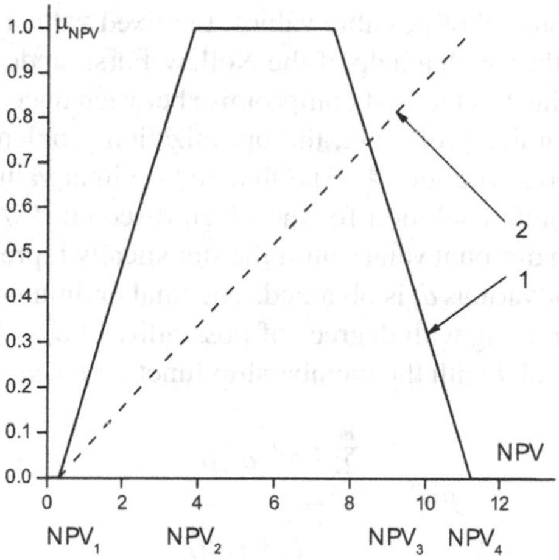

Fig. 4.39 Connection between the restriction and the local criterion: 1 - initial fuzzy interval of NPV (fuzzy restriction); 2 - desirability function $\mu_{NPV}(NPV)$.

The initial fuzzy P_t and KV_t are also considered as desirability functions μ_{P_1}, μ_{P_2}, ..., μ_{KV_1}, μ_{KV_2},... describing constraints on controlled input variables. It is clear that initial fuzzy numbers were already constructed in such a way that when they are interpreted as desirability functions, the more preferable values in the fuzzy intervals of P_t and KV_t appear to be those more realizable (possible). Since these desirability functions are connected with a possibility of realization of corresponding values of the variables P_t and KV_t, they implicitly describe the financial risk of the project.

As the result, the general criterion based on the set of all desirability functions has been defined as follows:

$$D(P_t, KV_t, d_t) = \mu_{NPV}^{\alpha_1}(NPV(P_t, KV_t, d_t)) \wedge (\mu_{P_1} \wedge \mu_{P_2} \wedge \ldots \wedge \mu_{KV_1}, \mu_{KV_2} \wedge \ldots)^{\alpha_2}, \quad (4.63)$$

where α_1 and α_2 are the ranks characterizing the relative importance of local criteria of profit maximization and risk minimization, $\wedge$ is the *min* operator, $\mu_{NPV}(NPV(P_t, KV_t, d_t))$ is the desirability function of NPV.

Many different forms of the general criterion are in use. As emphasized in [148], the choice of particular aggregating operator, usually called t-norm, is rather an application dependent problem. However, the choice of *min* operator in Eq.(4.63) is the most straightforward approach, when a compensation of small values of some criteria by the great values of others is not permitted [36, 106].

The problem is reduced to the search for crisp values of $PP_1, PP_2, \ldots, KKV_1, KKV_2, \ldots$ on corresponding fuzzy intervals $P_1, P_2, \ldots, KV_1, KV_2, \ldots$, maximizing the general criterion (4.63).

The problem is complicated by the fact that the discount d is a random parameter, distributed in a specific interval.

The solution was carried out as follows.

Firstly, from the interval of possible values, the fixed value of discount d_i is selected randomly. Further, with a help of the Nollaw-Furst random method the optimum solution is obtained as the best compromise between uncertainty of basic data and intention to maximize profit, i.e., the optimization problem reduces to maximization of the general criterion (4.63). Obtained optimal values PP_t^d and KKV_t^d present the local optimum solution for the given discount d. Above procedure is repeated with random discount values until the statistically representative sample of optimum solutions for various d_i is obtained. The final optimum values PP_t^0, KKV_t^0 are calculated by weighting with degrees of possibility of d_i, which are defined by the initial fuzzy interval d with the membership function $\mu_d(d_i)$:

$$PP_t^0 = \frac{\sum_{i=1}^{m} PP_t^d(d_i)\mu_d(d_i)}{\sum_{i=1}^{m} \mu_d(d_i)}, \quad (4.64)$$

where m is the number of random discount values used for the solution of the problem. Similarly, all KKV_t^0 can be calculated.

It is also possible to take into account the values of the general criterion in optimum points:

$$PP_t^0 = \frac{\sum_{i=1}^{m} \left(PP_t^d(d_i) \left(\mu_d^{\beta_1}(d_i) \wedge D^{\beta_2}(d_i) \right) \right)}{\sum_{i=1}^{m} \left(\mu_d^{\beta_1}(d_i) \wedge D^{\beta_2}(d_i) \right)}, \quad (4.65)$$

where β_1, β_2 are the corresponding weights.

The similar expression can be constructed for KKV_t^0. It is worth noting that the expression (4.64) provides an ability to take into account, apart from reliability of the values d_i, the degree of compatibility (in other words, the degree of consensus) for each of selected values of discount.

Obtained optimal PP_t^0 and KK_t^0 may be used for the final project's quality estimation. The results of calculation for the first example from the previous subsection (Table 4.22, project 1) are presented in Table 4.24.

Table 4.24 Results of optimization

	Expression (4.64)	Expression (4.65)
Years	PP_t^0, KK_t^0	PP_t^0, KK_t^0
0	0.00, 2.49	0.00, 2.50
1	0.00, 0.83	0.00, 0.79
2	8.05, 0.00	8.04, 0.00
3	7.12, 0.00	7.09, 0.00

Finally, with substituting PP_t^0, KK_t^0 and fuzzy interval d in the expression (4.51) the optimal fuzzy value of NPV was obtained.

For considered example we get the following results:
$NPV_1 = \{4.057293, 6.110165, 8.073906, 9.454419\}$ using (4.64)
and
$NPV_2 = \{4.065489, 6.109793, 8.064094, 9.436519\}$ using (4.65).

It is clear that there is no great deference between the results obtained using expressions (4.64) and (4.65) in this case.

In Fig. 4.40, the fuzzy NPV obtained using PP_t^0 and KK_t^0 is compared with the initial one, obtained with the use of initial fuzzy numbers P_t and KV_t, without optimization. It is obvious that in the optimal case the mean value of fuzzy interval NPV is greater.

Using the optimal PP_t^0 and KK_t^0 and the method described in Subsection 4.5.2, the degree of project risk may be also evaluated. This risk can be considered as the financial risk of a project as a whole.

For the needs of common accounting practice, it is possible to calculate the average weighted value of NPV using the following expression:

$$NPV = \frac{\sum_{i=1}^{m} NPV_i \cdot \mu_{NPV_i}}{\sum_{i=1}^{m} \mu_{NPV_i}}, \quad (4.66)$$

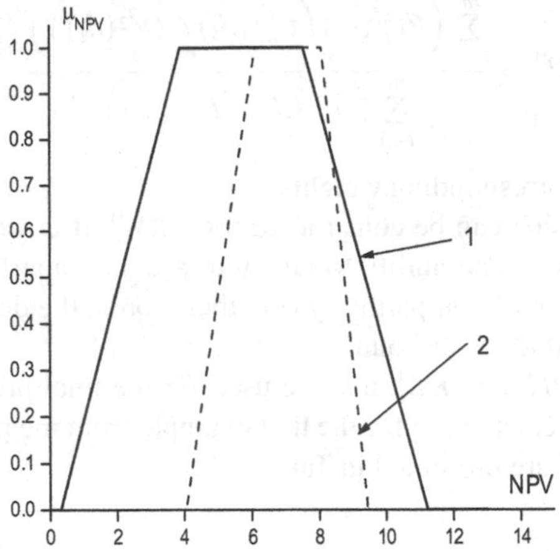

Fig. 4.40 Comparison of the initial and optimal fuzzy *NPV*: 1 - initial *NPV*; 2 - optimal *NPV*

For the considered example $NPV_1 = 6.8931$ and $NPV_2 = 6.8942$ were obtained.

4.6 Summary and Discussion

In this chapter, we have presented the generalized method for multiple criteria hierarchical evaluation of investments in the fuzzy setting. The key issue is the analyzing the familiar approaches to the aggregation of local criteria. The problems of ranked local criteria aggregation are analyzed and some new theoretical results which can be useful for proper choice of aggregation method are presented. It is proved that the most popular weighted sum method is the most unreliable one and can provide wrong results. As all known aggregation modes have their own advantages and drawbacks it seems impossible to choose the best one especially when dealing with complicated hierarchical problems. Therefore, a new approach which makes it possible to generalize the aggregating modes into some "ideal" criterion is developed using the mathematical tools of level–2 and type–2 fuzzy sets. It is shown that proposed method allows us to build in natural way the branched hierarchical structures of the investment project's local criteria. The theoretical consideration is illustrated using examples of investments. The main direction of future research will be generalization of the proposed approach using conception of hyperfuzzy sets [42] being the useful particular case of type–2 fuzzy sets for representation of verbally formulated local criteria and parameters of project's quality.

Using the developed methodology, a new method for the stock ranking based on the multiple criterion decision making and optimization has been proposed. Two general criteria are used in the analysis. The first of them is based on the financial indices and may be treated as the criterion of firm's "health" or its financial performance. The second one is the two-criteria performance of firm based on the stock prices. It represents the firm's market success. The method rests on the selection of the stocks with a great correlation of the firm's financial performance and its market success. The local criteria are built in the form of the membership function of corresponding fuzzy subsets. Two different strategies for the stock ranking and three most popular methods for the local criteria aggregation are compared. As the example, the values of financial rations and prices from the database comprising the data of 162 firms from the subsector of the biotechnology of US economy were used. It is shown that the proposed method makes it possible to select a small group of "good" stocks characterized by a great coincidence of firm's financial performance and its market success. The method rejects from the consideration all the "unsafe" firms, i.e., such ones that their market success is based rather on the public relations, rumors and other rather unreliable information. The method is addressed to those who prefer to select for a portfolio only the firms which demonstrate the closeness of their overall financial performance in the past year and success in the stock exchange in the following year.

Another application of the developed general approach to the solution of the *MCDM* problems presented in this chapter are the problems of calculation of *NPV* and *IRR* and investment project risk assessment in the fuzzy setting. It is shown that the straightforward way of project risk assessment is to consider this risk as a degree of fuzziness of the fuzzy *NPV*. Nevertheless, other methods for risk estimation based on the probability approach to interval and fuzzy values comparison can be relevant in the fuzzy budgeting as well. It is shown that the real values of *IRR* may be obtained as the solutions of the fuzzy equation and a set of new useful derivative parameters characterizing uncertainty of the problem may be obtained as an additional result.

The problem of the multiobjective optimization of a project in the mixed fuzzy and random environment is formulated in the form of compromise between local criteria of profit maximization and risk minimization. The numerical method for the solution of this problem is described and tested.

References

1. Abd El-Wahed, W.F., Lee, S.M.: Interactive fuzzy goal programming for multi-objective transportation problems. Omega 34, 158–166 (2006)
2. Ali, M.M., Torn, A.: Population set-based global algorithms: some modifications and numerical studies. Computers and Operations Research 31, 1703–1725 (2004)
3. Anderson, J.A.: Screening for investment gold. Black Enterprise 29, 93–97 (1998)
4. Babusiaux, D., Pierru, A.: Capital budgeting, project valuation and financing mix: Methodological proposals. Europian Journal of Operational Research 135, 326–337 (2001)
5. Bana, E., Costa, C.A. (eds.): Reading in multiple criteria decision aid. Springer, Berlin (1990)
6. Bana, E., Costa, C.A., Stewart, T.J., Vansnick, J.-C.: Multicriteria decision analysis: Some thoughts based on the tutorial and discussion session of the ESIGMA meetings. European Jurnal of Operational Research 99, 28–37 (1997)
7. Belletante, B., Arnaud, H.: Choisir ses investissements. Paris Chotard et Associes Editeurs (1989)

8. Beynon, M., Curry, B., Morgan, P.: The Dempster-Shafer theory of evidence: an alternative approach to multicriteria decision modeling. Omega 28, 37–50 (2000)
9. Beynon, M., Peel, M.J., Tang, Y.-C.: The application of fuzzy decision tree analysis in an exposition of the antecedents of audit fees. Omega 32, 231–244 (2004)
10. Biswas, A., Pal, B.B.: Application of fuzzy goal programming technique to land use planning in agricultural system. Omega 33, 391–398 (2005)
11. Bogle, H.F., Jehenck, G.K.: Investment Analysis: US Oil and Gas Producers Score High in University Survey. In: Proc. of Hydrocarbon Economics and Evaluation Symposium, Dallas, pp. 234–241 (1985)
12. Bollinger, D., Pictet, J.: Multiple criteria decision analysis of treatment and land-filling technologies for waste incineration residues. Omega 36, 418–428 (2008)
13. Borisov, A.N., Korneeva, G.V.: Linguistic approach to decision making model building under uncertainty. Methods of decision making under uncertainty, 4–6 (1980) (in Russian)
14. Brigham, E.F.: Fundamentals of Financial Management. The Dryden Press, New York (1992)
15. Buckley, J.J.: The fuzzy mathematics of finance. Fuzzy Sets and Systems 21, 257–273 (1987)
16. Chakraborty, M., Chandra, M.K.: Multicriteria decision making for optimal blending for beneficiation of coal: a fuzzy programming approach. Omega 33, 413–418 (2005)
17. Chan, F.T.S., Kumar, N.: Global supplier development considering risk factors using fuzzy extended AHP-based approach. Omega 35, 417–431 (2007)
18. Chanas, S., Delgado, M., Verdegay, J.L., Vila, M.A.: Ranking fuzzy interval numbers in the setting of random sets. Information Sciences 69, 201–217 (1993)
19. Chang, P.-T., Lee, E.S.: The Estimation of Normalized Fuzzy Weights. Computers and Mathematics with Applications 29, 21–42 (1995)
20. Chansa-ngavej, C., Mount-Campbell, C.A.: Decision criteria in capital budgeting under uncertainties: implications for future research. Int. J. Prod. Economics 23, 25–35 (1991)
21. Chen, C.T.: Extensions of the TOPSIS for group decision-making under fuzzy environment. Fuzzy Sets and Systems 114, 1–9 (2000)
22. Chen, C.T.: A fuzzy approach to select the location of the distribution center. Fuzzy Sets and Systems 118, 65–73 (2001)
23. Chen, S.: An empirical examination of capital budgeting techniques: impact of investment types and firm characteristics. Eng. Economist. 40, 145–170 (1995)
24. Chen, Y., Kilgour, D.M., Hipel, K.W.: A case-based distance method for screening in multiple-criteria decision aid. Omega 36, 373–383 (2008)
25. Chen, C., Klein, C.M.: An efficient approach to solving fuzzy MADM problems. Fuzzy Sets and Systems 88, 51–67 (1997)
26. Chiou, H.-K., Tzeng, G.-H., Cheng, D.-C.: Evaluating sustainable fishing development strategies using fuzzy MCDM approach. Omega 33, 223–234 (2005)
27. Chiu, C.Y., Park, C.S.: Fuzzy cash flow analysis using present worth criterion. Eng. Economist. 39, 113–138 (1994)
28. Chiu, C.Y., Park, C.S.: Fuzzy cash flow analysis using present worth criterion. Eng. Economist. 39, 113–138 (1994)
29. Choi, D.Y., Oh, K.W.: Asa and its application to multi-criteria decision making. Fuzzy Sets and Systems 114, 89–102 (2000)
30. Choo, E.U., Schoner, B., Wedley, W.C.: Interpretation of criteria weights in multicriteria decision making. Computers and Industrial Engineering 37, 527–541 (1999)

31. Choobineh, F., Behrens, A.: Use of intervals and possibility distributions in economic analysis. J. Oper. Res. Soc. 43, 907–918 (1992)
32. Chu, A., Kalaba, R., Springarn, R.: A Comparison of Two Methods for Determining the weights of Belonging to Fuzzy Sets. Journal of Optimization theory and applications 27, 531–538 (1979)
33. Delgado, M., Verdegay, J.L., Vila, M.A.: Linguistic decision making models. Internat. J. Intell. Systems 7, 479–492 (1992)
34. Demirtas, E.A., Üstün, Ö.: An integrated multiobjective decision making process for supplier selection and order allocation. Omega 36, 76–90 (2008)
35. Dimova, L., Sevastianov, D., Sevastianov, P.: Application of fuzzy sets theory, methods for the evaluation of investment efficiency parameters. Fuzzy Economic Review 5, 34–48 (2000)
36. Dimova, L., Sevastianov, P., Sevastianov, D.: MCDM in a fuzzy setting: investment projects assessment application. International Journal of Production Economics 100(2006), 10–29 (2006)
37. Doumpos, M., Kosmidou, K., Baourakis, G., Zopounidis, C.: Credit risk assessment using a multicriteria hierarchical discrimination approach: A comparative analysis. European Journal of Operational Research 1389, 392–412 (2002)
38. Dropsy, V.: Do macroeconomic factors help in predicting international equity risk premia? Journal of Applied Business Research 3, 120–132 (1996)
39. Dubois, D., Koenig, J.L.: Social choice axioms for fuzzy set aggregation. Fuzzy Sets and Systems 43, 257–274 (1991)
40. Dyckhoff, H.: Basic concepts for theory of evaluation: hierarchical aggregation via autodistributive connectives in fuzzy set theory. European J. Operation Research 20, 221–233 (1985)
41. Dymova, L.: A constructive approach to managing fuzzy subsets of type 2 in decision making. TASK Quarterly 7, 157–164 (2003)
42. Dymova, L., Rog, P., Sewastianow, P.: Hyperfuzzy estimations of financial parameters. In: Proceeding of the 2th International Conference on Mathematical Methods in Finance and Econometrics, pp. 78–84 (2002)
43. Freedman, J.D.: Behind the smoke and mirrors: gauging the integrity of investment simulations. Financial Analysts Journal 6, 26–31 (1992)
44. Gold, S.C., Lebowitz, P.: Computerized stock screening rules for portfolio selection. Financial Services Review 8, 61–70 (1999)
45. Gomes, C.F.S., Nunes, K.R.A., Xavier, L.H., Cardoso, R., Valle, R.: Multicriteria decision making applied to waste recycling in Brazil. Omega 36, 395–404 (2008)
46. Gottwald, S.: Set theory for fuzzy sets of higher level. Fuzzy Sets and Systems 2, 125–151 (1979)
47. Hauke, W.: Using Yager's t-norms for aggregation of fuzzy intervals. Fuzzy Sets and Systems 101, 59–65 (1999)
48. Helmer, O.H.: The Delphi Method for Systematizing Judgments about the Future, Institute of Government and Public Aairs, University of California (1966)
49. Herrera, F., Herrera-Vieda, E., Verdegay, J.L.: Direct approach processes in group decision making using linguistic OWA operators. Fuzzy Sets and Systems 79, 175–190 (1996)
50. Huang, C.-C., Chu, P.-Y., Chiang, Y.-H.: A fuzzy AHP application in government-sponsored R&D project selection. Omega 36, 1038–1052 (2008)

51. Investopedia,
`http://www.investopedia.com/terms/o/outstandingshares.asp`
52. Investorwords,
`http://www.investorwords.com/4316/ROI.html`
53. Jaulin, L., Kieffir, M., Didrit, O., Walter, E.: Applied Interval Analysis. Springer, London (2001)
54. Kahraman, C.: Fuzzy versus probabilistic benefit/cost ratio analisis for public work projects. Int. J. Appl. Math. Comp. Sci. 11, 705–718 (2001)
55. Kahraman, C., Ruan, D., Tolga, E.: Capital budgeting techniques using discounted fuzzy versus probabilistic cash flows. Information Sciences 142, 57–76 (2002)
56. Kahraman, C., Tolga, E., Ulukan, Z.: Justification of manufacturing technologies using fuzzy benefit/cost ratio analysis. Int. J. Product Econom. 66, 45–52 (2000)
57. Kahraman, C., Ulukan, Z.: Continuous compounding in capital budgeting using fuzzy concept. In: Proc. of the 6th IEEE International Conference on Fuzzy Systems, pp. 1451–1455 (1997)
58. Kahraman, C., Ulukan, Z.: Fuzzy cash flows under inflation. In: Proc. of the Seventh International Fuzzy Systems Association World Congress (IFSA 1997), vol. 4, pp. 104–108 (1997)
59. Karnik, N.N., Mendel, J.M.: Application of Type-2 Fuzzy Logic Systems to Forecasting of Time-Series. Information Sciences 120, 89–111 (1999)
60. Kaufmann, A., Gupta, M.: Introduction to fuzzy-arithmetic theory and applications. Van Nostrand Reinhold, New York (1985)
61. Kiang, M.Y., Chi, R., Tam, K.Y.: DKAS: a distributed knowledge acquisition system in a DSS. Journal of Management Information Systems 4, 59–82 (1993)
62. Kosko, B.: Fuzzy entropy and conditioning. Information Science 30, 165–174 (1986)
63. Krishnapuram, R., Keller, J.M., Ma, Y.: Quantitative analysis of properties and spatial relations of fuzzy image regions. IEEE Trans. Fuzzy Systems 1, 222–233 (1993)
64. Kryzanowski, L., Galler, M., Wright, D.: Using artificial neural network to pick stocks. Financial Analysts Journal 1, 21–27 (1993)
65. Kuchta, D.: Fuzzy capital budgeting. Fuzzy Sets and Systems 111, 367–385 (2000)
66. Kundu, S.: Min-transitivity of fuzzy leftness relationship and its application to decision making. Fuzzy Sets and Systems 86, 357–367 (1997)
67. Kundu, S.: Preference relation on fuzzy utilities based on fuzzy leftness relation on interval. Fuzzy Sets and Systems 97, 183–191 (1998)
68. Lam, M.: Neural network techniques for financial performance prediction: integrating fundamental and technical analysis. Decision Support Systems 37, 567–581 (2004)
69. Lee, H.: Group decision making using fuzzy sets theory for evaluating the rate of aggregative risk in software development. Fuzzy Sets and Systems 80, 261–271 (1996)
70. Leigh, W., Purvis, R., Ragusa, J.M.: Forecasting the NYSE composite index with technical analysis, pattern recognizer, neural network, and genetic algorithm: a case study in romantic decision support. Decision Support Systems 32, 361–377 (2002)
71. Li Calzi, M.: Towards a general setting for the fuzzy mathematics of finance. Fuzzy Sets and Systems 35, 265–280 (1990)
72. Li, Q., Sterali, H.D.: An approach for analyzing foreign direct investment projects with application to China's Tumen River Area development. Computers & Operations Research 3, 1467–1485 (2003)
73. Liang, P., Song, F.: Computer-aided risk evaluation system for capital investment. Omega 22, 391–400 (1994)

74. Lin, F.C., Lin, M.: Analysis of financial data using neural nets. AI Expert 2, 36–41 (1993)
75. Liu, D., Stewart, T.J.: Object-oriented decision support system modeling for multicriteria decision making in natural resource managment. Computers and Operations Research 31, 985–999 (2004)
76. Longerstaey, J., Spenser, M.: RiskMetric-Technical document. RiskMetric Group, J.P. Morgan, New York (1996)
77. Lootsma, F.A.: Performance evaluation of non-linear optimization methods via multicriteria decision analysis and via linear model analysis. In: Powell, M.J.D. (ed.) Nonlinear Optimization, pp. 419–453 (1981)
78. Lopes, M.D.S., Flavel, R.: Project appraisal-a framework to assess non-financial aspects of projects during the project life cycle. International Journal of Project Management 16, 223–233 (1998)
79. Masaharu, M., Kokichi, T.: Fuzzy sets of type II under algebraic product and algebraic sum. Fuzzy Sets and Systems 5, 277–290 (1981)
80. McIvor, R.T., McCloskey, A.G., Humphreys, P.K., Maguire, L.P.: Using a fuzzy approach to support financial analysis in the corporate acquisition process. Expert Systems with Applications 27, 533–547 (2004)
81. Migdalas, A., Pardalos, P.M.: Editorial: hierarchical and bilevel programming. J. Global Optimization 8, 209–215 (1996)
82. Mikhailov, L.: Fuzzy analytical approach to partnership selection information of virtual enterprises. Omega 30, 393–401 (2002)
83. Mikhailov, L.: Deriving priorities from fuzzy pairwise comparison judgments. Fuzzy Sets and Systems 134, 365–385 (2003)
84. Miller, G.A.: The magical number seven plus or minus two: some limits on our capacity for processing information. Psychological Review 63, 81–97 (1956)
85. Milner, P.M.: Physiological psychology. Holt, New York (1970)
86. Mitra, G.: Mathematical Models for Decision Support. Springer, Berlin (1988)
87. Mohamed, S., McCowan, A.K.: Modelling project investment decisions under uncertainty using possibility theory. International Journal of Project Management 19, 231–241 (2001)
88. Moore, R.E.: Interval analysis. Prentice-Hall, Englewood Cliffs (1966)
89. Nakamura, K.: Preference relations on set of fuzzy utilities as a basis for decision making. Fuzzy Sets and Systems 20, 147–162 (1986)
90. Nedosekin, A., Kokosh, A.: Investment risk estimation for arbitrary fuzzy factors of investment project. In: Proc. of Int. Conf. on Fuzzy Sets and Soft Computing in Economics and Finance, St. Petersburg, pp. 423–437 (2004)
91. Pardalos, P.M., Siskos, Y., Zopounidis, C.: Advances in multicriteria analysis. Kluwer Academic Publishers, Dordrecht (1995)
92. Peneva, V., Popchev, I.: Properties of the aggregation operators related with fuzzy relations. Fuzzy Sets and Systems 139, 615–633 (2003)
93. Perrone, G.: Fuzzy multiple criteria decision model for the evaluation of AMS. Comput. Integrated Manufacturing Systems 7, 228–239 (1994)
94. Racine, J.: On the nonlinear predictability of stock returns using financial and economic variables. Journal of Business and Economic Statistics 19, 380–382 (2001)
95. Reilly, F.K., Brown, K.C.: Investment Analysis and Portfolio Management. South-Western College Pub. (2002)

96. Reinganum, M.R.: The anatomy of a stock market winner. Financial Analysts Journal 1, 16–28 (1988)
97. Ribeiro, R.A.: Fuzzy multiple attribute decision making: a review and new preference elicitation techniques. Fuzzy Sets and Systems 78, 155–181 (1996)
98. Roubens, M.: Fuzzy sets and decision analysis. Fuzzy Sets and Systems 90, 199–206 (1997)
99. Roy, B.: Methodologie Multicriterie d'Aide a la Decision (1985); Economica, Paris, English edn. Multicriteria Methodology for Decision Aiding. Kluwer Academic Publlishers, Boston (1996)
100. Saaty, T.: A scaling method for priorities in hierarchical structures. Journal of Mathematical Psychology 15, 234–281 (1977)
101. Saaty, T.: Mathematical Methods of Operations Research. Dover Pub., New York (2004)
102. Schwager, J.D.: The New Market Wizards: Conversations with America's Top Traders. John Wiley and Sons, NY (1995)
103. Sengupta, A., Pal, T.K.: On comparing interval numbers. European Journal of Operational Research 127, 28–43 (2000)
104. Shih, H.S., Lee, E.S.: Compensatory fuzzy multiple level decision making. Fuzzy Sets and Systems 114, 71–87 (2000)
105. Chen, S.-M.: A new method for tool steel materials selection under fuzzy environment. Fuzzy sets and systems 92, 265–274 (1997)
106. Sevastianov, P., Dimova, L., Zhestkova, E.: Methodology of the multicriteria quality estimation and its software realizing. In: Proc. of the Fourth International Conference on New Information Technologies NITe', vol. 1, pp. 50–54 (2000)
107. Sewastianow, P., Jonczyk, M.: Bicriterial fuzzy selection. Operations research and decisions 4, 149–165 (2003)
108. Sevastjanov, P., Figat, P.: Aggregation of aggregating modes in MCDM, Synthesis of Type 2 and Level 2 fuzzy sets. Omega 35, 505–523 (2007)
109. Sewastianow, P., Rog, P.: A probabilistic approach to fuzzy and interval ordering. Task Quarterly, Special Issue Artificial and Computational Intelligence 7, 147–156 (2002)
110. Sewastianow, P., Rog, P.: Fuzzy modeling of manufacturing and logistic systems. Mathematics and Computers in Simulation 63, 569–585 (2003)
111. Sewastianow, P., Rog, P.: Two-objective method for crisp and fuzzy interval comparison in Optimization. Computers & Operations Research 33, 115–131 (2006)
112. Sewastianow, P., Rog, P., Venberg, A.: The Constructive Numerical Method of Interval Comparison. In: Wyrzykowski, R., Dongarra, J., Paprzycki, M., Waśniewski, J. (eds.) PPAM 2001. LNCS, vol. 2328, pp. 756–761. Springer, Heidelberg (2002)
113. Sevastianov, P., Sevastianov, D.: Risk and capital budgeting parameters evaluation from the fuzzy sets theory position. Reliable software 1, 10–19 (1997)
114. Sevastianov, P., Tumanov, N.: Multi-criteria identification and optimization of technological processes. Science and Engineering (1990) (in Russian)
115. Silva, C.G., Figueira, J., Lisboa, J., Barman, S.: An interactive decision support system for an aggregate production planning model based on multiple criteria mixed integer linear programming. Omega 34, 167–177 (2006)
116. Silvert, W.: Ecological impact classification with fuzzy sets. Ecological Moddeling (1997)
117. Steuer, R.E.: Multiple criteria optimisation: theory, computation and application. Wiley, New York (1986)

118. Steuer, R.E., Na, P.: Multiple criteria decision making combined with finance. A categorical bibliographic study. European Jurnal of Operational Research 150, 496–515 (2003)
119. Stewart, T.J.: A critical survey on the status of multiple criteria decision making. OriON 5, 1–23 (1989)
120. Sugeno, M.: Industrial applications of fuzzy control. Elsevier Science Publishing Company, Amsterdam (1985)
121. Sugeno, M., Kang, G.T.: Structure identification of fuzzy model. Fuzzy Sets and Systems 28, 15–33 (1988)
122. Sugeno, M., Yasukawa, T.: A fuzzy-logic-based approach to qualitative modelling. IEEE Transactions on Fuzzy Systems 1, 7–31 (1993)
123. Tiryaki, F., Ahlatcioglu, M.: Fuzzy stock selection using a new fuzzy ranking and weighting algorithm. Applied Mathematics and Computation 170, 144–157 (2005)
124. Thomsett, M.C.: Mastering Technical Analysis. A Kaplan Professional, Chicago (1999)
125. Tong, M., Bonissone, P.P.: A linguistic approach to decision making with fuzzy sets. IEEE Trans. Systems Man Cybernet. 10, 716–723 (1980)
126. Torn, A., Zilinskas, A.: Global optimization. Springer, Berlin (1989)
127. Tre, G., Caluwe, R.: Level-2 fuzzy sets and their usefulness in object-oriented database modeling. Fuzzy Sets and Systems 140, 29–49 (2003)
128. Valls, A., Torra, V.: Using classification as an aggregation tool in *MCDM*. Fuzzy Sets and Systems 115, 159–168 (2000)
129. Wadman, D., Schneider, M., Schnaider, E.: The use of interval mathematics in fuzzy expert system. International Journal of Intelligent Systems 9, 241–259 (1994)
130. Wagenknecht, M., Hartmann, K.: On fuzzy rank ordering in polyoptimisation. Fuzzy Sets and Systems 11, 253–264 (1983)
131. Walczak, S.: An empirical analysis of data requirements for financial forecasting with neural networks. Journal of Management Information Systems 17, 203–222 (2001)
132. Wang, J., Hwang, W.-L.: A fuzzy set approach for R&D portfolio selection using a real options valuation model. Omega 35, 247–257 (2007)
133. Wang, X., Kerre, E.E.: Reasonable properties for the ordering of fuzzy quantities (I) (II). Fuzzy Sets and Systems 112, 375–385, 387–405 (2001)
134. Wanga, M.-J., Chang, T.-C.: Tool steel materials selection under fuzzy environment. Fuzzy Sets and Systems 72, 263–270 (1995)
135. Ward, T.L.: Discounted fuzzy cash flow analysis. In: Proceeding of the 1985 Fall Industrial Engineering Conference, pp. 476–481 (1985)
136. Weck, M., Klocke, F., Schell, H., Rüenauver, E.: Evaluating alternative production cycles using the extended fuzzy AHP method. European Journal of Operational Research 100, 351–366 (1997)
137. Yager, R.: On the measure of fuzziness and negation. Part 1. Membership in the Unit. Interval Int. J. Gen. Syst. 5, 221–229 (1979)
138. Yager, R.: Multiple objective decision-making using fuzzy sets. International Journal of Man-Machine Studies 9, 375–382 (1979)
139. Yager, R.R.: A foundation for a theory of possibility. Journal of Cybernetics 10, 177–209 (1980)
140. Yager, R.R.: Fuzzy subsets of type II in decisions. Journal of Cybernetics 10, 137–159 (1980)
141. Yager, R.R.: On ordered weighted averaging aggregation operators in multicriteria decision making. IEEE Trans. Systems Man and Cybern. 18, 183–190 (1988)

142. Yager, R.R., Detyniecki, M., Bouchon-Meunier, B.: A context-dependent method for ordering fuzzy numbers using probabilities. Information Sciences 138, 237–255 (2001)
143. Zadeh, L.A.: Fuzzy sets. Inf. Control 8, 338–353 (1965)
144. Zadeh, L.A.: Quantitative fuzzy semantics. Information Sciences 3, 177–200 (1971)
145. Zadeh, L.A.: Fuzzy logic and its application to approximate reasoning. Information Processing 74, 591–594 (1974)
146. Zadeh, L.A.: The Concept of linguistic Variable and its Application to approximate Reasoning- I. Information Sciences 8, 199–249 (1975)
147. Zimmerman, H.J.: Fuzzy Sets, Decision-Making and Expert Systems. Kluwer Academic Publishers, Dordrecht (1987)
148. Zimmermann, H.J., Zysno, P.: Latest connectives in human decision making. Fuzzy Sets and Systems 4, 37–51 (1980)
149. Zimmermann, H.J., Zysno, P.: Decision and evaluations by hierarchical aggregation of information. Fuzzy Sets and Systems 104, 243–260 (1983)
150. Zollo, G., Iandoli, L., Cannavacciuolo, A.: The performance requirements analysis with fuzzy logic. Fuzzy Economic Review 4, 35–69 (1999)

Chapter 5
Interval and Fuzzy Arithmetic in Logistic

This chapter deals with the so-called distribution problem, which belong to the wide class of the logistic problems.

It is known that distribution and transportation problems have similar mathematical structures and are usually treated as particular cases of the general linear programming problem.

There are many effective algorithms for the solution of transportation and distribution problems proposed in the scientific literature and in the textbooks. So we can say that these problems in the case of real valued parameters are, generally, solved. Nevertheless, in practice, we often meet different kinds of uncertainty when the parameters of these optimization problems are presented by intervals or fuzzy values.

The known approaches to the solution of fuzzy transportation and distribution problems are usually based on some restrictions imposed on the form of membership functions. These restrictions make it possible, using analytical procedures, to transform the initial fuzzy problem to the set of usual linear programming problems with real valued parameters. Nevertheless, in practice, membership functions representing the parameters of the problem may have substantially complicated forms and analytical procedures can not be used.

Therefore, in this chapter a new approach to the solution of fuzzy distribution problem is developed. In the framework of this approach, all parameters and variables may be fuzzy values without any additional restrictions.

It is important that real-world distribution problems are usually multiple criteria ones.

In this chapter, the results obtained as the solution of fuzzy single criterion distribution problem are used as the base for the formalization and solution of multiple criteria fuzzy distribution problem.

L. Dymowa: Soft Computing in Economics and Finance, ISRL 6, pp. 187–206.
springerlink.com

5.1 Fuzzy Linear Programming Approach to the Distribution Planning Problem

A new numerical approach to the solution of the fuzzy distribution problem based on the direct fuzzy extension of the simplex method is developed. The fuzzy extension is based on the fuzzy arithmetic rules and the method for fuzzy values comparison based on the probabilistic approach (see Chapters 3 and 4). In the framework of proposed approach, all parameters and variables may be fuzzy values without any additional restrictions. The α-cut representation of all fuzzy parameters and variables is used and any additional assumption regarding their form is not needed. The implementation of the method is made using the object-oriented programming technique.

The advantages of the proposed approach are illustrated with the use of case study, where the fuzzy solution of the fuzzy distribution problem is compared with that obtained using the Monte-Carlo method.

5.1.1 The Methods for the Solution of Fuzzy Linear Programming Problem

The transportation and distribution problems have similar mathematical structures and are usually treated as particular cases of the general linear programming problem. The first efficient algorithm for the solution of the transportation problem was proposed in 1979 by Isermann [28]. In 1987, Ringuest i Rinks [47] developed two iterative algorithms for the solution of the linear multiple criteria transportation problem.

There are many effective algorithms for the solution of transportation and distribution problems proposed in the scientific literature and in the textbooks. So we can say that these problems in the case of real valued parameters are, generally, solved. Nevertheless, in practice, we often meet different kinds of uncertainty when the parameters of these optimization problems are presented by intervals or fuzzy values. Zimmermann first proposed the formulation of fuzzy linear programming problem (*FLPP*) and its approximate solution in [60]. Steuer [51], Tong [52], Chanas and Kuchta [12] proposed the solutions of the linear programming problem with the interval target function. The generalization of these solution was presented by Kuchta in [33]. Similarly, in [16] the authors proposed the procedure for the solution of transportation problem with interval parameters of the objective function and constraints.

Chanas [9] and Chanas and Kuchta [11, 13] developed an approach to the solution of *FLPP*. This approach has been generalized in [21, 22]. It is based on some restrictions imposed on the form of membership functions. These restrictions make it possible, using analytical procedures, to transform the initial fuzzy problem to the set of usual linear programming problems with real valued parameters.

Nevertheless, in practice, membership functions representing the parameters of the problem may have substantially complicated forms and analytical procedures can not be used. In [36], the fuzzy transportation problem has been reduced to the pair of optimization tasks for the lower and upper bounds of the fuzzy target function (fuzzy cost of transportation).

The literature analysis allows us to mark out two main approaches to the solution of *FLPP*. In the most general form, this problem may be presented as follows:

$$\max/\min \hat{Z} = \sum_{j=1}^{n} \hat{c}_j \hat{x}_j,$$

$$\sum_{j=1}^{n} \hat{a}_{ij} \hat{x}_j \leq \hat{b}_i, i = 1,2,...,m,$$

$$\hat{x}_j \geq 0,$$

where $\hat{X} = \{\hat{x}_j\}$ is the vector of fuzzy variables, $\hat{C} = \{\hat{c}_j\}$, $\hat{B} = \{\hat{b}_i\}$ and $\hat{A} = \{\hat{a}_{ij}\}$ are the vectors and the matrix of fuzzy parameters characterizing the objective function and constraints.

In the framework of the first approach, the above problem is solved in the assumption that X is the vector of real valued variables. Hence, the non-fuzzy solution of the fuzzy problem is obtained. In [24, 29, 31, 32, 35, 39, 46, 48], the set of real valued solutions of the above problem obtained using different assumptions regarding the form of fuzzy or interval parameters is presented. Similarly, in [2, 30, 36, 57] the set of approximate solutions of the fuzzy transportation problem has been obtained assuming that the variables $X = \{x_{ij}\}$ are real values.

The second approach is free of the above assumption that variables X are real values. Hence, the solution of *FLPP* should be obtained in the fuzzy form ($\hat{X}$). Obviously, such approach seems to be more natural since all the model's parameters are fuzzy values. However, this approach leads to the more general formulation of the problem and to additional mathematical problems that complicate the obtaining of exact solutions. Therefore, only a few papers represent this approach.

In [25, 40, 41, 42, 44], the approximate fuzzy solutions $\hat{X} = \{\hat{x}_j\}$ of *FLPP* were obtained in the assumption that the elements of matrix A and vector C are real values. Such an assumption makes it possible to obtain an approximate solution when the elements of the vector $\hat{B}$ are trapezoidal fuzzy values [40]. In [7], it is assumed that the elements of $\hat{X}$, $\hat{B}$, $\hat{C}$ and $\hat{A}$ are fuzzy values, but instead of fuzzy objective function $\hat{Z}$ its non-fuzzy representation was used. Three additional real valued parameters (local criteria) were used to represent $\hat{Z}$. After such a simplification of the initial problem, these local criteria were aggregated and finally the problem was formulated as the multiple criteria optimization with fuzzy constraints. The similar approach has been used in [37], where the additional restriction was introduced: all parameters and variables of *FLPP* are triangular fuzzy valuess. These fuzzy triangular values were approximated by their nearest symmetric triangular fuzzy values, with the assumption that all decision variables are symmetrical triangles.

Summarizing we can say that now there are no such general approaches that make it possible to obtain the solution of $FLPP$, fuzzy transportation or fuzzy distribution problems without additional restrictions on the form of fuzzy parameters and variables. Obviously, such restrictions substantially limit the ability of known methods to solve practically important problems when additional restrictions disturb the initial structure of the problem.

Therefore, in the current section we propose a new numerical approach to the solution of fuzzy distribution problem (FDP) based on the direct fuzzy extension of the simplex method. This extension is based on the fuzzy arithmetic rules and the method for fuzzy values comparison.

The α-cut representation of all fuzzy parameters and variables is used and any additional assumption regarding their form is not needed. The implementation of this extension is made using object-oriented programming technique.

5.1.2 The Direct Fuzzy Extension of the Simplex Method

In the framework of the proposed approach to the fuzzy extension of the simplex method, all the steps of the developed algorithm are similar to those of the standard simplex method with only one difference: the usual arithmetical operations and operation of comparison are replaced by the corresponding operations on the fuzzy values. Therefore, we call this approach "the direct fuzzy extension of the simplex method". Obviously, the choice of appropriate operations on fuzzy values, especially the operation of fuzzy values comparison, plays a pivotal role in the development of the proposed method.

Here we shall use the methos of applied interval analysis as the basis of fuzzy arithmetic presented in Section 3.

When dealing with the distribution problem, we not only minimize the transportation costs, but in addition we maximize the distributor's profits.

Suppose the distributor deals with M wholesalers and N consumers (see Fig.5.1). Let a_i, i=1 to M, be the maximal quantities of goods that can be proposed by wholesalers and b_i, j=1 to N, be the maximal good requirements of consumers. The fuzzy profit $\hat{z}_{ij}$ obtained as the result of delivering of a good unit from ith wholesaler to jth consumer can be calculated as $\hat{z}_{ij} = c_j - c_i - \hat{t}_{ij}$, where c_j is the price of selling, c_i is the price of buying, $\hat{t}_{ij}$ is the total fuzzy transportation cost of delivering of a good unit from ith wholesaler to jth consumer. In accordance with the signed contracts, distributor must buy at least p_i good units at price of c_i monetary units for unit of good from each ith wholesaler and to sell at least q_j good units at price of c_j monetary units for unit of good to each jth consumer. These constraints p_i, q_j limit only the lower bounds for the possible optimal quantities of goods which can be bought and sold. Therefore, they can be negotiated and hereinafter we shall treat them as the fuzzy constraints denoted as $\hat{p}_i$, $\hat{q}_j$. Therefore, the problem is to find such optimal good quantities $\hat{x}_{ij}$ (i=1,...,M;j=1,...,N) delivered from ith wholesaler to jth consumer which maximize the distributor's total fuzzy profit $\hat{D}$ under fuzzy constraints:

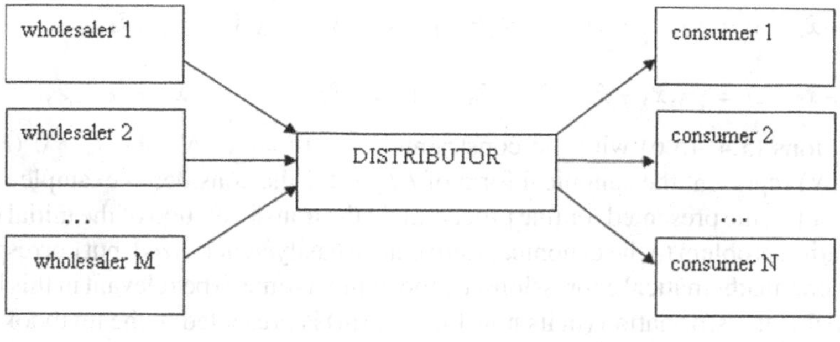

Fig. 5.1 The distributor's activity

$$\hat{D} = \sum_{i=1}^{M} \sum_{j=1}^{N} (\hat{z}_{ij} \cdot \hat{x}_{ij}) \rightarrow \max, \tag{5.1}$$

$$\sum_{j=1}^{N} \hat{x}_{ij} \leq a_i \; (i = 1..M), \; \sum_{i=1}^{M} \hat{x}_{ij} \leq b_j \; (j = 1..N), \tag{5.2}$$

$$\sum_{j=1}^{N} \hat{x}_{ij} \geq \hat{p}_i \; (i = 1..M), \; \sum_{i=1}^{M} \hat{x}_{ij} \geq \hat{q}_j \; (j = 1..N). \tag{5.3}$$

In the above model, only the parameters a_i and b_j are real valued as they represent the maximal quantities of goods proposed by wholesalers and the maximal good requirements of consumers that in common practice usually can not be negotiated.

To transform the model (5.1)-(5.3) into its canonical form, we substitute the two-index representation of this model for the single-index one.

To illustrate this routine procedure, let us consider the case of N=M=2. Then introducing $\hat{x}_1 = \hat{x}_{11}$, $\hat{x}_2 = \hat{x}_{12}$, $\hat{x}_3 = \hat{x}_{21}$, $\hat{x}_4 - \hat{x}_{22}$ and $\hat{z}_1 = \hat{z}_{11}$, $\hat{z}_2 = \hat{z}_{12}$, $\hat{z}_3 = \hat{z}_{21}$, $\hat{z}_4 = \hat{z}_{22}$ we rewrite (5.1) as follows:

$$\hat{D} = \sum_{i=1}^{f} \hat{z}_i \cdot \hat{x}_i \rightarrow \max, \tag{5.4}$$

where in our case $f = M \cdot N$=4.

Introducing the variable $\hat{g}_i$ (i=1 to $2N + 2M$) such that $g_1 = a_1$, $g_2 = a_2$, $g_3 = b_1$, $g_4 = b_2$, $\hat{g}_5 = \hat{p}_1$, $\hat{g}_6 = \hat{p}_2$, $\hat{g}_7 = \hat{q}_1$, $\hat{g}_8 = \hat{q}_2$,
from (5.2) and (5.3) we get $\hat{x}_1 + \hat{x}_2 \leq g_1$, $\hat{x}_3 + \hat{x}_4 \leq g_2$, $\hat{x}_1 + \hat{x}_3 \leq g_3$,$\hat{x}_2 + \hat{x}_4 \leq g_4$, $\hat{x}_1 + \hat{x}_2 \geq \hat{g}_5$, $\hat{x}_3 + \hat{x}_4 \geq \hat{g}_6$, $\hat{x}_1 + \hat{x}_3 \geq \hat{g}_7$, $\hat{x}_2 + \hat{x}_4 \geq \hat{g}_8$.

The simplex algorithm requires the linear programming problem to be in augmented form, so that the inequalities are replaced by equalities [15]. Therefore, the next step is the presentation of the above inequalities in the canonic form. Introducing the so-called slack variables $\hat{s}_i$, i=1 to r ($r = 2M + 2N$), we transform these inequalities to the set of equalities in the canonical form:

$$\hat{x}_1+\hat{x}_2+\hat{s}_1=g_1, \hat{x}_3+\hat{x}_4+\hat{s}_2=g_2, \hat{x}_1+\hat{x}_3+\hat{s}_3=g_3, \hat{x}_2+\hat{x}_4+\hat{s}_4=g_4, \quad (5.5)$$

$$\hat{x}_1+\hat{x}_2-\hat{s}_5=\hat{g}_5, \hat{x}_3+\hat{x}_4-\hat{s}_6=\hat{g}_6, \hat{x}_1+\hat{x}_3-\hat{s}_7=\hat{g}_7, \hat{x}_2+\hat{x}_4-\hat{s}_8=\hat{g}_8. \quad (5.6)$$

Expressions (5.4)-(5.6) with the constraints $\hat{x}_i \geq 0$ (i=1 to $N \cdot M$), $\hat{s}_i \geq 0$ (i=1 to $2M+2N$) represent the canonical form of $FLPP$ for the considered example.

Of course, the presented routine procedure of the transformation of the initial fuzzy distribution problem to the canonical form can be easily generalized, but corresponding general mathematical expressions are too cumbersome to be relevant in this book. Indeed, this transformation (in its non-fuzzy form) is presented in the textbooks. All the following steps of the developed approach are similar to those of standard simplex method with only one difference: usual arithmetical operations and the operation of comparison are replaced by the corresponding operations on fuzzy values.

To implement this approach, the methods of object -oriented programming have been used. To do that, the special class "Fuzzy value" was built using the language C++ . This class contains the overloaded operators representing the operations on fuzzy values. Since all the fuzzy parameters and variables are presented by objects of this class, the algebraic structure of "fuzzy extended simplex method" formally inherits the algorithm of the usual simplex method.

Let us consider an illustrative example.

We first solve the simplest distribution problem with N=2, M=2 and real valued parameters: z_{11}=3, z_{12}=5, z_{21}=6, z_{22}=4, a_1=20, a_2=30, b_1=25, b_2=25, p_1=20, p_2=25, q_1=20, q_2=25.

Its solution (in the two-index form) is x_{11}=0, x_{12}=20, x_{21}=25, x_{22}=5 and for the total optimal profit we have obtained D=270.

The second step is the fuzzy extension of the above example such that all fuzzy parameters are centered around the corresponding real valued parameters of this distribution problem. To perform the fuzzy extension, we have used trapezoidal fuzzy values so that the fuzzy parameters were presented by real values and quadruples as follows:

a_1=20, a_2=30, b_1=25, b_2=25, $\hat{z}_{11}$=[2,2.5,3.5,4], $\hat{z}_{12}$=[4,4.5,5.5,6],
$\hat{z}_{21}$ =[5,5.5,6.5,6], $\hat{z}_{22}$=[3,3.5,4.5,5], $\hat{p}_1$ =[19,19.5,20.5,21], $\hat{p}_2$ =[24,24.5,25.5,26],
$\hat{q}_1$=[19,19.5,20.5,21], $\hat{q}_2$=[24,24.5,25.5,26].

Using the developed algorithm for the direct fuzzy extension of the simplex method, the following results (in the two-index form) were obtained:
$\hat{x}_{11}$=[0,0,0,0], $\hat{x}_{12}$ =[10,14,26,30], $\hat{x}_{21}$ =[12,17,28,33], $\hat{x}_{22}$ =[2,3.5,6.5,8],
$\hat{D}$=[220,240,300,320].

It is easy to see that these results are centered around those obtained using real valued version of the considered distribution problem. This may be treated as an evidence in favor of the method's correctness. On the other hand, we can see that the relative widths of the results $\hat{x}_{11}$, $\hat{x}_{12}$, $\hat{x}_{21}$, $\hat{x}_{22}$ are greater than those of the initial fuzzy parameters. This phenomenon is well known in interval analysis as the "access width effect" and will be analyzed in the following subsection on the base of comparison of the fuzzy solution with that obtained using Monte-Carlo method.

5.1.3 Numerical Studies

To perform the proposed method, we compare the results of *FDP* solution with those obtained from (5.1)-(5.3) when all the uncertain parameters are considered as normally distributed random values.

The standard Monte-Carlo procedure was used, i.e., for each set of randomly chosen real valued parameters the real valued solution of problem (5.1)-(5.3) was obtained. Finally, repeating this procedure the results were presented in the form of probability density functions of optimal x_{ij} and D.

To make the results obtained using the fuzzy and probability approaches comparable, the special simple method for the transformation of probability density distributions into fuzzy values without drastic loss of useful information was used. This method makes it possible to achieve the comparability of uncertain initial data in the fuzzy and the random cases.

For the sake of simplicity, we have used the simplest normally distributed probability density functions, exhaustively represented be their averages m and standard deviations σ. This method consist of two steps.

At the first step, using initial probability density function $f(x)$, the cumulative distribution function $F(x)$ is obtained as follows: $F(x) = \int_{-\infty}^{x} f(x)dx$.

At the second step, the function $F(x)$ is used to obtain a trapezoidal fuzzy number. We ask the decision-makers (experts) for the four values $F(x_i)$, i=1,...,4, which define the mapping of $F(x)$ on X in such a way that they provide the bottom and upper α-cuts of the trapezoidal fuzzy number.

Let us consider the example presented in Fig.5.2. Since the probability that x lies in the interval $[a,b]$ is equal to $F(b) - F(a)$, the intervals [95, 105] and [78, 120] (in the considered example) correspond to the 30% and 90% confidence intervals. Obviously, we place these intervals in such a way that they are centered around the center of the cumulative distribution $F(x)$.

Therefore, the resulting trapezoidal fuzzy number μ is presented by the quadruple as follows μ= [78,95,105,120].

Is easy to see that accuracy of the proposed transformation depends only on the expert's subjective opinion about suitability and correctness of chosen upper and button confidence intervals. Of course, this subjectivity is the source of additional uncertainty. Nevertheless, taking into account that the transformation of a probability density function into a fuzzy value leads inevitable to the loss of some information, we can expect that the choice of 30% and 90% confidence intervals will provide at least the satisfactory results of transformation.

Let us consider the example of the distribution problem (5.1)-(5.3) with N=3, M=3. To compare the results of fuzzy programming with those obtained when using the Monte-Carlo method, all the uncertain parameters were previously represented by normally distributed probability density functions. As the parameters a_i and b_j represent the maximal quantities of goods proposed by wholesalers and the maximal good requirements of consumers and in common practice usually can not be negotiated, they were presented by the real constant values a_1=460, a_2=460, a_3=610,

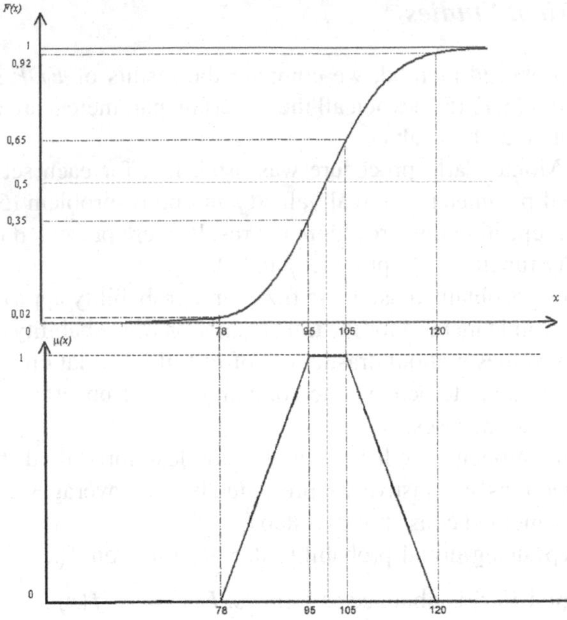

Fig. 5.2 The transformation of cumulative function into a fuzzy value

b_1=410, b_2=510, b_3=610. The other parameters were represented by normally distributed probability density functions with the following averages:

p_1 =440, p_2=440, p_3=590, q_1=390, q_2=490, q_3=590, z_{11}=300, z_{12} =480, z_{13}=490, z_{21} =400, z_{22}=580, z_{23}=290, z_{31}=300, z_{32}=380, z_{33} =600.

For simplicity, all the standard deviations σ were equal to 10. Using described above method for the transformation of probability distribution function into a fuzzy value, the following trapezoidal fuzzy parameters of the problem (5.1)-(5.3) have been obtained:

$\hat{p}_1$=[417,435,444,459], $\hat{p}_2$=[417,435,444,459], $\hat{p}_3$=[567,585,594,609],
$\hat{q}_1$=[367,385,394,409], $\hat{q}_2$=[467,485,494,509], $\hat{q}_3$ =[567,585,594,609],
$\hat{z}_{11}$=[277,295,304,319], $\hat{z}_{12}$=[457,475,484,499], $\hat{z}_{12}$=[467,485,494,509],
$\hat{z}_{21}$=[377,395,304,319], $\hat{z}_{22}$=[561,579,588,603], $\hat{z}_{23}$=[272,290,299,314],
$\hat{z}_{31}$=[377,395,304,319], $\hat{z}_{32}$=[561,579,588,603], $\hat{z}_{33}$=[272,290,299,314].

Some results we have obtained using the fuzzy optimization method and the Monte-Carlo method (usual linear programming with real valued, but random parameters) are presented in Fig.5.3-Fig.5.6. All the probability density functions in Fig.5.3-Fig.5.5 were obtained using Monte-Carlo method with 1 000 000 random steps. It is easy to see that the Monte-Carlo method sometimes provides two-extreme resulting probability density functions. Obviously, it is difficult to interpret these results,

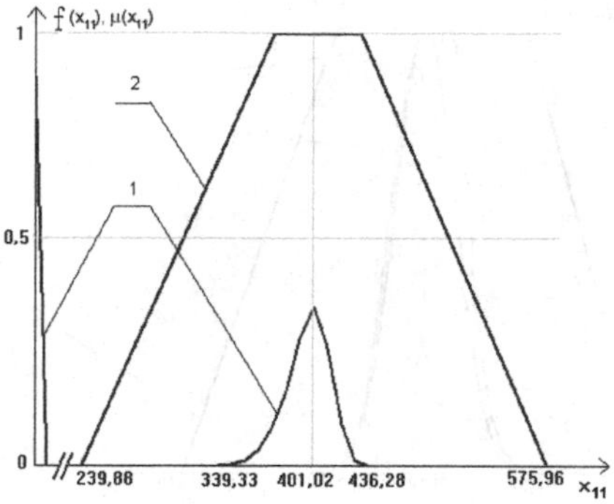

Fig. 5.3 The probability density function f -(1) and the fuzzy value μ -(2) for optimal x_{11}.

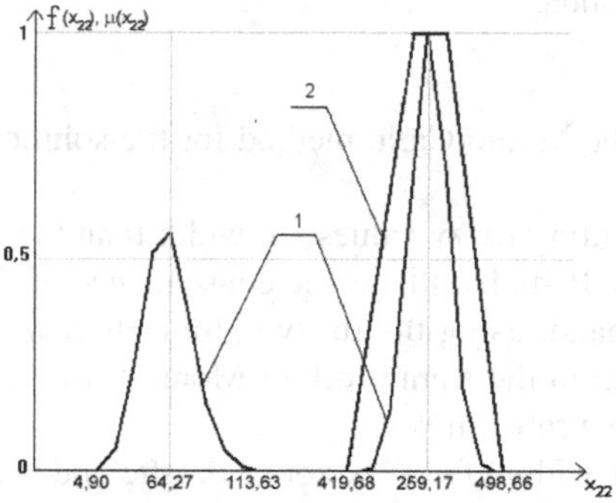

Fig. 5.4 The probability density function f -(1) and the fuzzy value μ -(2) for optimal x_{22}.

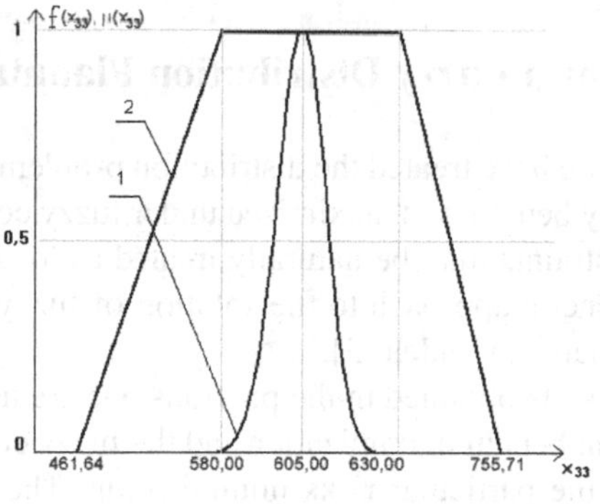

Fig. 5.5 The probability density function f -(1) and the fuzzy value μ -(2) for optimal x_{33}.

whereas when using the fuzzy optimization we have no such a problem since the results are always presented by trapezoidal fuzzy values. It is worth noting that to obtain the smooth resulting probability density function of the benefit D, too much of random steps (about 1 000 000) are needed (see Fig.5.6). Therefore, it seems

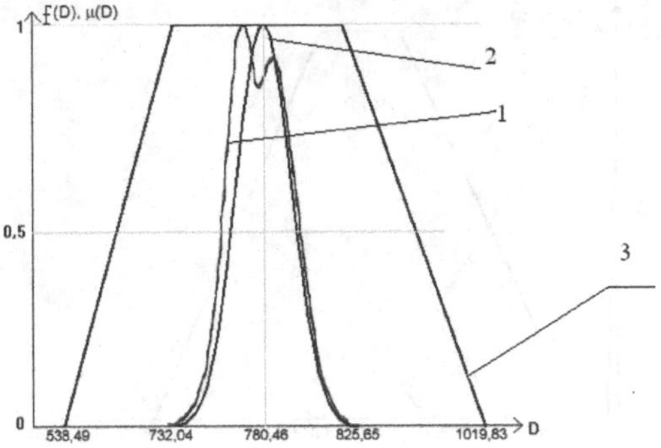

Fig. 5.6 The probability density function f and the fuzzy number μ for optimized benefit D: 1 - Monte-Carlo method with 10 000 random steps, 2 - Monte-Carlo method with 1 000 000 random steps, 3 - fuzzy solution.

rather senseless to use the Monte-Carlo method for the solution of the distribution problem in practice.

It is seen that the resulting fuzzy values are wider than the corresponding probability density functions. Partially, this is the consequence of the "access width effect", but on the other hand, using the fuzzy optimization we implicitly take into account the events which in the framework of Monte-Carlo method are treated as those with extremely low probability.

The observed "access width effect" is not so drastic and it does not prevent the use of the developed direct fuzzy extension of the simplex method for the solution of the fuzzy distribution problem.

5.2 Multiple Criteria Fuzzy Distribution Planning Problem

In the previous section, we have treated the distribution problem as a single criterion task. Only the total fuzzy benefit was maximized under fuzzy constraints. Nevertheless some of these constraints may be naturally treated as local criteria. This is in compliance with the general approach to the solution of fuzzy optimization problems proposed by Bellman and Zadeh [3].

In this section, the results obtained in the previous one are used to formulate the local criterion of the total benefit maximization and the fuzzy constraints are treated as the local criteria of the particular risks minimization. The general criterion is formulated as an aggregation of aggregation modes using level-2 fuzzy sets. These modes are the different types of aggregations of local criteria characterizing the total benefit and risks of breach of contracts. The developed method is illustrated with the use of numerical example.

The mathematical model of the fuzzy multi-objective transportation problem ($FMOTP$) [21, 22, 38] generally can be presented as follows:

$$\min \hat{F}^k(x) = \sum_{i=1}^{m} \sum_{j=1}^{n} \hat{C}_{ij}^k \cdot x_{ij},$$

subject to

$\sum_{j=1}^{n} x_{ij} = \hat{a}_i$, i=1 to m, $\sum_{i=1}^{m} x_{ij} = \hat{b}_j$, j=1 to n, $x_{ij} \geq 0$, i=1 to m, j=1 to n,

where $\hat{F}^k(x) = \{\hat{F}^1(x), \hat{F}^2(x), ..., \hat{F}^K(x)\}$ is a fuzzy vector of k objective functions, the superscript on both $\hat{F}^k(x)$ and fuzzy penalties $\hat{C}_{ij}^k$ are used to identify the number of objective functions (k = 1 to K), m and n are the numbers of fuzzy sources and destinations, respectively. In [21] and [38], the fuzzy approaches that make it possible to get the compromise solution of $FMOTP$ were developed and studied. Some shortcomings of using fuzzy set theory in solving such $MOTP$ were noted. In [21], it is shown that the use of fuzzy programming for solving $MOTP$ changes the standard form of the well known transportation problem. In addition, Lushu and Lai [38] proved that the use of *min*-operator does not guarantee an efficient solution. The other researches in this realm are presented in [4, 5, 6, 8, 10, 11, 20, 23, 34]. It should be emphasized here that in all cases the authors considered only the linear form of $FMOTP$.

In the current section, we propose a new approach to the formulation of multiple criteria fuzzy distribution problem based on the treating of fuzzy constraints as the local criteria of risks minimization. To aggregate these local criteria with the criterion of the total profit maximization, different aggregation modes are used. Finally, these modes are aggregated using the level 2 fuzzy sets. Therefore, the proposed general approach to the solution of multiple criteria fuzzy distribution problem is realized in two stages. At the first of them, the direct numerical method for solving the single-criterion fuzzy distribution problem is used. Using obtained results, the local criteria needed in the second stage are formulated. As the result we get the fuzzy multiple criteria nonlinear distribution problem which is solved using numerical methods.

5.2.1 The Problem Formulation

Let us consider the parameters $\hat{p}_i$, $\hat{q}_i$, in the fuzzy constraints (5.3). They represent the lower fuzzy bounds for the possible optimal quantities of goods which can be bought and sold. Hence, they can be negotiated and the choice of the real values $p_i \in \hat{p}_i$, $q_i \in \hat{q}_i$, has the strong influence on the total benefit.

Therefore, the fuzzy values $\hat{p}_i$, $\hat{q}_i$ may be treated as local criteria.

For example, let us consider $\hat{p}_i$ (the fuzzy bounds for quantities of goods which can be bought from the wholesaler). Generally, it can be presented by a trapezoidal fuzzy number (see Fig.5.7). The interval $[p_{i1}, p_{i4}]$ can be treated as the fuzzy interval of acceptable values of p_i with the membership function $\mu_i(p_i)$. On the other hand, in the interval $[p_{i1}, p_{i2}]$, the lowering of $\mu_i(p_i)$ with the decreasing of p_i may be naturally interpreted as follows: the risk that the contracts signed with the customers will be unfilled is rising with the decreasing of p_i. Similarly, the lowering of $\mu_i(p_i)$ with the increasing of p_i in the interval $[p_{i3}, p_{i4}]$ leads to the rising of the overbuying

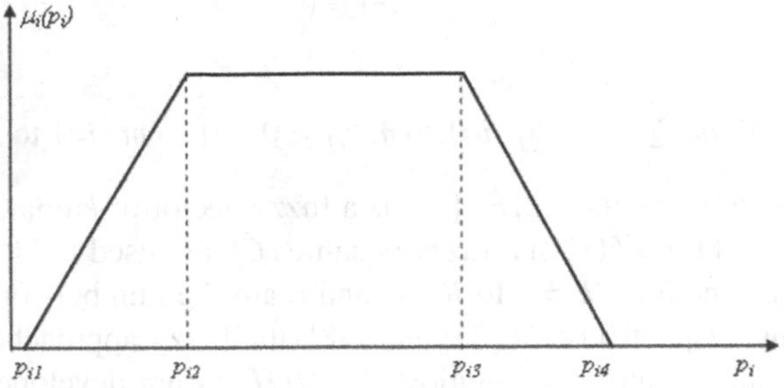

Fig. 5.7 The trapezoidal fuzzy constraint $\hat{p}_i$.

risk (the risk that the part of the bought goods could not be sold to the customers). In the interval $[p_{i2}, p_{i3}]$ we have no above risks and therefore $\mu_i(p_i) = 1$. Obviously, in a such interpretation, the function $\mu_i(p_i)$ represents the risk ranging from 0 to 1 and calculated as $1\text{-}\mu_i(p_i)$.

Using the similar reasoning, it can be shown that the membership function of q_j, i.e., $\mu_j(q_j)$ represents the corresponding risk ranging from 0 to 1 and calculated as $1\text{-}\mu_j(q_j)$.

Hence, the distribution problem can be treated as the multiple criteria optimization task including the local criteria of total benefit maximization and particular risks minimization. It is seen that the local criteria of particular risks minimization can be formulated directly using the fuzzy values $\hat{p}_i$, $\hat{q}_i$, but an explicit mathematical formulation of the total benefit maximization criterion needs an additional analysis. To formulate this criterion we can use the solution obtained in the previous stage in Section 5.1 in the framework of the direct fuzzy extension of simplex method in application to the fuzzy distribution problem. Since this approach is free of any additional restriction on the form of fuzzy parameters and variables and is based on the total benefit maximization under fuzzy constraints, the obtained optimal fuzzy benefit $\hat{D}$ can be treated as the fuzzy interval of achievable real valued benefits. Therefore, we can use the support of $\hat{D}$ to formulate the local criterion $\lambda(D)$ performing the distributor's intention to maximize the total benefit as it is shown in Fig.5.8. Of course, this criterion doesn't reflect the "possibility" to gain the benefit that implicitly performs the fuzzy benefit $\hat{D}$ obtained taking into account the constraints defined by the fuzzy parameters $\hat{p}_i$, $\hat{q}_i$, since in the multiple criteria distribution problem we use these parameters to formulate the local criteria of risk minimization. It is seen that to calculate the value of $\lambda(D)$, the real valued D are needed.

It is worthy to note that in practice, all the bought and sold good quantities p_i, q_j and optimal good quantities x_{ij} delivered from ith wholesaler to jth customer in the signed contracts should be presented in the form of real values. Hence, some simplifications of the initial fuzzy problem can be justified. Therefore, we propose to use in (5.1) instead of fuzzy $\hat{z}_{ij}$ their real valued representations z_{ij}. For example,

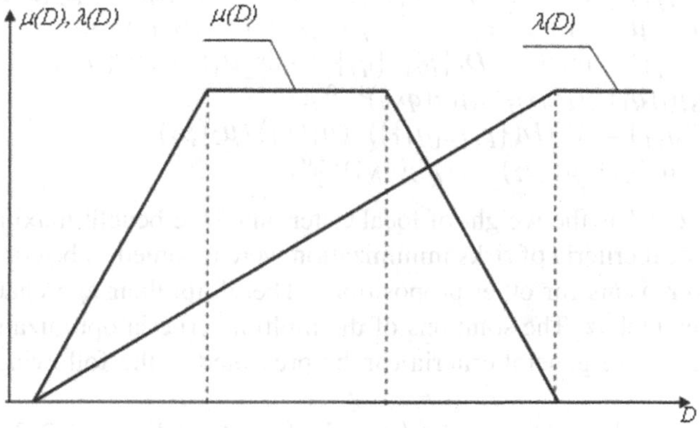

Fig. 5.8 The local criterion of total benefit maximization $\lambda(D)$.

in the case of symmetrical trapezoidal fuzzy $\hat{z}_{ij}$ (as in our examples in Section 5.1), the geometric centers of such trapezes will be used. Such a simplification allows us to obtain from (5.1)-(5.3) for the real valued $p_i \in \hat{p}_i$, $q_i \in \hat{q}_i$, i=1 to N, j=1 to M, the real valued profit $D(\{p_i\},\{q_j\})$ needed for the calculation of $\lambda(D(\{p_i\},\{q_j\}))$ on the consequent steps of the developed algorithm for the solution of multiple criteria distribution problem. Since the minimization of local risks is equivalent to the maximization of $\mu_i(p_i)$, $\mu_j(q_j)$, the solution of the multiple criteria distribution problem should be the optimal $\{p_i\}_{opt} \in \{\hat{p}_i\}$,$\{q_j\}_{opt} \in \{\hat{q}_j\}$ maximizing some general criterion that aggregates all the considered local criteria $\lambda(D(\{p_i\},\{q_j\}))$, $\mu_i(p_i)$, $\mu_j(q_j)$ taking into account their weights.

Under such conditions, the optimal good quantities x_{ij} delivered from ith wholesaler to jth customer are finally obtained from (5.1)-(5.3) as the solution of this problem for the real valued z_{ij} and optimal $\{p_i\}_{opt} \in \{\hat{p}_i\}$, $\{q_j\}_{opt} \in \{\hat{q}_j\}$, i=1 to N, j=1 to M.

5.2.2 The Solution of Multiple Criteria Fuzzy Distribution Problem Using the Aggregation of Aggregation Modes

To formulate the general criterion, we propose here to aggregate the criteria of local risks into the aggregated criterion of risk minimization and aggregate it with the local criterion of profit maximization. There are many approaches to the aggregation of local criteria proposed in the literature and the choice of an appropriate one is an application dependent problem [61]. Nevertheless, the weighted sum, Yager's [55] and multiplicative aggregation modes can be marked out as the most popular [17]. In our case, they are presented as follows:

$$F_1(\{p_i\},\{q_j\}) = \alpha \cdot \lambda(D(\{p_i\},\{q_j\})) + (1-\alpha)\cdot(\mu_1(p_1)+\mu_2(p_2)+...$$
$$... + \mu_M(p_M) + \mu_1(q_1) + \mu_2(q_2) + ... + \mu_N(q_N))/(2*(N+M)),$$
$$F_2(\{p_i\},\{q_j\}) = \min(\lambda^{\alpha}(D(\{p_i\},\{q_j\})), \min(\mu_1(p_1),\mu_2(p_2),$$
$$..., \mu_M(p_M), \mu_1(q_1), \mu_2(q_2), ..., \mu_N(q_N))^{1-\alpha}),$$
$$F_3(\{p_i\},\{q_j\}) = \lambda^{\alpha}(D(\{p_i\},\{q_j\}))\cdot(\mu_1(p_1)\cdot\mu_2(p_2)\cdot$$
$$...\cdot\mu_M(p_M)\cdot\mu_1(q_1)\cdot\mu_2(q_2)\cdot...\cdot\mu_N(q_N))^{1-\alpha},$$

where $0 \le \alpha \le 1$ is the weight of local criterion of the benefit maximization. The weights of local criteria of risks minimization were assumed to be equal ones since there are no reasons for other propositions. Therefore, their aggregation have the common weight 1-α. The solutions of the multiple criteria optimization problems based on the above general criteria can be presented in the following generalized form:

$$(\{p_i\},\{q_j\})_{k,opt} = \arg(\max(F_k(\{p_i\},\{q_j\}))), k = 1,2,3.$$

The results of theoretical analysis and practical experience make it possible to state that the most reliable aggregation mode is the Yager's type aggregation (F_2), the multiplicative mode (F_3) appears to be somewhat less reliable and, finally, the weighted sum (F_1) may be considered as unreliable and insensitive when choosing an alternative in Pareto-region [17](see also Chapter 4).

On the other hand, it is known that Yager's type aggregation sometimes does not comply with intuitive concepts of the decision maker about optimality [18]. Therefore, when dealing with a complex problem characterized by a great number of local criteria, it seems reasonable to use all relevant types of aggregation. If the results obtained using different aggregation methods are similar then we can say that that they are rather optimal ones. In the opposite case an additional analysis of local criteria and their weights should be carried out.

The natural consequence of the described problems is a rising interest in the methods for the aggregation of aggregation modes [49]. The different approaches to the aggregation of aggregation operators were proposed in [14, 18, 19, 27, 43, 45, 56, 62]. These approaches are based on the use of weighted sum, *min* operator, multiplicative aggregation modes and their combinations for the aggregation of aggregation modes. The common restriction of these approaches is that they do not allow us to aggregate all possible aggregation operators.

Therefore, here we shall use a simple, but intuitively obvious and mathematically strong approach proposed in [17, 50] which is free of this restriction. This method for the aggregation of aggregation modes is based on the level-2 fuzzy sets [26, 54, 58].

As proposed by Zadeh in [59], the level-2 fuzzy sets is such a fuzzy set, of which membership grades assigned to the elements of the universal set are ordinary fuzzy sets. The approach proposed in [17, 50] was developed for the solution of the decision making problems. Therefore, here it is adapted to the use for the solution of the multiple criteria optimization problems.

The solution of our problem can be obtained only using numerical methods. Hence, when realizing such method we shall deal with the discrete finite set of alternative $v_k = (\{p_i\},\{q_j\})_k$, k=1 to L, where L is the number of steps of an algorithm which is searching for the optimal alternative $v_{Lopt} = (\{p_i\}, \{q_j\})_{Lopt}$ in the domain $p_i \in \hat{p}_i$, $q_j \in \hat{q}_j$, i=1 to N, j=1 to M.

Therefore, let us suppose we have K different aggregation modes F_l, l=1 to K, and L alternatives $v_k = (\{p_i\}, \{q_j\})_k$, k=1 to L. Let the membership function $\mu(F_l)$, l=1 to K, represents expert's opinion about the closeness of considering aggregation operator F_l to some perfect type of aggregation, which can be treated as the "best" one or "ideal" method for aggregation. The values of such membership function may be treated as weights of aggregation modes. Then such "ideal" method F_{ideal} can be represented by its membership function and by the set of compared aggregation modes F_l in the form of following fuzzy set:

$$F_{ideal} = \left\{ \frac{\mu(F_l)}{F_l} \right\}, l = 1\ to\ K. \tag{5.7}$$

In turn, each F_l can be represented formally by the set of compared alternatives $v_k = (\{p_i\}, \{q_j\})_k$, k=1 to L, on which it is factually defined and for which the values of $F_l(v_k)$ are calculated. It is clear that the value of $F_l(v_k)$ may be treated as a degree to which the alternative v_k satisfies the aggregated criterion F_l or as an extent to which the alternative v_k belongs to a set of alternatives satisfying F_l . Then

$$F_l = \left\{ \frac{F_l(v_k)}{v_k} \right\}, k = 1\ to\ L. \tag{5.8}$$

Substituting (5.8) into (5.7) we obtain the F_{ideal} in the form of level-2 fuzzy set and using the operations on such fuzzy sets proposed in [59] we have finally:

$$F_{ideal} = \left\{ \frac{\mu_{ideal}(v_k)}{v_k} \right\}, k = 1\ to\ L, \tag{5.9}$$

$$\mu_{ideal}(v_k) = \max_l(\mu(F_l) \cdot F_l(v_k)). \tag{5.10}$$

It is clear that the best alternative can found as $v_{opt} = \arg\max_k \mu_{ideal}(v_k)$.

The well known direct random search method [53] has been adopted to develop the numerical algorithm for the solution of the above optimization problem. Of course, any other modern optimization method, e.g., genetic algorithm can be used as well. Nevertheless in [1], it is shown that "when the optimizing function is non-linear, non-differentiable and non-smooth, direct search methods are the methods of choice".

The algorithm of direct random search method has been realized as follows. On the kth step of the random searching, the alternative $v_k = (\{p_i\}, \{q_j\})_k$ has been randomly chosen in the domain $p_i \in \hat{p}_i$, $q_j \in \hat{q}_j$. For the chosen $(\{p_i\}, \{q_j\})_k$ and the real valued representations of $\hat{z}_{ij}$ from (5.1)-(5.3) the benefit D_k and the good quantities $\{x_{ij}\}_k$ are obtained. This makes it possible to calculate the value of the local criterion $\lambda(D_k)$ and the values of aggregation modes $F_l(v_k)$, l=1 to K. In our case, the weighted sum (F_1), Yager's aggregation (F_2) and multiplicative mode (F_3) have been used.

Finally, from (5.10) the value of $\mu_{ideal}(v_k)$ is calculated. If $\mu_{ideal}(v_k) > \mu_{ideal}(v_{k-1})$, we treat the kth step of random search method as the successful one. Repeating the random steps, we gradually converge to the maximal (optimal) $\mu_{ideal}(v_{opt})$.
Let us consider a numerical example.

We shall use the same example as in Section 5.1 using instead of fuzzy $\hat{z}_{ij}$ their real valued representations z_{ij}. In the considered case of symmetrical trapezoidal fuzzy $\hat{z}_{ij}$ the geometric centers of such trapezes were used.

In Table 5.1, the solution of multiple criteria fuzzy distribution problem obtained using the aggregation of aggregating modes is presented in comparison with those obtained with the use of weighted sum (F_1), Yager's (F_2) and multiplicative (F_3) aggregation modes in the described above procedure of direct random search method. According to the recommendation justified in [17], the following weights of aggregation modes were used: $\mu(F_1) = 0.05$, $\mu(F_2) = 0.7$, $\mu(F_3) = 0.25$. It is seen that the solution obtained using the aggregation of the aggregation modes can be treated as the compromise one since it lies in the range of solutions we get with the use of considered aggregation modes solely.

Table 5.1 The comparison of the obtained solutions of multiple criteria fuzzy distribution problem

	$\mu_{ideal}(v_{opt})$	$F_1(v_{opt})$	$F_2(v_{opt})$	$F_3(v_{opt})$
D_{opt}	853062	859701	853706	854259
x_{11}	421	467	421	411
x_{12}	52	44	50	54
x_{13}	8	0	4	0
x_{21}	0	0	0	0
x_{22}	516	516	518	516
x_{23}	0	0	0	0
x_{31}	0	0	0	4
x_{32}	0	0	0	0
x_{33}	662	663	666	665
p_1	476	511	476	465
p_2	518	516	518	505
p_3	666	663	666	669
q_1	421	467	421	415
q_2	568	569	568	560
q_3	670	663	670	665

5.3 Summary and Discussion

A two stage approach to the solution of multiple criteria fuzzy distribution problem is developed. At the first stage, the direct numerical method for the solution of single-criterion fuzzy distribution problem is used. The method is based on the α-cut representation of fuzzy numbers and probability estimation of the fact that given fuzzy value is greater/equal than another one. The proposed approach makes it possible to accomplish the direct fuzzy extension of usual simplex method without restrictions on the form of fuzzy variables. The results of case studies with the use of fuzzy optimization method and Monte-Carlo method (usual linear programming with real valued, but random parameters) show that the fuzzy approach have considerable advantages in comparison with Monte-Carlo method, especially from the computational point of view.

At the next stage the results obtained at the first one, are used to formulate the local criterion of the total benefit maximization and the fuzzy constraints are treated as the local criteria of the particular risks minimization. The general criterion is formulated as an aggregation of aggregation modes using level-2 fuzzy sets. These modes are the different types of aggregations of local criteria characterizing the total benefit and risks of breach of contracts. With the use of numerical example it is shown that the solution obtained using the aggregation of the aggregation modes can be treated as the compromise one since it lies in the range of solutions we get with the use of considered aggregation modes solely.

References

1. Ali, M.M., Törn, A.: Population set-based global algorithms: some modifications and numerical studies. Computers and Operations Research 31, 1703–1725 (2004)
2. Ammar, E.E., Youness, E.A.: Study on multiobjective transportation problem with fuzzy numbers. Applied Mathematics and Computation 166, 241–253 (2005)
3. Bellman, R., Zadeh, L.: Decision-making in fuzzy environment. Management Science 17, 141–164 (1970)
4. Bit, A.K., Biswal, M.P., Alam, S.S.: Fuzzy programming approach to multicriteria decision making transportation problem. Fuzzy Sets and Systems 50, 35–41 (1992)
5. Bit, A.K., Biswal, M.P., Alam, S.S.: Fuzzy programming approach to multiobjective solid transportation problem. Fuzzy Sets and Systems 57, 183–194 (1993)
6. Bit, A.K., Biswal, M.P., Alam, S.S.: An additive fuzzy programming model for multiobjective transportation problem. Fuzzy Sets and Systems 57, 13–19 (1993)
7. Buckley, J.J., Feuring, T.: Evolutionary algorithm solution to fuzzy problems: Fuzzy linear programming. Fuzzy Sets and Systems 109, 35–53 (2000)
8. Challam, G.A.: Fuzzy goal programming (FGP) approach to a stochastic transportation problem under budgetary constraints. Fuzzy Sets and Systems 66, 293–299 (1994)
9. Chanas, S.: The use of parametric programming in fuzzy linear programming. Fuzzy Sets and Systems 11, 243–251 (1983)
10. Chanas, S., Kolodziejckzy, W., Machaj, A.A.: A fuzzy approach to the transportation problem. Fuzzy Sets and Systems 13, 211–221 (1984)
11. Chanas, S., Kuchta, D.: Fuzzy integer transportation problem. Fuzzy Sets and Systems 98, 291–298 (1998)
12. Chanas, S., Kuchta, D.: Multiobjective programming in optimization of interval objective functions - A generalized approach. European Journal of Operational Research 94, 594–598 (1996)

13. Chanas, S., Kuchta, D.: A concept of the optimal solution of the transportation problem with fuzzy cost coefficients. Fuzzy Sets and Systems 3, 299–305 (1996)
14. Choi, D.Y., Oh, K.W.: Asa and its application to multi-criteria decision making. Fuzzy Sets and Systems 114, 89–102 (2000)
15. Dantzig, G.B.: Linear Programming and Extensions. Princeton University Press, Princeton (1963)
16. Das, S.K., Goswami, A., Alam, S.S.: Multiobjective transportation problem with interval cost, source and destination parameters. European Journal of Operational Research 117, 100–112 (1999)
17. Dimova, L., Sevastjanov, P., Sevastjanov, D.: MCDM in a Fuzzy Setting: Investment Projects Assessment Application. Int. Journal of Production Economy 100, 10–29 (2006)
18. Dubois, D., Koenig, J.L.: Social choice axioms for fuzzy set aggregation. Fuzzy Sets and Systems 43, 257–274 (1991)
19. Dyckhoff, H.: Basic concepts for theory of evaluation: hierarchical aggregation via autodistributive connectives in fuzzy set theory. European Journal of Operation Research 20, 221–233 (1985)
20. Ehrgott, M., Verma, R.A.: Note on solving multicriteria transportation-location problems by fuzzy programming. Asia-Pacific Operational Research 18, 149–164 (2001)
21. El-Wahed, W.F.A.: A multi-objective transportation problem under fuzziness. Fuzzy Sets and Systems 117, 27–33 (2001)
22. El-Wahed, W.F.A., Lee, S.M.: Interactive fuzzy goal programming for multi-objective transportation problems. Omega 34, 158–166 (2006)
23. El-Wahed, W.F.A., Abo-Sinna, M.A.: A hybrid fuzzy-goal programming approach to multiple objective decision making problems. Fuzzy Sets and Systems 119, 71–78 (2001)
24. Galperin, E.A., Ekel, P.Y.: Synthetic Realization Approach to Fuzzy Global optimization via Gamma Algorithm. Mathematical and Computer Modelling 41, 1457–1468 (2005)
25. Ganesan, K., Veeramani, P.: Fuzzy linear programming with trapezoidal fuzzy numbers. Ann. Oper. Res. 143, 305–315 (2006)
26. Gottwald, S.: Set theory for fuzzy sets of higher level. Fuzzy Sets and Systems 2, 125–151 (1979)
27. Hauke, W.: Using Yager's t-norms for aggregation of fuzzy interval. Fuzzy Sets and Systems 101, 59–65 (1999)
28. Isermann, H.: The enumeration of all efficient solution for a linear multiple-objective transportation problem. Naval Research Logistics Quarterly 26, 123–139 (1979)
29. Iskander, M.G.: A computational comparison between two evaluation criteria in fuzzy multiobjective linear programs using possibility programming. Computers and Mathematics with Applications 55, 2506–2511 (2008)
30. Islam, S., Roy, T.K.: A new fuzzy multi-objective programming: Entropy based geometric programming and its application of transportation problems. European Journal of Operational Research 173, 387–404 (2006)
31. Jiménez, M., Arenas, M., Bilbao, A., Rodríguez, M.V.: Linear programming with fuzzy parameters: An interactive method resolution. European Journal of Operational Research 177, 1599–1609 (2007)
32. Kuchta, D.: A generalisation of an algorithm solving the fuzzy multiple choice knapsack problem. Fuzzy Sets and Systems 127, 131–140 (2002)
33. Kuchta, D.: A generalisation of a solution concept for the linear programming problem with interval coefficients. Operations Research and Decisions 4, 115–123 (2003)

34. Lai, Y., Hwang, C.: Fuzzy multiple objective decisions making: methods and applications. Springer, Berlin (1996)
35. Li, G., Guo, R.: Comments on Formulation of fuzzy linear programming problems as four-objective constrained optimization problems. Applied Mathematics and Computation 186, 941–944 (2007)
36. Liu, S.-T.: Fuzzy total transportation cost measures for fuzzy solid transportation problem. Applied Mathematics and Computation 174, 927–941 (2006)
37. Lotfi, F.H., Allahviranloo, T., Jondabeh, M.A., Alizadeh, L.: Solving a full fuzzy linear programming using lexicography method and fuzzy approximate solution. Applied Mathematical Modelling 33, 3151–3156 (2009)
38. Lushu, L., Lai, K.K.: A fuzzy approach to the multiobjective transportation problem. Computers and Operational Research 27, 43–57 (2000)
39. Maeda, T.: On characterization of fuzzy vectors and its applications to fuzzy mathematical programming problems. Fuzzy Sets and Systems 159, 3333–3346 (2008)
40. Mahdavi-Amiri, N., Nasseri, S.H.: Duality results and a dual simplex method for linear programming problems with trapezoidal fuzzy variables. Fuzzy Sets and Systems 158, 1961–1978 (2007)
41. Maleki, H.R.: Ranking functions and their applications to fuzzy linear programming. Far East J. Math. Sci (FJMS) 4, 283–301 (2002)
42. Maleki, H.R., Tata, M., Mashinchi, M.: Linear programming with fuzzy variables. Fuzzy Sets and Systems 109, 21–33 (2000)
43. Migdalas, A., Pardalos, P.M.: Editorial: hierarchical and bilevel programming. Journal of Global Optimization 8, 209–215 (1996)
44. Mishmast, N.H., Maleki, H.R., Mashinchi, M.: Multiobjective linear programming with fuzzy variables. Far East J. Math. Sci (FJMS) 5, 155–172 (2002)
45. Mitra, G.: Mathematical Models for Decision Support. Springer, Berlin (1988)
46. Ramík, J.: Duality in fuzzy linear programming with possibility and necessity relations. Fuzzy Sets and Systems 157, 1283–1302 (2006)
47. Ringuest, J.L., Rinks, D.B.: Interactive solutions for the linear multiobjective transportation problem. European Journal of Operational Research 32, 96–106 (1987)
48. Rommelfanger, H.: A general concept for solving linear multicriteria programming problems with crisp, fuzzy or stochastic values. Fuzzy Sets and Systems 158, 1892–1904 (2007)
49. Roubens, M.: Fuzzy sets and decision analysis. Fuzzy Sets and Systems 90, 199–206 (1997)
50. Sevastjanov, P., Figat, P.: Aggregation of aggregating modes in MCDM, Synthesis of Type 2 and Level 2 fuzzy sets. Omega 35, 505–523 (2007)
51. Steuer, R.E.: Algorithm for linear programming problems with interval objective function co-efficient. Mathematics of Operations Research 6, 333–348 (1981)
52. Tong, S.: Interval number and fuzzy number linear programming. Fuzzy Sets and Systems 66, 301–306 (1994)
53. Törn, A., Žilinskas, A.: Global optimization. Springer, Berlin (1989)
54. Tre, G., Caluwe, R.: Level-2 fuzzy sets and their usefulness in object-oriented database modeling. Fuzzy Sets and Systems 140, 29–49 (2003)
55. Yager, R.: Multiple objective decision-making using fuzzy sets. International Journal of Man-Machine Studies 9, 375–382 (1979)
56. Yager, R.: On ordered weighted averaging aggregation operators in multicriteria decision making. IEEE Transactions on Systems Man and Cybernetics 18, 183–190 (1988)

57. Yang, L., Liu, L.: Fuzzy fixed charge solid transportation problem and algorithm. Applied Soft Computing 7, 879–889 (2007)
58. Zadeh, L.A.: Quantitative fuzzy semantics. Information Sciences 3, 177–200 (1971)
59. Zadeh, L.A.: Fuzzy logic and its application to approximate reasoning. Information Processing 74, 591–594 (1974)
60. Zimmermann, H.J.: Fuzzy programming and linear programming with several objective functions. Fuzzy Sets and Systems 1, 45–55 (1978)
61. Zimmermann, H.J., Zysno, P.: Latest connectives in human decision making. Fuzzy Sets Systems 4, 37–51 (1980)
62. Zimmermann, H.J., Zysno, P.: Decision and evaluations by hierarchical aggregation of information. Fuzzy Sets and Systems 104, 243–260 (1983)

Chapter 6
The Synthesis of Fuzzy Logic and DST in Stock Trading Decision Support Systems

Modern computerized stock trading systems (mechanical trading systems) are based on the simulation of the decision making process and generate advice for traders to buy or sell stocks or other financial tools taking into account the price history, technical analysis indicators, accepted rules of trading and so on. There are many approaches to building stock trading systems proposed in the literature. The applications of the methods of soft computing in this field of researches are analysed in Chapter 2. It is noted that the source of many failures when building really profitable stock trading systems is the ignoring of human factor. It was recognized in [32], after obtaining a negative result that "The trading system loses money and gets a negative Sharpe Ratio. We believe that if expert's experience is available, it will generate more promising results".

We can say that the last statement is the pivotal idea on which the methods presented in this chapter are based. We believe that the wisdom accumulated by generations of traders in the form of well-known trading rules of technical analysis are an adequate base on which it is possible to build optimal fuzzy expert systems for stock trading. Our starting point was the paper [9], where the authors presented an expert system based on the fuzzy logic representation of technical analysis trading rules which are usually used by traders for decision making. Since technical analysis provides indicators used by experts to predict stock price movements, the method proposed in [9] maps these indicators into new inputs that can be used in a fuzzy logic system. This chapter generalizes our experience in building stock trading systems. Some results we have obtained are partially presented in [10, 30].

Here we present and compare three different expert systems for stock trading based on the synthesis of fuzzy logic and technical analysis. The first one is the special adaptation of classical Mamdani's approach. Another method is based on the so-called "logic-motivated fuzzy logic operators" [29]. The third system that will be presented is based on the synthesis of fuzzy logic and the Dempster-Shafer theory.

L. Dymowa: Soft Computing in Economics and Finance, ISRL 6, pp. 207–240.
springerlink.com

6.1 Stock Trading Systems Based on Conventional Fuzzy Logic

In this section, two different expert systems for stock trading based on the synthesis of fuzzy logic and technical analysis are presented and compared. The first one is, in essence, a special adaptation of classical Mamdani's approach. Another method is based on the recognition of the fact that Mamdani's approach was developed for fuzzy logic controllers, not for solving decision making problems. Therefore, the so-called "logic-motivated fuzzy logic operators" based on the other mathematical representation of t-norm and Yager's implication rule were used in the expert system. The efficiency of such expert systems is naturally measured by comparing system outputs versus stock price movement. The preliminary results obtained using the real data from extremely different markets (NYSE and Warsaw Stock Exchange) allow us to say that optimized expert system based on the "logic-motivated fuzzy logic operators" framework provides substantially greater benefits and is more reliable than the expert system based on the Mamdani's approach. Moreover, the developed optimal investment strategy makes it possible to get a profit even in conditions of trading in the direction opposite to the downtrend of stock prices.

6.1.1 Modern Approaches to Building Stock Trading Systems

During the last two decades, powerful mathematical methods have been employed to find a way to predict stock prices accurately, but they have produced less than successful results in practice [13]. In [19], it was shown that numerous studies addressing stock price prediction have generally employed time series analysis techniques [17] and multiple regression models. Recently, artificial intelligence techniques such as artificial neural networks (*ANNs*) and genetic algorithms (*GAs*) have been applied in this area. However, the above-mentioned concern still exists [2, 20, 22].

In [18], *ANNs* had some limitations when learning the patterns since stock price data have tremendous noise and complex dimensionality. In [19], it is pointed out that numerous factors such as macro-economical and political events may have a major influence on stock prices. As noted in [39], "in recent times interest has turned to the use of neural networks for this task, but had less than successful results". Therefore, the growing interest among researchers in the application of rough sets theory [24] for trading rules extraction can be observed [32, 39].

In [12], neuro-fuzzy (*NF*) systems were compared with rough sets (*RS*) based methods applied in medicine and finance. It was shown that the decisions generated by *NF* systems are not transparent. In *RS* based methods, the knowledge discovered during the process of classifier generation is represented in a transparent form and so provides a better understanding of the problem under consideration and a better explanation of the circumstances behind decisions. *RS* methods "...ensure relatively high accuracy using smallest numbers of rules. It is related to their design philosophy to take into account only key relationships between attributes and decisions that are specific for the objects from the whole universe" [12].

As it was pointed out in [36], "One of the most important problems on rule induction methods is that they cannot extract rules, which plausibly represent expert's decision processes. On one hand, rule induction methods induce probabilistic rules, the description length of which is too short, compared with the expert's rules. On the other hand, construction of Bayesian networks generates too lengthy rules".

Thus, the rough sets theory is not a universal remedy for the real-world problems of stock trading. It was recognized in [32], after obtaining a negative result that "The trading system loses money and gets a negative Sharpe Ratio. We believe that if expert's experience is available, it will generate more promising results".

We can say that the last statement was the pivotal idea on which the approaches presented in this chapter are based. We believe that the wisdom accumulated by generations of traders in the form of well-known trading rules of technical analysis are an adequate base on which it is possible to build optimal fuzzy expert systems for stock trading. Our starting point was the paper [9], where the authors presented an expert system based on the fuzzy logic representation of technical analysis trading rules which are usually used by traders for decision making. Since technical analysis provides indicators used by experts to predict stock price movements, the method proposed in [9] maps these indicators into new inputs that can be used in the fuzzy logic system. Past sequences (history) of stock prices are used to calculate these indicators. This method relies on fuzzy logic to generate a decision when certain price movements or certain price formations occur. The main idea is to use technical indicators and fuzzy logic to create a new fuzzy indicator that recommends the buying or selling of a stock. This method avoids over-reliance on quantitative data. It consists of a few inputs (e.g., rate of change (*ROC*), stochastic and support-resistance indicators), one output variable (e.g., level of confidence to take a certain action), and a few fuzzy rules expressing the relationships between financial indicators.

To build the stock trading expert system, the authors of [9] directly used the classical Mamdani's multi-input-single-output [21] general form of fuzzy rules. However, it should be noted that the applied Mamdani's approach was developed for the use in automatic control systems, not for solving decision making problems. As a consequence, it often produces somewhat artificial non-transparent systems of fuzzy rules when dealing with the application of human reasoning. In [29], a new system of the so-called "logic-motivated fuzzy logic operators" ($LMFL$) is proposed to accommodate better the specificity of human reasoning in decision making processes. This system is based on the modified mathematical representation of t-norm and Yager's implication rule [37]. In our opinion, a proper choice of the basic concept of a fuzzy logic system plays a pivotal role in the building of stock trading systems. In [30], we demonstrated that $LMFL$ approach can be split into two different frameworks for expert systems building, which are called in this chapter $Yager_{max}$ and $Yager_{ave}$, respectively. Using data from the Warsaw Stock Exchange we have shown that $Yager_{ave}$ method yields significantly better results than $Yager_{max}$ and Mamdani's approaches. It allows us (after optimization during a teaching period) to make high profits even in the risky case of trading long positions against a dominating downtrend, whereas in [9] only some reduction in the total loss is indicated in such cases.

6.1.2 Technical Analysis Indicators and Their Fuzzy Representation

To present the merits of our optimization-based approach more transparently, we have used nearly the same set of technical analysis indicators as in [9]. However, it should be noted that we have used only triangular membership functions in the fuzzy logic systems as there is no need to consider more complicated membership functions in the technical analysis setting.

Firstly, the historical data $R(\tau)$ representing the closing prices of a stock at consequent moments τ were used to obtain a set of technical indicators (inputs). As in [9], we shall use the notation $\tau = nT$, where T is the employed time period (e.g., T=30 min, T=1 hour, T=1 day ...), n is the number of consequent time periods. The closing price and the following five well known technical indicators (see [1]) were used in the analysis:

- Rate of change momentum indicator $ROC(nT) = R(nT) - R((n-r)T)$, where r is the depth of analysis ($n > r$). It is worth noting here that r plays an important role in further analysis as it is one of the model's parameters which is optimized.
- Stochastic indicators

$$\%K(nT) = \left(\frac{R(nT) - R_{\min}(nT)}{R_{\max}(nT) - R_{\min}(nT)}\right) \cdot 100,$$

where $R_{min}(nT)$=min$(R(nT), R((n-1)T), ..., R((n-r)T))$,

$R_{max}(nT)$=max$(R(nT), R((n-1)T), ..., R((n-r)T))$.

$$\%D(nT) = \sum_{(n=3)}^{n} \frac{\%K(nT)}{3}, \quad n \geq 3.$$

- Support level SL=$Avg(nT) - 2\sigma(nT)$,
- Resistance level RL=$Avg(nT) + 2\sigma(nT)$,

where $\sigma(nT) = \sqrt{\frac{\sum_{n-g}^{n} (R(nT)-Avg(nT))^2}{g}}$, $Avg(nT) = \frac{\sum_{n-g}^{n} R(nT)}{g}$ is the so-called moving average, g is the number of days for averaging.

In line with the conventional technical analysis practice [1], we assume g=20.

The next step is to transform the initial technical indicators into the modified set of seven indicators which can be directly used in the fuzzy logic system [9]:

$$Y_1 = Y_{ROC} = \frac{R(nT)-R((n-r)T)}{R((n-30)T)},$$

$$Y_2 = Y_1((n-2)T) - Y_1(nT) = Y_{ROC}((n-2)T) - Y_{ROC}(nT), n \geq 2,$$

$$Y_3 = \%D(nT), Y_4 = \%K(nT) - Y_3(nT), Y_5 = Avg(nT) + 2\sigma - R(nT),$$

$$Y_6 = R(nT) - (Avg(nT) - 2\sigma), Y_7 = R(nT) - Avg(nT).$$

These indicators represent the signals that a trader can take into account in the decision making process. Nevertheless, the trader's attitude to the values of these indicators is usually expressed in linguistic (fuzzy) form, for example: If Y_1 is large, in other words, if the *ROC* indicator is large, then the price is likely to rise. If Y_6 is large, in other words, if the price is close to the resistance level, then the price is likely to fall.

Fuzzy logic provides a method for quantifying such fuzzy concepts. This is achieved using a membership function, which is a mapping from the domain of the input value to a real value ranging from 0 to 1. This mapping represents the degree of membership to a class defined by the user for a given application. For the technical analysis application, four classes were selected [9]- *small*, *medium*, *big* and *large* - to represent the four levels of quantification for each input value range. Numerous methods for representing membership functions are proposed in the literature. In [9], Gaussian type membership functions were used. In this chapter we shall use only triangular membership functions in fuzzy logic systems since there is no need to introduce more complicated membership functions in the technical analysis setting. Thus, all the results presented in this chapter were obtained using triangular membership functions shown in Fig. 6.1. As all these membership functions were presented as triangles of the same width of supports (see Fig. 6.1), only the minimal $Inf(Y_j)$ and maximal $Sup(Y_j)$ values of used indicators $Y_1 - Y_7$ need to be defined. Of course, they can be treated as the adaptive parameters of a system, which should be found when teaching (optimizing) the system using teaching time series. However, for the sake of simplicity, we have used the minimal $Inf(Y_j)$ and maximal $Sup(Y_j)$ values of Y_j on the considered teaching time series. In our case, such an approach seems to be sufficiently justified. Moreover, as we shall compare several different stock trading simulating systems, such an approach allows us to compare and contrast them on the equal footing since the values of $Inf(Y_j)$ and $Sup(Y_j)$ do not depend on the simulating systems used.

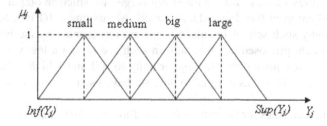

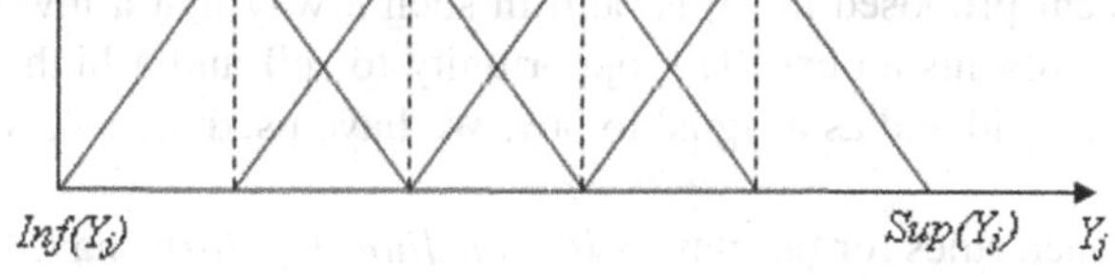

Fig. 6.1 Triangular membership functions: $Inf(Y_j), Sup(Y_j)$ are minimal and maximal values of the input Y_j in the considered time series

A more general representation of the triangular membership functions in our case is:

$$\mu_{jv} = \begin{cases} 0, & Y_j \leq a_{jv} \\ \frac{Y_j - a_{jv}}{b_{jv} - a_{jv}}, & a_{jv} < Y_j \leq b_{jv} \\ \frac{c_{jv} - Y_j}{c_{jv} - b_{jv}}, & b_{jv} < Y_j < c_{jv} \\ 0, & Y_j \geq c_{jv} \end{cases}$$

where j is the number of the fuzzy variable, v is the number of used classes or linguistic terms $(small, medium, big, large)$.

Of course, this approach offers many opportunities to tune the fuzzy logic system via an appropriate (optimal) selection of the values of the parameters a_{jv}, b_{jv}, c_{jv}, but this is one of many directions for future elaboration of our expert systems and is outside the scope of this book.

The next step is the formulation of the fuzzy rules that will govern trading. This collection of fuzzy rules approximately represents the human thinking in the decision making process. These rules, in the case of multi-input-single-output systems (*MISO*) [21], can be presented as follows:

$$IF\ Y_j\ is\ A_1, ..., AND/OR\ Y_m\ is\ A_m\ THEN\ C\ is\ A_L, \tag{6.1}$$

where (*IF* Y_1 *is* A_1,..., *AND/OR* Y_m *is* A_m) are preconditions (antecedents) and A_L are postconditions, Y_1 and Y_m are input variables, C is the output variable, A_1 is the class defined on Y_1, A_m is the class defined on Y_m, and A_L is a class defined on C.

The antecedent (rule's premise) describes to what degree a rule is applied, while the conclusion (rule's consequent) assigns a membership function to the output variable.

As mentioned in Subsection 6.1.1, there are problems associated with representing the decision making process using conventional fuzzy logic. It is for this reason that in both [9] and this chapter we consider an approach to the formulation of consequents of fuzzy rules as only an approximation of the decision making process. As in [9], four fuzzy classes - $low, medium, big, large$ - to which an output variable can be assigned are used (see Fig. 6.1). As in [9], we assume *Inf*(Y)=0, *Sup*(Y)=100, though usually such sets of linguistic terms are represented by the interval [0,1]. The rule system proposed in [9] is built in such a way that a low value of C, i.e., "C *is low*" represents an excellent opportunity to sell and a high value of C, i.e., "C *is big*" is considered as a signal to buy. We have used the rule system proposed in [9].

The combined rules for the classes $low, medium, big, large$ have been formulated as follows:

$$\begin{aligned} &IF\ \{(\mathrm{Y}_1\ is\ \text{big})\ and\ (\mathrm{Y}_2\ is\ \text{big})\}\ or\ \{(\mathrm{Y}_2\ is\ \text{large})\ and\ (\mathrm{Y}_3\ is\ \text{large})\} \\ &or\ (\mathrm{Y}_7\ is\ \text{large})\ THEN\ C\ is\ \text{large}, \end{aligned} \tag{6.2}$$

$$\begin{aligned} &IF\ (\mathrm{Y}_2\ is\ \text{large})\ or\ (\mathrm{Y}_6\ is\ \text{large})\ or\ (\mathrm{Y}_3\ is\ \text{large})\ or\ (\mathrm{Y}_5\ is\ \text{low})\ THEN \\ &C\ is\ \text{big}, \end{aligned} \tag{6.3}$$

$$IF\,(Y_4\ is\ \text{big})\,or\{(Y_7\ is\ \text{medium})\ or\,(Y_4\ is\ \text{big})\}THEN\ C\ is\ \text{medium}, \tag{6.4}$$

$$\begin{aligned}&IF\,\{(Y_1\ is\ \text{low})\,and\,(Y_2\ is\ \text{medium})\}\,or\,\{(Y_6\ is\ \text{low})\,and\,(Y_7\ is\ \text{low})\}\\&THEN\ C\ is\ \text{low}.\end{aligned} \tag{6.5}$$

Of course, this set of rules (6.2)-(6.5) is not a complete and exhaustive one. It may be modified and extended to incorporate other trading strategies. Moreover, the rule "*IF* (Y_3 *is large*) *THEN C is large*" is rather controversial, since according to the technical analysis theory [1] a large value of Y_3 (*%D*) means overbuying, i.e., it is a signal to sell. Indeed, using the technical analysis theory and fuzzy logic we can build more accurate, but somewhat more complicated rules, e.g., "*IF* (Y_3 *is large*) *and* ($Y_3(nT)$ *is smaller than* $Y_3((n-2)T)$) *and* (Y_4 *is substantially smaller than* 0) *THEN C is low*".

However, we do not intend to criticize here the set of rules proposed in [9], since a robust trading system should work well even if a few of its rules are not correct. It is clear that the intension to build a complete or exhaustive set of rules seems to be a senseless endeavor since, in practice, there may be thousands of rules proposed by different traders. Thus, any rule based trading strategy will omit some probably good opportunities to generate buying or selling signals that would be generated by another set of rules. Obviously, in such cases, the system generates nothing and this can be naturally treated as the generation of the "hold" signal.

Indeed, the aim of our work is to develop an adequate methodological framework for building mechanical trading systems (*MTS*), not to create them. Therefore, we avoid such typical trading attributes as Stop Loss Orders, Take Profit Orders and so on in our analysis.

6.1.3 Stock Trading System Based on the Mamdani's Approach

Here we present an adaptation of Mamdani's approach to the specificity of decision making in the stock trading to build the optimized trading system.

Let μ_{iv} denotes the fuzzy membership grade of input Y_j to a class v. For example, "Y_2 is *big*" denotes the membership of the fuzzy input 2 to the class *big*. Its value is computed using the membership grade μ_{23}. Then the value of the antecedent of rule (6.5) which is denoted in [9] as a_0 may be calculated as follows:

$$a_0 = \max\{\min(\mu_{22},\mu_{11}),\min(\mu_{61},\mu_{71})\},$$

where the "and" and "or" operators are replaced by *min* and *max* operators, respectively. Similarly all other antecedents are calculated:

$$a_1 = \max\{\mu_{21},\min(\mu_{43},\mu_{72})\},$$

$$a_2 = \max\{\mu_{24},\mu_{64},\mu_{34},\mu_{51}\},$$

$$a_3 = \max\{\min(\mu_{23},\mu_{13}),\min(\mu_{24},\mu_{34}),\mu_{74}\}.$$

Mamdani's fuzzy implication method is used to combine the rules and calculate the outputs: for class *low*

$$\mu_{C-0}(\bullet) = a_0 \wedge \mu_{c-0}(\bullet), \tag{6.6}$$

for class *medium*

$$\mu_{C-1}(\bullet) = a_1 \wedge \mu_{c-1}(\bullet), \tag{6.7}$$

for class *big*

$$\mu_{C-2}(\bullet) = a_2 \wedge \mu_{c-2}(\bullet), \tag{6.8}$$

for class *large*

$$\mu_{C-3}(\bullet) = a_3 \wedge \mu_{c-3}(\bullet), \tag{6.9}$$

where $\mu_{c-0}(\bullet)$, $\mu_{c-1}(\bullet)$, $\mu_{c-2}(\bullet)$ and $\mu_{c-3}(\bullet)$ are membership functions corresponding to the classes "low", "medium", "big" and "large" of output variable C, respectively and symbol $\wedge$ denotes the so-called clipping operation [21] that produces an output membership function $\mu_{C-i}(\bullet)$ clipped off at a height equal to a_i. Finally, the Mamdani's process produces

$$\begin{aligned} &\mu_C(C) = \mu_{C-0}(C) \vee \mu_{C-1}(C) \vee \mu_{C-2}(C) \vee \mu_{C-3}(C) = \\ &(a_0 \wedge \mu_{c-0}(C)) \vee (a_1 \wedge \mu_{c-1}(C)) \vee (a_2 \wedge \mu_{c-2}(C)) \vee (a_3 \wedge \mu_{c-3}(C)), \end{aligned} \tag{6.10}$$

where $\vee$ is the *max* operator. Since the resulting C is a fuzzy value, its real value representation C_r is needed. To get the real value estimation C_r, the standard center of area (*COA*) defuzzification method [21] was used. Obviously, when C_r for analyzed session is close to 100 (high end), the stock is a strong buy. On the other hand, when C_r is close to 0 (low end), the stock is a strong sell.

It is shown in [9] that a proper choice of some critical values $C_{\max}$ and $C_{\min}$ such that fuzzy logic system generates advice for buying when $C_r > C_{\max}$ and for selling when $C_r < C_{\min}$ makes it possible to create an effective investment strategy. Of course, the presented trading system will be optimized using the teaching procedure on the base of real historical data (the history of stock prices) and the results will be shown in Subsection 6.1.5.

6.1.4 Expert System Based on Logic-Motivated Fuzzy Logic Operators

As mentioned above, the Mamdani's approach to fuzzy modeling was developed for the use in automatic control, not for solving decision making problems.

However, trading is a process of decision making. The simplest fuzzy rule in a trading system can be presented as follows:

$$if\ x = A\ then\ y = B,$$

where A and B belong to one of the classes defined above, x is the value of an indicator used in the technical analysis. For example, let B = *large*. As assumed above, in this case the trader decides to buy. In other words, the term *large* in our trading rule is equivalent to the buy decision, whereas A may be naturally treated as an argument in favor of such a decision. It is easy to see that factually we are dealing with an implication $A \rightarrow B$. The complete set of formal conditions that a fuzzy implication should satisfy was formulated by Fodor [11]. Mamdani's type implication does not satisfy these conditions. Moreover, in his foreword to the recent book [27], L. Zadeh wrote "A source of confusion is that Mamdani and Assilian used this interpretation in their seminal 1974 paper, but referred to it as implication, which it is not, rather than as a joint constraint". Therefore, the so-called logical-type implication modes satisfying all or at least most of conditions defined in [11] seems to be more suitable for the solution of decision making problems.

Another problem is the choice of an appropriate type of aggregation for the particular antecedents to the resulting one. As this problem is very similar to the multiple criteria decision making problem, we think that as it was pointed out in [52], the aggregation is a context dependent problem. From this point of view, the classical Mamdani's non-compensative *min* operator can not correctly reflect the trader's reasoning as he/she usually "compensates in mind" a small value of one technical indicator by large values of others. Different Logical-type implication modes were proposed by Lukasiewicz, Fodor, Zadeh, Yager, Willmott, Dubois and Prade. In addition, a lot of methods were developed for aggregation: usual t-norms and t-conorms [26], parameterized triangular norms, soft fuzzy norms (see [27] for more detail). It seems unrealistic to test all possible combinations of implication and aggregation modes. Therefore, in line with "context dependence" strategy, we have selected an approach based on the so-called "logic-motivated fuzzy logic operators" [29]. Although this approach is based on the well known interpretation of *and* and *or* operators and Yager's implication rule, we have chosen it because the method for inferring the final mathematical expressions (for implication and aggregation as well) within the framework of this approach is similar to the trader's reasoning. It is important that in our case, the implication $A \rightarrow B$ has a natural interpretation as the "sum" of arguments (represented by A) in favor of action B (buying or selling). As a consequence, the value of membership function representing implication $\mu_{A\rightarrow B}(x,y) = I(\mu_A(x), \mu_B(y))$ should be treated as the degree of certainty in statements such as "if $x = A$ then Buy/Sell". Similar reasoning is used in the approach based on the "logic-motivated fuzzy logic operators" [29].

Let us consider the implication operation $f_{\rightarrow}(a,b)$. According to [29], the purpose of an implication operation is, given our degrees of certainty $a = d(A)$ and $b = d(B)$ in statements A and B, to estimate our degree of certainty $d(A \rightarrow B)$ in the composite statement "A implies B". From the viewpoint of the logic-motivated idea [29], the fact that our degree of certainty in statement A is equal to $d(A)$ means that we have $d(A)$ arguments in favor of A. Similarly, the fact that our degree of certainty in statement B is equal to $d(b)$ means that we have $d(b)$ arguments in favor of B. If we have an argument in favor of the implication "A implies B", then by combining each argument in favor of A with an argument in favor of the implication we obtain a transformation of each argument in favor of A into an argument in favor of B. In mathematical terms, we thus have a function mapping the set $S(A)$ of arguments in favor of A into the set $S(B)$ of arguments in favor of B. Thus, the number of arguments in favor of $A \to B$ coincides with the number of functions from the set $S(A)$ to the set $S(B)$. The number of such functions is known to be equal to $d(B)^{d(A)}$. Hence, logic motivates the use of $f_{\to}(a,b) = b^a$ as an implication operation. This operation was first introduced by R.Yager and is called *Yager's* implication. An alternative justification for the use of *Yager's* implication comes from the requirement that several natural properties of classical implication, such as

$$(A \to B)\&(A \to C) \equiv (A \to (B\&C)) \text{ and } (A \to (B \to C)) \equiv (A\&B) \to C$$

hold for the fuzzy implication as well [37]. As it was shown in [29], logic motivates the use of an algebraic product $f_{\&}(a,b) = a \cdot b$ as the *and*-operation (t-norm) and algebraic sum $f_{\&}(a,b) = a + b - a \cdot b$ as the *or*-operation (t-conorm).

So, when using "logic-motivated fuzzy logic operators" all *and*-operators in expressions (6.2)-(6.5) were represented by algebraic products and similarly all *or*-operators in the preconditions of rules (6.2)-(6.5) were represented by algebraic sums.

Let μ_{jv} denote a fuzzy membership grade of an input Y_j in a class V. For example, "Y_2 *is big*" denotes a membership of the fuzzy input Y_2 in the class *big*. Its value is computed using the membership grade μ_{23}. Then the value of the antecedent of the rule (6.5) denoted in [9] as a_0 may be calculated as follows

$a_0 = \mu_{11} \cdot \mu_{22} \oplus \mu_{61} \cdot \mu_{71}$.

The values of all the other antecedents are calculated in a similar way:

$a_1 = \mu_{21} \oplus \mu_{43} \cdot \mu_{72}$,

$a_2 = \mu_{24} \oplus \mu_{64} \oplus \mu_{34} \oplus \mu_{51}$,

$a_3 = \mu_{23} \cdot \mu_{13} \oplus \mu_{24} \cdot \mu_{34} \oplus \mu_{74}$,

where $\oplus$ denotes the algebraic sum (*or* -operation).

Since the values of a_v, in general, may be more than 1, they were normalized using the expression $a'_v = \frac{a_v}{\sum_{i=0}^{3} a_i}$. Obviously, $\sum_{i=0}^{3} a'_i = 1$.

So instead of Mamdani's outputs [9], we obtain the following implications:

$$\mu_{C-i}(C) = (\mu_{c-i}(C))^{a'_i}, i = 0 \text{ to } 3, \tag{6.11}$$

where $\mu_{c-0}(\bullet), \mu_{c-1}(\bullet), \mu_{c-2}(\bullet)$ *and* $\mu_{c-3}(\bullet)$ are the membership functions corresponding to the classes *low, medium, big, large* of the output variable C.

Finally, the result of aggregation is presented as follows:

$$\mu_C(C) = \mu_{C-0}(C) \oplus \mu_{C-1}(C) \oplus \mu_{C-2}(C) \oplus \mu_{C-3}(C). \tag{6.12}$$

Since the resulting C is a fuzzy value, its real value representation C_r is needed. To obtain the real value C_r, an appropriate defuzzification method is needed. There is an important point concerning expressions (6.11) which needs to be clarified before a proper defuzzification method is chosen. It is easy to see that when treating the involution in (6.11) as an operation on a fuzzy set [51], i.e., as the involution of values of its membership function, the critical points of the initial triangular fuzzy value could not be changed (see Fig. 6.2). Obviously, such an operation is of no practical importance in our case, since the defuzzification of the fuzzy sets represented by the membership functions $\mu_{C-i}(C)$ and $(\mu_{C-i}(C))^{a_i'}$ cannot provide different results.

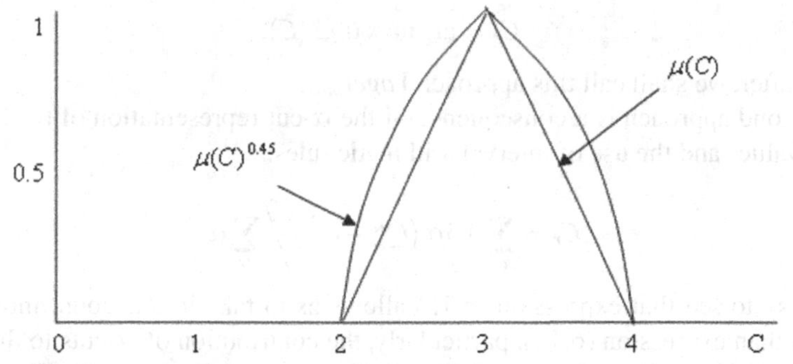

Fig. 6.2 Involution of fuzzy values in the fuzzy set sense

On the other hand, the treatment of expression (6.11) in the fuzzy arithmetic sense seems to be reasonable (see Fig. 6.3).

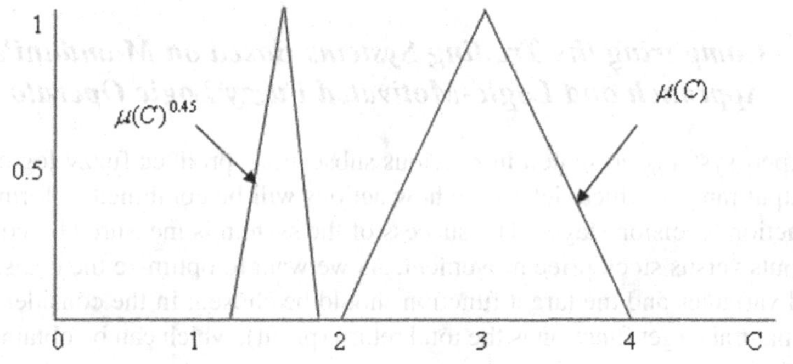

Fig. 6.3 Involution of fuzzy values in the fuzzy arithmetic sense

To implement the last approach, the α -cut representation [16] of the analyzed fuzzy values has been used. As a result, the fuzzy values $\widetilde{C}-i$ represented by the membership functions $\mu_{C-i}(C)$ were obtained:

$$\widetilde{C}-i=\bigcup_{\alpha}\left[\underline{C},\overline{C}\right]_{\alpha}^{a_i}=\bigcup_{\alpha}\left[\underline{C}^{a_i},\overline{C}^{a_i}\right]_{\alpha}, \tag{6.13}$$

where $\left[\underline{C},\overline{C}\right]_{\alpha}$ are crisp intervals such that $\mu_{c-i}(C)\geq\alpha$, $C\in\left[\underline{C},\overline{C}\right]_{\alpha}$.

To get a real value estimation of a fuzzy number $\widetilde{C}-i$, two different approaches may be used. The first is based on the results of [29] where the following logic-motivated expression was proposed:

$$C_r = \arg\max \mu_{C-i}(C). \tag{6.14}$$

Hereinafter, we shall call this approach $Yager_{max}$.
The second approach is a consequence of the α-cut representation of the analyzed fuzzy values and the use of interval arithmetic rules:

$$C_r = \sum_{\alpha} 0.5\alpha \left(\underline{C}^{\alpha_i} + \overline{C}^{\alpha_i}\right) \Big/ \sum_{\alpha} \alpha. \tag{6.15}$$

It is easy to see that expression (6.15) allows us to take into account more information than expression (6.14), particularly, the contribution of α-cuts to the generalized estimation. Hereafter, we shall call this approach $Yager_{ave}$. Finally, we have developed an expert system based on logic-motivated fuzzy logic operators, which provides the real-valued parameter C_r indicating the conditions for buying and selling.

Since the Mamdani's type expert system described in Subsection 6.1.3 works in the same way as the expert system based on logic-motivated fuzzy logic operators, it is possible to compare the results obtained using Mamdani's, $Yager_{ave}$ and $Yager_{max}$ approaches.

6.1.5 Comparing the Trading Systems Based on Mamdani's Approach and Logic-Motivated Fuzzy Logic Operators

The expert systems presented in previous subsections produce fuzzy logic outputs and output ranges, which determine how actions will be combined to form an executed action (decision stage). The success of the system is measured by comparing its outputs versus stock price movement. As we want to optimize the decisions, the control variables and the target function should be chosen. In the considered situation, a natural target function is the total return (profit), which can be obtained using our optimized strategy during some control time period.

Since the value C_r may serve as indicator of good opportunity to buy or sell a stock, it seems natural to introduce the top C_{max} and the bottom C_{min} , such that if $C_r > C_{max}$ then the system buys and if $C_r < C_{min}$ then the system sells.

In our optimization task, the parameters C_{min}, C_{max} were considered as control variables. A third control variable was the number r of stock exchange sessions used in calculation of initial technical indicators inputs (see Subsection 6.1.2).

The optimization task was formulated as the maximization of the total return R obtained during teaching period:

$$(C_{\min}, C_{\max}, r)_{opt} = \arg\max R(C_{\min}, C_{\max}, r).$$

The direct random search method [35] was used for the optimization in the teaching time period, and transaction costs were taken into account since they can significantly affect the total return in the real-world situations.

We have used the obtained optimal C_{max}, C_{min} and r in the simulation of decision making process during the following time period (testing period) to get the predicted total returns.

The use of optimization procedure provides us much more benefits in the testing period than the method proposed in [9]. Moreover, the developed optimal investment strategy employing technical analysis and fuzzy logic makes it possible to get a profit even in conditions of downtrends (dropping stock prices) that is opposite to results of [9] where only some reduction of total loss is indicated in such a case.

The results obtained using Mamdani's, $Yager_{ave}$ and $Yager_{max}$ approaches were compared with those obtained on the base of simplest trend dependent (buy and hold) strategy, i.e., when we buy some stocks at the beginning and sell them only at the end of analyzed time period.

The typical results we got for stocks of the polish company COMARCH (Warsaw Stock Exchange) are presented in Table 6.1, Fig. 6.4 and Fig. 6.5. In all cases the initial investments of 10000 PLN were simulated at the beginning of teaching and testing periods.

Table 6.1 Returns (PLN) obtained with use of different approaches for COMARCH, Warsaw Stock Exchange

Method	Teaching time period (one month)	Testing time period (one month)
Mamdani's	590	-514
$Yager_{max}$	600	146
$Yager_{ave}$	1100	200
Buy and hold strategy	80	-429

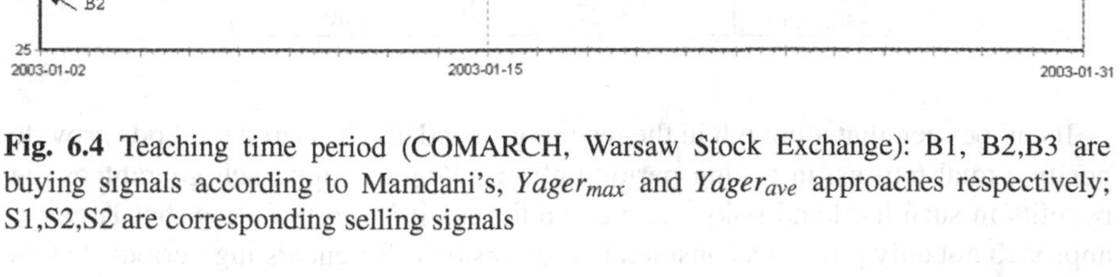

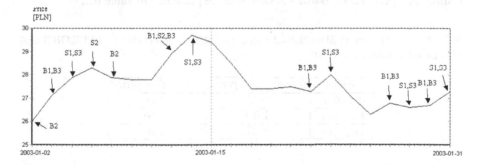

Fig. 6.4 Teaching time period (COMARCH, Warsaw Stock Exchange): B1, B2,B3 are buying signals according to Mamdani's, $Yager_{max}$ and $Yager_{ave}$ approaches respectively; S1,S2,S2 are corresponding selling signals

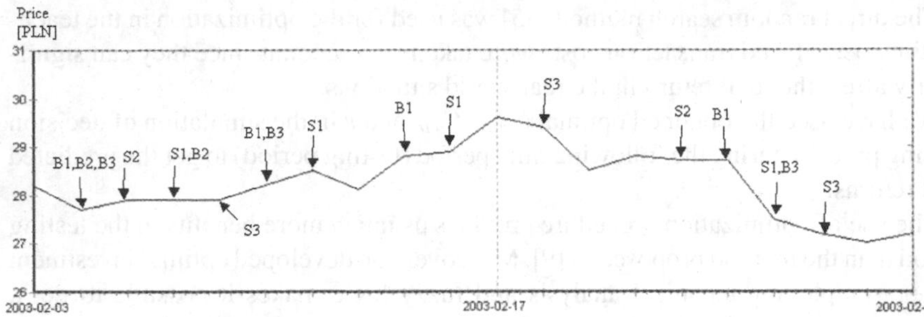

Fig. 6.5 Testing time period (COMARCH, Warsaw Stock Exchange): B1, B2,B3,.. are buying signals according to Mamdani's, $Yager_{max}$ and $Yager_{ave}$ approaches respectively; S1,S2,S2,... are corresponding selling signals

Analyzing the results presented in Table 6.1 we can say that $Yager_{ave}$ approach seems to be more beneficial and reliable. On the other hand, Mamdani's approach can produce a negative profit (losses) in testing periods if they are characterized by dropping stock prices. The same result for Mamdani's approach was indicated in [9]. Finally, the passive trend dependent strategy (buy and hold) can provide some lower losses than Mamdani's approach when dropping trend takes place, but this strategy is absolutely not paying in the rising trends in comparison with any considered optimized strategy.

To estimate the robustness of the developed optimised trading systems, they were tested in most difficult for trading and risky conditions - in the case of operating by long position (buying) only in the direction opposite to a dominating downtrend. For this purpose the quotations of Legal Play Entertainment Inc (NYSE) were used. In all cases at the beginning of teaching and testing periods, the initial investments of 10000 USD were simulated with corresponding transaction costs.

The typical results for the teaching time period 2003-10-31 - 2003-12-05 and the testing time period 2004-04-02 - 2004-05-25 are presented in Table 6.2.

Table 6.2 Total returns (USD) obtained with use of different approaches for LPLE.OB (Legal Play Entertainment Inc), NYSE

Method	Teaching time period	Testing time period
Mamdani's	1412	-545
$Yager_{max}$	1420	-2664
$Yager_{ave}$	3327	6286

It can be seen that although in the teaching period all compared methods provide positive total returns, in testing period only the $Yager_{ave}$ approach guarantees the benefits in such hard and risky trading conditions. It is worth noting that $Yager_{ave}$ approach not only produces considerable returns on different testing periods, but its returns on teaching period are at least two times greater than those obtained using Mamdani's and $Yager_{max}$ approaches (see Table 6.2).

Summarizing, we can say that $Yager_{ave}$ approach is the most adaptable, profitable and robust method for building the fuzzy logic based stock trading systems.

It is seen in Fig. 6.4 and Fig. 6.5 that the optimized strategy does not always generate only the best from common sense trading decisions. Nevertheless, we can see that optimization with the use of fuzzy expert systems makes it possible to generate effective investment strategy providing positive final profits even in the case of risky trading by long positions opposite to the downtrend.

6.2 The Stock Trading System Based on Fuzzy Logic and Evidential Reasoning

The synthesis of fuzzy logic and methods of the Dempster-Shafer theory (the so-called rule-base evidential reasoning) is proved to be a powerful tool for building expert and decision making systems. Nevertheless, there are two limitations of such approaches that reduce their ability to deal with uncertainties the decision makers often meet in practice.

The first limitation is that in the framework of known approaches to the rule-base evidential reasoning, a degree of belief can be assigned only to a particular hypothesis, not to a group of them, whereas an assignment of a belief mass to a group of events is a key principle of the Dempster-Shafer theory (the basics of this theory are presented in Chapter 3).

The second limitation is concerned with the observation that in many real-world decision problems we deal with different sources of evidence and the combination of them is needed. The known methods for the rule-base evidential reasoning do not provide a technique for the combination of evidence from different sources.

In this section, an approach free of these limitations is presented. The advantages of this approach are demonstrated using simple numerical examples and the developed stock trading expert system optimized and tested on the real data from Warsaw Stock Exchange.

6.2.1 Experts Systems Based on Rule-Base Evidential Reasoning

It is well known that the best solutions of real-world problems can be usually obtained using the synthesis of some modern powerful methods.

Therefore, during the last two decades, the rising interest of researchers has been observing in searching for an appropriate synthesis of fuzzy sets theory and the Dempster-Shafer theory (*DST*) methods. Several researchers have investigated relationships between fuzzy sets and *DST* and suggested different approaches to integrate them.

The integration of fuzzy sets and *DST* methods within symbolic rule-based models primarily has been used for solving control and classification problems [4, 5, 15, 43, 50]. These models combine these theories in a synergic way, preserving their strengths while avoiding disadvantages they present when used as monostrategy approaches: a capacity for the representation of fuzzy classifiers is enhanced by introducing the measure of ambiguity; limitations of *DST* in providing effective procedures to draw inferences from belief functions are softened by integrating the rule of propagation of evidence within the fuzzy deduction paradigm.

Generally, such a rule-base evidential reasoning system may be presented as in [5]:

$$IF\ ((A\ is\ L)\ and\ (B\ is\ M))\ THEN\ C\ is\ m_0,$$
$$IF\ ((A\ is\ H)\ and\ (B\ is\ L))\ THEN\ C\ is\ m_1,$$

where m_0 and m_1 are two credibility structures with two focal elements and variable C is defined in the universe of discourse which usually is a set of classes to deal with in considered classification problem.

In the above example adopted from [5], the credibility structures were presented as follows:

$$m_0\colon D_{00} = \left\{\frac{\mu_{00}^0}{y_0}, \frac{\mu_{01}^0}{y_1}\right\}, m_0(D_{00}), D_{01} = \left\{\frac{\mu_{01}^1}{y_1}\right\}, m_0(D_{01}),$$
$$m_1\colon D_{10} = \left\{\frac{\mu_{11}^0}{y_1}, \frac{\mu_{12}^0}{y_2}\right\}, m_1(D_{10}), D_{11} = \left\{\frac{\mu_{10}^1}{y_0}, \frac{\mu_{11}^1}{y_1}, \frac{\mu_{12}^1}{y_2}\right\}, m_1(D_{11}),$$

where D_{00}, D_{01}, D_{10}, D_{11} are fuzzy subsets in $Y = (y_0, y_1, y_2)$, μ_{00}^0, μ_{01}^0, μ_{01}^1, μ_{11}^0, μ_{12}^0, μ_{10}^1, μ_{11}^1, μ_{12}^1 are the corresponding membership grades, $m_0(D_{00})$, $m_0(D_{01})$, $m_1(D_{10})$, $m_1(D_{11})$ are the basic probability values associated with fuzzy subsets $D_{00}, D_{01}, D_{10}, D_{11}$.

The output of the system is obtained in [5] with use of *COA* method [44] in the form of defuzzified value $\bar{y}$.

This approach seems to be justified when universe of discourse $\{y_i\}$ is presented by real values. On the other hand, if we deal with expert or decision support systems, the elements of universe of discourse $\{y_i\}$ can be only the names or labels of corresponding actions or decisions, e.g., *Buy*, *Sell* and *Hold* in trading systems or the names of medical diagnoses. It is clear that in such cases, the methods based on conventional fuzzy logic, developed for the controlling can not be used at least directly.

A more suitable for building expert and decision support systems seems to be the methodology proposed by Yang, Liu, Sii, and Wang [46, 47] based on the evidential reasoning approach [45, 48, 49]. In the belief rule system, each possible consequent of a rule is associated with a belief degree. Such a rule base is capable to capture more complicated and continuous causal relationships between different factors than traditional *IF-THEN* rules. Therefore, the traditional *IF-THEN* rules may be treated as a special cases of the more general belief rule systems [14, 23, 34].

In the framework of rule-base inference methodology, using the evidential reasoning (*RIMER*) approach [46] a belief *IF-THEN* rule, e.g., the kth rule R_k, is expressed as follows:

$$IF(X_1 \ is \ A_1^k) \wedge (X_2 \ is \ A_2^k) \wedge ... \wedge (X_{T_k} \ is \ A_{T_k}^k)$$
$$THEN(D_1,\beta_{1k}),\ (D_2,\beta_{2k}),\ ...,\ (D_N,\beta_{Nk}), \tag{6.16}$$

with rule weights θ_k, $k = 1$ to L, and attribute weights $\delta_1, \delta_2, ..., \delta_{T_k}$, where $A_i^k, i = 1$ to T_k is the referential value of the ith antecedent attribute, T_k is the number of antecedent attributes used in the kth rule, β_{ik}, i=1 to N, is the belief degree to which D_i is believed to be the consequent of kth antecedent, L is the number of rules in the rule-base, $\wedge$ denotes t-norm.

If $\sum_{i=1}^{N} \beta_{ik} = 1$, the kth rule is said to be complete; otherwise, it is incomplete. The case of $\sum_{i=1}^{N} \beta_{ik} = 0$ corresponds to the total ignorance about the output given the input in the kth rule. The rule (6.16) is also referred to as a belief rule. In the framework of *RIMER* approach, the final outcome obtained as the aggregation of the rules (6.16) is presented as $O = \{(D_j, \beta_j)\}$, where β_j, $j = 1$ to N, is the aggregated degree of belief in the decision (hypothesis, action, diagnosis) D_j.

Therefore, the decision characterized by the maximal aggregated degree of belief is the best choice. So the *RIMER* approach can be used for building expert and decision support systems.

Nevertheless, there are two restrictions in the *RIMER* approach that reduce its ability to deal with uncertainties the decision makers often meet in practice. The first restriction is that in the framework of *RIMER* approach, a degree of belief can be assigned only to a particular hypothesis, not to a group of them, whereas the assignment of a belief mass to a group of events is a key principle of *DST*. The second restriction is concerned with the observation that in many real-world decision problems we deal with different sources of evidence and the combination of them is needed. The *RIMER* approach does not provide a technique for the combination of evidence from different sources. In this section, a new approach free of these restrictions is proposed.

It is important that usually the advantages of the approach based on the rule-base evidential reasoning were demonstrated using only simple numerical examples. Only a few examples of solving real-world problems using this approach were found in the literature. In [5], the neural model for fuzzy Dempster-Shafer classifiers was developed and tested. The *RIMER* approach has been successfully used to build the belief-rule-base expert system for pipeline leak detection in [41]. In [33], the synthesis of fuzzy logic and *DST* is used to build the expert system for medical diagnosis.

Therefore, the validity and capability of belief-rule-base systems in dealing with more practical and complicated problems need to be examined. In this section, we present a method for building fuzzy belief-rule-base systems and apply it to build the stock trading expert system.

A brief review of the papers devoted to the building stock trading expert systems has been presented in Section 6.1.

6.2.2 A Modern Approach to the Rule-Base Evidential Reasoning

Here we present a modern approach to the rule-based evidential reasoning free of the above mentioned restrictions of *RIMER* method.

To present this approach in more transparent form, we shall use the simple example which makes it possible to expose the features of the proposed approach and avoid here the complicated general formulas.

Suppose a doctor has to diagnose the diseases for a patient who has such outward symptoms as high temperature, cough, heavy breathing, headache and so on, which are typical for cold and flu.

There are two different groups of symptoms (different sources of evidence) available to make a diagnostic conclusion: outward symptoms based on the direct observation (high temperature, cough, heavy breathing, headache) and symptoms based on the laboratory analysis of components of blood.

It is clear that part of them may be treated as arguments in favor of cold , other as arguments in favor of flu and there may be symptoms of cold and flu simultaneously. There are two possible types of reasoning in the analyzed situation: symptoms from both groups (outward symptoms and symptoms based on the laboratory analysis) are treated as having common source of evidence; each group of symptoms is treated as a separate source of evidence.

It is clear that the first type of reasoning is only the special case of the second one. Therefore, to represent our approach in a more general form, our further analysis will be based on assumption that there are different sources of evidence in a rule-base evidential reasoning system.

Let as denote $A_j^k = (x_1\ is\ A_{j1}^k) \wedge (x_2 is A_{j2}^k) \wedge \ldots \wedge (x_{T_j}\ is\ A_{jT_j}^k)$, where A_j^k, $j = 1$ to N, $k = 1$ to M, is the antecedent of the jth rule for kth source of evidence, M is the number of sources of evidence, N is the number of rules, A_{ji}^k, $i = 1$ to T_j is the referential value of the ith antecedent attribute, T_j is the number of antecedent attributes used in the jth rule, $x_1, x_2, ..., x_{T_j}$ are the values of parameter used in the jth rule, e.g., if x_1 is the temperature, then one of the antecedent attribute can be presented as (x_1 is $High$), where the linguistic term $High$ is presented by the corresponding fuzzy subset.

Let us return to our example. Denote D_1 =*cold*, D_2 =*flu*. Suppose the rule-base evidential reasoning system is presented as follows:

$$IF\ A_1^1\ THEN\ D_1, \tag{6.17}$$

$$IF\ A_2^1\ THEN\ (D_1, D_2), \tag{6.18}$$

$$IF\ A_3^1\ THEN\ D_2, \tag{6.19}$$

$$IF\ A_1^2\ THEN\ D_1, \tag{6.20}$$

$$IF\ A_2^2\ THEN\ (D_1, D_2), \tag{6.21}$$

$$IF\ A_3^2\ THEN\ D_2, \tag{6.22}$$

with rule weights θ_i^k, $i = 1$ to 3 , $k = 1, 2$.

It is important that there are no implications in the sense of conventional fuzzy logic in the above rules. In these rules, $THEN$ denotes the assignment of some decision (action, diagnosis) to the conclusion part of a rule. For example, D_1 in the rule (6.17) is not any real or fuzzy value. It is only a label denoting the action which a rule-base evidential reasoning system proposes if the value presented by the precondition part of the rule is great enough. The similar treatment of rules in a rule-base evidential reasoning system has been proposed in [33].

As an antecedent (rule's premise) describes to what degree a rule is applied, we can treat its value as the measure of belief that a decision (action or diagnosis) in the conclusion part of a rule will be a good choice. In other words, we propose to treat the value of rule's antecedent as a focal element of some basic probability assignment bpa (see Chapter 3). For example, the rules (6.17), (6.19),(6.20), (6.22) taking into account their weights can be rewritten as $m_1^*(D_1) = \theta_1^1 A_1^1$, $m_1^*(D_2) = \theta_3^1 A_3^1$, $m_2^*(D_1) = \theta_1^2 A_1^2$, $m_2^*(D_2) = \theta_3^2 A_3^2$.

The more problems we have with the rules (6.18), (6.21). In these cases, a doctor hesitates over the choice of an unambiguous diagnostic decision, as the observed symptoms are typical for both possible diseases.

As in these cases a doctor cannot choose more probable diagnosis, the simplest solution of the problem seems to be the assignment of the probability 50% to each diagnosis. Such a solution is in the spirit of $RIMER$ type approaches and conforms to the Bayesian "principle of insufficient reason": if there is no reliable information about the probabilities of events, they are treated as equally probable [3].

Nevertheless, in the considered example, such approach seems to be somewhat artificial one since the doctor's hesitation over the choice of diagnosis is caused just by the lack of reliable information of the probabilities of possible diagnosis to be true. Moreover, in the considered case, both the diagnoses can be true simultaneously. The problem of the Bayesian approach is that the probability can be assigned only to a single event, whereas in the framework of DST, the probability of a group of events can be analyzed too. Therefore, in context of DST and taking into account the rule weights, the rules (6.18), (6.21) can be rewritten as follows:
$m_1^*(D_1,D_2) = \theta_2^1 A_2^1$, $m_2^*(D_1,D_2) = \theta_2^2 A_2^2$.

Thus, we obtain two belief structures based on two sources of evidence:
$m_1^*(D_1) = \theta_1^1 A_1^1$, $m_1^*(D_2) = \theta_3^1 A_3^1$, $m_1^*(D_1,D_2) = \theta_2^1 A_2^1$ and $m_2^*(D_1) = \theta_1^2 A_1^2$, $m_2^*(D_2) = \theta_3^2 A_3^2$, $m_2^*(D_1,D_2) = \theta_2^2 A_2^2$.

Generally, these belief structures are not normalized, i.e., the conditions
$m_1^*(D_1) + m_1^*(D_2) + m_1^*(D_1,D_2) = 1$, $m_2^*(D_1) + m_2^*(D_2) + m_2^*(D_1,D_2) = 1$ are not verified.

Therefore, to obtain the conventional basic assignment function, we have used the following normalization procedure:
$m_1(D_1) = m_1^*(D_1)/S_1$, $m_1(D_2) = m_1^*(D_2)/S_1$, $m_1(D_1,D_2) = m_1^*(D_1,D_2)/S_1$,
$m_2(D_1) = m_2^*(D_1)/S_2$, $m_2(D_2) = m_2^*(D_2)/S_2$, $m_2(D_1,D_2) = m_2^*(D_1,D_2)/S_2$,
where $S_1 = m_1^*(D_1) + m_1^*(D_2) + m_1^*(D_1,D_2)$, $S_2 = m_2^*(D_1) + m_2^*(D_2) + m_2^*(D_1,D_2)$.

The next step is obtaining the combined basic assignment function using the Dempster's rule (3.81) (see Chapter 3). As the result we get:

$$m_{12}(D_1) = \frac{m_1(D_1)m_2(D_1) + m_1(D_1)m_2(D_1,D_2) + m_2(D_1)m_1(D_1,D_2)}{1-K},$$

$$m_{12}(D_2) = \frac{m_1(D_2)m_2(D_2) + m_1(D_2)m_2(D_1,D_2) + m_2(D_2)m_1(D_1,D_2)}{1-K},$$

$$m_{12}(D_1,D_2) = \frac{m_1(D_1,D_2)m_2(D_1,D_2)}{1-K},$$

$K = m_1(D_1)m_2(D_2) + m_1(D_2)m_2(D_1)$.

The obtained combined basic assignment function is not suitable enough for choosing the best decision among D_1 and D_2. Therefore, using the expressions (3.78), (3.80) (see Chapter 3), the belief and plausibility measures have been obtained as follows:

$Bel(D_1) = m_{12}(D_1)$, $Pl(D_1) = m_{12}(D_1) + m_{12}(D_1,D_2)$,
$Bel(D_2) = m_{12}(D_2)$, $Pl(D_2) = m_{12}(D_2) + m_{12}(D_1,D_2)$.

This makes it possible to present the competing decisions in the form of belief intervals:

$BI(D_1) = [Bel(D_1), Pl(D_1)]$, $BI(D_2) = [Bel(D_2), Pl(D_2)]$.

It is known that a belief interval can be interpreted as imprecision of the "true probability". In our case, this can be treated as that $BI(D_1)$ encloses the true probability of D_1 to be the best choice and the corresponding probability of D_2 lies in $BI(D_2)$. Therefore, to choose the best decision among D_1 and D_2, it is enough to compare the intervals $BI(D_1)$, $BI(D_2)$.

It is easy to see that in such interval-probabilistic context we deal with, the use of the probabilistic approach to the interval comparison presented in Subsection 3.3.3 (see Chapter 3) seems to be the best choice.

Obviously, the presented method can be used in the case of more sources of evidence. The attribute weights may be also used in the way similar to that developed in the framework of the *RIMER* approach. The proposed approach can be adopted to the case when there exists some additional information of the probabilities of the decisions D_1 and D_2 in the rules such as (6.18) and (6.21). For example, these rules may be presented as follows:

$$IF\ A_2^1\ THEN\ ((\beta_{21}^1 D_1), (\beta_{22}^1 D_2)), \tag{6.23}$$

$$IF\ A_2^2\ THEN\ ((\beta_{21}^2 D_1), (\beta_{22}^2 D_2)), \tag{6.24}$$

where β_{ij}^k, $(i, j, k = 1, 2)$ is the belief degree to which D_j is believed to be the consequent of the antecedent A_i^k. In context of the proposed approach, the rules (6.23), (6.24) can be rewritten as follows:

$m_1^*((\beta_{21}^1 D_1), (\beta_{22}^1 D_2)) = \theta_2^1 A_2^1$, $m_2^*((\beta_{21}^2 D_1), (\beta_{22}^2 D_2)) = \theta_2^2 A_2^2$.

Finally, after normalization we get two belief structures based on two sources of evidence:

$m_1(D_1)$, $m_1(D_2)$, $m_1((\beta_{21}^1 D_1), (\beta_{22}^1 D_2))$;
$m_2(D_1)$, $m_2(D_2)$, $m_2((\beta_{21}^2 D_1), (\beta_{22}^2 D_2))$.

Here we propose the following way to utilize an additional information of D_1 and D_2 presented by β_{ij}^k.

Firstly, let us consider the calculation of *Bel* and *Pl* measures using two following examples.

Example 6.1. Suppose, there is no an additional information of D_1, D_2 and
$m_1(D_1) = 0.2, m_1(D_2) = 0.5, m_1(D_1, D_2) = 0.3,$
$m_2(D_1) = 0.2, m_2(D_2) = 0.3, m_2(D_1, D_2) = 0.5.$
Then from these sources of evidence we get respectively:
$BI_1(D_1) = [0.2, 0.5], BI_1(D_2) = [0.5, 0.8],$
$BI_2(D_1) = [0.2, 0.7], BI_2(D_2) = [0.3, 0.8].$
Using the Dempster's rule of combination we obtain:

$$m_{12}(D_1) = 0.238, m_{12}(D_2) = 0.583, m_{12}(D_1, D_2) = 0.179. \tag{6.25}$$

Example 6.2. Suppose, there is the additional information of D_1 and D_2, presented by β_{ij}^k:

$$\begin{aligned} m_1(D_1) = 0.2, m_1(D_2) = 0.5, m_1((\beta_{21}^1 D_1), (\beta_{22}^1 D_2)) = 0.3, \\ m_2(D_1) = 0.2, m_2(D_2) = 0.3, m_2((\beta_{21}^2 D_1), (\beta_{22}^2 D_2)) = 0.5. \end{aligned} \tag{6.26}$$

We propose to utilize the additional information of D_1 and D_2, presented by β_{ij}^k in such a way that β_{ij}^k will reduce the values of focal elements representing hesitations of the decision makers concerned with the probabilities of D_1 and D_2 to be best decisions in the rules such as (6.23), (6.24). The reason behind this proposition is that any additional reliable information used in a mathematic model should reduce the result's uncertainty. Therefore, for the calculations of the *Bel* and *Pl* measures we propose the following method for utilizing the additional information of D_1 and D_2:

$$\begin{aligned} Bel_1(D_1) = m_1(D_1), Pl_1(D_1) = m_1(D_1) + \beta_{21}^1 m_1((\beta_{21}^1 D_1), (\beta_{22}^1 D_2)), \\ Bel_1(D_2) = m_1(D_2), Pl_1(D_2) = m_1(D_2) + \beta_{22}^1 m_1((\beta_{21}^1 D_1), (\beta_{22}^1 D_2)), \\ Bel_2(D_1) = m_2(D_1), Pl_2(D_1) = m_2(D_1) + \beta_{21}^2 m_2((\beta_{21}^2 D_1), (\beta_{22}^2 D_2)), \\ Bel_2(D_2) = m_2(D_2), Pl_2(D_2) = m_2(D_2) + \beta_{22}^2 m_1((\beta_{21}^2 D_1), (\beta_{22}^2 D_2)). \end{aligned} \tag{6.27}$$

Let $\beta_{21}^1 = 0.4, \beta_{22}^1 = 0.6, \beta_{21}^2 = 0.3, \beta_{22}^2 = 0.7$. Then from (6.26) and (6.27) we get:

$BI_1(D_1) = [0.2, 0.32], BI_1(D_2) = [0.5, 0.68],$
$BI_2(D_1) = [0.2, 0.35], BI_2(D_2) = [0.3, 0.65].$

Comparing these results with those obtained in the Example 1, we can see that utilizing the additional information concerned with the probabilities of D_1 and D_2 to be best decisions in the rules such as (6.23), (6.24) leads to the reduction of the lengths of all calculated belief intervals, i.e., to the reduction of the overall result's uncertainty. It is important that although in both examples the interval relations $BI_1(D_2) > BI_1(D_1)$ and $BI_2(D_2) > BI_2(D_1)$ are observed, the differences between

$BI_1(D_2)$ and $BI_1(D_1)$, $BI_2(D_2)$ and $BI_2(D_1)$ are greater in the Example 2 than in the Example 1. Moreover, opposite to the case in the Example 1, there is no intersection of intervals $BI_1(D_2)$ and $BI_1(D_1)$ in the Example 2.

To explain this fact, let us consider the average belief degrees to which D_1, D_2 are believed to be the consequents of the antecedents in the rules (6.23) and (6.24): $\overline{\beta}_1 = \frac{1}{2}(\beta_{21}^1 + \beta_{21}^2)$, $\overline{\beta}_2 = \frac{1}{2}(\beta_{22}^1 + \beta_{22}^2)$. Since $\overline{\beta}_1 = 0.35$ and $\overline{\beta}_2 = 0.65$, it is quite natural for the decision D_2 to be more preferable in the Example 2 than in the Example 1.

Thus, the results of the above analysis are in compliance with common sense.

Therefore, we propose here to use the described above method for utilizing the additional information of D_1 and D_2 in the Dempster's combination rule which we propose to modify as follows:

$$\begin{aligned}
m_{12}^*(D_1) = (m_1(D_1)m_2(D_1) + m_1(D_1)\beta_{21}^2 m_2((\beta_{21}^2 D_1),(\beta_{22}^2 D_2)) + \\
+ m_2(D_1)\beta_{21}^1 m_1((\beta_{21}^1 D_1),(\beta_{22}^1 D_2)))/Nf, \\
m_{12}^*(D_2) = (m_1(D_2)m_2(D_2) + m_1(D_2)\beta_{22}^2 m_2((\beta_{21}^2 D_1),(\beta_{22}^2 D_2)) + \\
+ m_2(D_2)\beta_{22}^1 m_1((\beta_{21}^1 D_1),(\beta_{22}^1 D_2)))/Nf, \\
m_{12}^*(D_1,D_2) = (\beta m_1((\beta_{21}^1 D_1),(\beta_{22}^1 D_2))m_2((\beta_{21}^2 D_1),(\beta_{22}^2 D_2)))/Nf,
\end{aligned} \quad (6.28)$$

where $Nf = 1 - K$ is the normalization factor, $K = m_1(D_1)m_2(D_2) + m_1(D_2)m_2(D_1)$, $\beta = \overline{\beta}_1\overline{\beta}_2$, $\overline{\beta}_1 = \frac{1}{2}(\beta_{21}^1 + \beta_{21}^2)$, $\overline{\beta}_2 = \frac{1}{2}(\beta_{22}^1 + \beta_{22}^2)$.

It is easy to see that always $\overline{\beta}_1 + \overline{\beta}_2 = 1$.

Thus, the maximal value of β equal to 0.25 we have when $\overline{\beta}_1 = \overline{\beta}_2 = 0.5$. The value of β is reduced from 0.25 to zero with increasing the difference between of $\overline{\beta}_1$ and $\overline{\beta}_2$.

In the extreme case, when, e.g., $\beta_{21}^1 = \beta_{21}^2 = 0$, $\beta_{22}^1 = \beta_{22}^2 = 1$, $\beta = 0$, from (6.28) we get:

$m_{12}^*(D_1) = m_1(D_1)m_2(D_1)/(1-K)$, $m_{12}^*(D_1,D_2) = 0$,
$m_{12}^*(D_2) = (m_1(D_2)m_2(D_2) + m_1(D_2)m_2((\beta_{21}^2 D_1),(\beta_{22}^2 D_2)) +$
$+ m_2(D_2)m_1((\beta_{21}^1 D_1),(\beta_{22}^1 D_2)))/Nf$.

Thus, in this case when decision maker possesses an additional information that in the rules (6.23), (6.24) only D_2 is the best decision, there is no the focal element corresponding to the hesitation (i.e., $m_{12}^*(D_1,D_2) = 0$) in the combined *bpa* and the values of the rest of focal elements are redistributed in favor of D_2. Therefore, we can say that when $\beta = 0$, we deal with the most certain case with minimal hesitation. In the opposite situation, when $\beta = 0.25$ (e.g., $\beta_{22}^1 = \beta_{22}^2 = 0.5$, $\beta_{22}^1 = \beta_{22}^2 = 0.5$), we deal with the maximally uncertain case since there is fifty-fifty chance for both D_1 and D_2 to be the best decision. Therefore, the value of β can be treated as the measure of uncertainty concerned with the probabilities of D_1 and D_2 to be best decisions.

Let us consider some numerical examples illustrating this assertion.

For $\beta_{21}^1 = 0.4$, $\beta_{22}^1 = 0.6$, $\beta_{21}^2 = 0.3$, $\beta_{22}^2 = 0.7$ ($\overline{\beta}_1 = 0.35$, $\overline{\beta}_2 = 0.65$, $\beta = 0.2275$), using the modified Dempster's rule of combination (6.28) and the subsequent normalization, from *bpa*s (6.26) we get:

$$m_{12}(D_1) = 0.184, m_{12}(D_2) = 0.741, m_{12}(D_1, D_2) = 0.075. \qquad (6.29)$$

For $\beta_{21}^1 = 0.5$, $\beta_{22}^1 = 0.5$, $\beta_{21}^2 = 0.5$, $\beta_{22}^2 = 0.5$ ($\overline{\beta}_1 = 0.5$, $\overline{\beta}_2 = 0.5$, $\beta = 0.25$), we obtain:

$$m_{12}(D_1) = 0.258, m_{12}(D_2) = 0.686, m_{12}(D_1, D_2) = 0.056. \qquad (6.30)$$

For $\beta_{21}^1 = 0$, $\beta_{22}^1 = 1$, $\beta_{21}^2 = 0$, $\beta_{22}^2 = 1$ ($\overline{\beta}_1 = 0$, $\overline{\beta}_2 = 1$, $\beta = 0$), we obtain:

$$m_{12}(D_1) = 0.0476, m_{12}(D_2) = 0.580, m_{12}(D_1, D_2) = 0. \qquad (6.31)$$

It is seen that the proposed modified Dempster's rule of combination allows us to utilize an additional information of D_1 and D_2 in such a way that the obtained results are in good accordance with common sense. From (3.79), (3.80) and (6.28) after the normalization procedure, the competing decisions are presented in the form of belief intervals $BI(D_1) = [Bel(D_1), Pl(D_1)]$, $BI(D_2) = [Bel(D_2), Pl(D_2)]$. Finally, to choose the best decision among D_1 and D_2, it is enough to compare these intervals. A proper method for interval comparison is presented in Subsection 3.3.3 (see Chapter 3).

6.2.3 Stock Trading Expert System

Modern computerized stock trading systems (or "mechanical trading systems") are based on the simulation of the decision making process and generate advices for the trader to buy or sell some stocks or other financial tool he/she deals with taking into account the price history, technical analysis indicators, accepted rules of trading and so on.

Generally, the stock trading expert systems ($STES$) are based on the analysis of charts such us as shown in Fig. 6.6. The objects presented by $High$, Low, $Open$ and $Close$ stock prices on a chosen time frame are called "Bars". In Fig. 6.6, the 1-hour time frame (1h-Bars) is performed.

In the most of $STES$s, different trend following strategies are used which are based on the technical analysis, i.e., on methods for evaluating securities by analyzing statistics generated by market activity, such as past prices and volumes (number of transactions during a unit of a time frame).

Usually moving averages (e.g., averages of closing prices in the last n *Bars*) are used to indicate the direction of a trend and to smooth out price and volume fluctuations or "noise" that can confuse interpretation. Typically, upward momentum is confirmed when a short-term average crosses above a longer-term average (see Fig. 6.7). Downward momentum is confirmed when a short-term average crosses below a long-term average.

While some suggest the use of simple moving average rules [6, 7] others have considered more complex indicators such as momentum and exponential moving averages [25]. The seeking for proper trading rules, in general, involves learning good indicators, as well as combinations of these in defining good rules. It is known

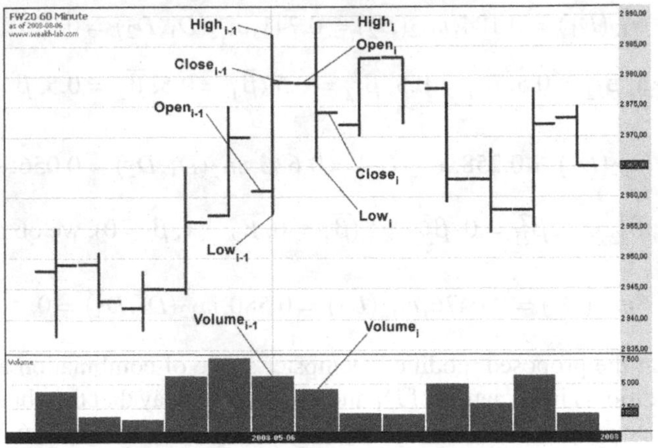

Fig. 6.6 Prices and volumes of the futures contract FWIG20 (WIG20 is the main index of Warsaw Stock Exchange)

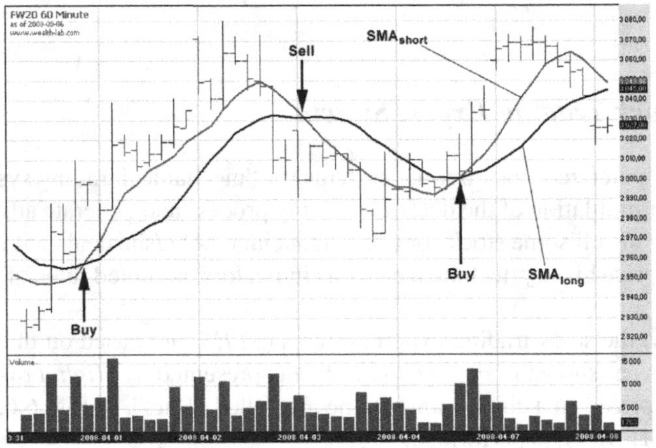

Fig. 6.7 Simplest trend following strategy: *SMA*-simple moving average

that "trend-following methods typically utilize moving averages of closing price data for buy and sell signals. Often, the signals turn out to be false due to short-term market fluctuations" [40]. This drawback of moving averages is especially important when short time frames are used on the market with high volatility or in the times of financial crisis. On the other hand, the methods based on moving averages can also be used in the development of other technical indicators.

To weaken the undesirable consequences of the above mentioned deficiency of usual moving averages, we propose here to use instead of prices and volumes the values of their changes on the nearest Bars. Finally, the short and long-term moving averages based on these changes are used to advice *Buy* , *Sell* or *Hold* decisions like in usual trend-following strategies.

6.2.3.1 Trading Rules

In the proposed trading system, the changes of prices and volumes are used to build the basic parameters of the system as follows:

- the change of the close price on *i*th *Bar*: $\Delta C_i = Close_i - Close_{i-1}$;
- the change of price movement presented by *Bar*s taken as a whole. Generally, this parameter should be presented by interval subtraction $\Delta B_i = B_i - B_{i-1}$, where B_i =$[Low_i, High_i]$ is the interval of prices observed in *i*th *Bar*. For the sake of simplicity, this price movement is approximately presented here by the parameter $\varepsilon_i = \frac{(Low_i+High_i)-(High_{i-1}+Low_{i-1})}{2}$, i.e., by the change of interval's means on the nearest *Bar*s; - the change of the volume on *i*th *Bar*: $\Delta V_i = V_i - V_{i-1}$;
- the change of the Bollinger band width $\Delta WBB(n)_i$ =$WBB(n)_i$-$WBB(n)_{i-1}$, where $WBB(n)_i$ is equal to two standard deviations away from a simple moving average calculated on *i*th *Bar* using *n* previous *Bar*s. Because standard deviation is a measure of volatility, Bollinger bands adjust themselves to the market conditions. Thus, $\Delta WBB(n)_i$ represents the change of volatility.

Traders usually make decisions using subjective assessments of parameters such as introduced above and explain possible actions in the linguistic form, e.g., if (ΔC_i is *Big* and ΔV_i is *Big* and $\Delta WBB(n)_i$ is *Medium*) then *Buy*. Since the linguistic terms such as *Low*, *Big*, *Medium* are the classes representing some fuzzy concepts, the use of fuzzy logic for building stock trading systems seems to by quite natural. In the proposed system, only three fuzzy classes *Low*, *Big* and *Medium* are used. They are represented by corresponding triangular membership functions.

Since in the case of $\Delta C_i > 0$ or $\varepsilon_i > 0$, the *Buy* signal should be generated and if $\Delta C_i < 0$ or $\varepsilon_i < 0$, the *Sell* signal should be the best choice, we have introduced the classes Low_B^C, $Medium_B^C$ and Big_B^C for $\Delta C_i > 0$ and Low_S^C, $Medium_S^C$ and Big_S^C for $\Delta C_i < 0$. Similarly the classes Low_B^ε, $Medium_B^\varepsilon$, Big_B^ε, Low_S^ε, $Medium_S^\varepsilon$ and Big_S^ε, have been introduced for ε_i. The membership functions $\mu^C_{LOW_B}$, $\mu^C_{MEDIUM_B}$, $\mu^C_{BIG_B}$, $\mu^C_{LOW_S}$, $\mu^C_{MEDIUM_S}$, $\mu^C_{BIG_S}$ of the classes for ΔC_i are presented in Fig. 6.8.

In the same way, the membership functions $\mu^\varepsilon_{LOW_B}$, $\mu^\varepsilon_{MEDIUM_B}$, $\mu^\varepsilon_{BIG_B}$, $\mu^\varepsilon_{LOW_S}$, $\mu^\varepsilon_{MEDIUM_S}$, $\mu^\varepsilon_{BIG_S}$ of classes for assessment of ε_i have been built. Since the change of the volume ΔV_i does not generate the *Buy* or *Sell* signal, but can be treated as the strength of signals generated by the parameters ΔC_i and ε_i, the membership functions μ^V_{LOW}, μ^V_{MEDIUM}, μ^V_{BIG} of the classes Low^V, $Medium^V$, Big^V have been built as presented in Fig. 6.9.

The membership functions μ^{WBB}_{LOW}, μ^{WBB}_{MEDIUM}, μ^{WBB}_{BIG} of the classes μ^{WBB}_{LOW}, μ^{WBB}_{MEDIUM}, μ^{WBB}_{BIG} based on the $\Delta WBB(n)_i$ were built similarly, since the values of this parameter can only weaken or intensify the *Buy* and *Sell* signals. The optimal values of parameters ΔC_{Big}, ε_{Big}, ΔV_{Big}, ΔWBB_{Big} used to define the membership functions (see Fig. 6.8, Fig. 6.9) and the optimal period *n* for the calculation of $WBB(n)_i$ were found in the stage of teaching the trading system using historical data.

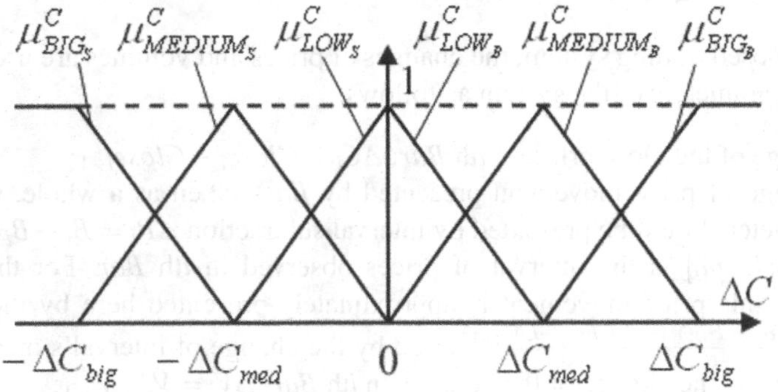

Fig. 6.8 Membership functions of the fuzzy classes based on ΔC

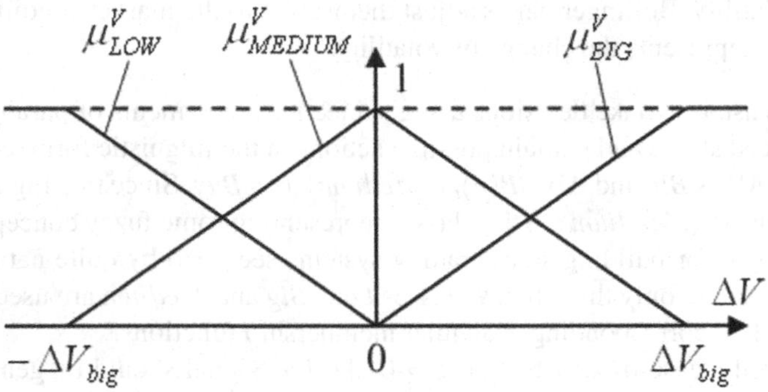

Fig. 6.9 Membership functions of the fuzzy classes based on ΔV

There are only three possible decisions *Buy*, *Sell* and *Hold* in any trading situations, but in practice, there are more complicated situations when a trader hesitates over the choice of an unambiguous decision. In such cases, a trader usually formulates his/her opinion in the imprecise form, e.g., as (*Buy* or *Hold*) or (*Sell* or *Hold*). Hereinafter, such situations will be denoted as (*Buy*, *Hold*) and (*Sell*, *Hold*), respectively. Basing on the introduced fuzzy classes for assessment of market parameters, the set of 108 rules generating possible trader's decisions including the cases of hesitation on each *i*th *Bar* were built. For illustration, some of them are presented below (here and hereinafter the index *i* is omitted for simplicity of notation):

$IF\ (\Delta C\ is\ Big^C_B) \wedge (\Delta V\ is\ Big^V) \wedge (\Delta WBB\ is\ Big^{WBB})\ THEN\ Buy,$
$IF\ (\Delta C\ is\ Big^C_B) \wedge (\Delta V\ is\ Big^V) \wedge (\Delta WBB\ is\ Low^{WBB})\ THEN\ (Buy, Hold),$
$IF\ (\Delta C\ is\ Big^C_B) \wedge (\Delta V\ is\ Medium^V) \wedge (\Delta WBB\ is\ Low^{WBB})\ THEN\ Hold,$
$IF\ (\Delta C\ is\ Big^C_S) \wedge (\Delta V\ is\ Big^V) \wedge (\Delta WBB\ is\ Big^{WBB})\ THEN\ Sell,$
$IF\ (\Delta \varepsilon\ is\ Big^\varepsilon_B) \wedge (\Delta V\ is\ Big^V) \wedge (\Delta WBB\ is\ Medium^{WBB})\ THEN\ Buy,$

$IF\ (\Delta\varepsilon\ is\ Big^{\varepsilon}_{B}) \wedge (\Delta V\ is\ Big^{V}) \wedge (\Delta WBB\ is\ Low^{WBB})\ THEN\ (Buy, Hold),$
$IF\ (\Delta\varepsilon\ is\ Big^{\varepsilon}_{S}) \wedge (\Delta V\ is\ Big^{V}) \wedge (\Delta WBB\ is\ Medium^{WBB})\ THEN\ Sell,$
$IF\ (\Delta\varepsilon\ is\ Big^{\varepsilon}_{S}) \wedge (\Delta V\ is\ Big^{V}) \wedge (\Delta WBB\ is\ Low^{WBB})\ THEN\ (Sell, Hold).$

It is worth noting that the proposed set of 108 rules performs the reasoning of rather cautious (and experienced) trader participating in our research. All rules are based on the parameters ΔV and ΔWBB, but the half of them (54 rules) include fuzzy assessments of ε and the next half include fuzzy assessments of ΔC. Since at the same moment (*Bar*) a positive change $\Delta C > 0$ generating *Buy* signal and a negative change $\varepsilon < 0$ generating *Sell* signal can appear simultaneously, the rules based on ΔC and ε can be treated as those originated from two different sources of evidence.

The *min* operator has been used (as the connective "and") to aggregate attributes in antecedents as it is the most frequently used t-norm in the fuzzy logic. This operator is in compliance with requirements of the cautious trader, although the sets of parameterized families of more flexible t-norms and s-conorms were proposed in the literature [27, 28]. All the rules were assumed to be of equal importance, i.e, for the rule weights we have $\theta_k = 1, k = 1$ to 108, and no reliable additional information of probabilities of the decisions *Buy*, *Sell* and *Hold* in the rules with the ambiguous consequents $(Buy, Hold)$ and $(Sell, Hold)$ has been reveled. On the other hand, according to the trader's opinion, the different weights of attributes in antecedents should be introduced taking into account that attributes based on ΔC and ε are of equal importance. Therefore, the weights w_C=w_ε, w_V, w_{WBB} were introduced in the system of rules and their values were obtained in the stage of teaching the trading system using historical data under the condition w_ε+ w_V+ w_{WBB}=1 . Then in the framework of the approach presented in Chapter 4, the above eight illustrative rules can be rewritten as follows:

$$m^C_1(Buy) = \min((\mu^C_{BIG_B})^{w_C}, (\mu^V_{BIG})^{w_V}, (\mu^{WBB}_{BIG})^{w_{WBB}}),$$
$$m^C_1(Buy, Hold) = \min((\mu^C_{BIG_B})^{w_C}, (\mu^V_{BIG})^{w_V}, (\mu^{WBB}_{LOW})^{w_{WBB}}),$$
$$m^C_1(Hold) = \min((\mu^C_{BIG_B})^{w_C}, (\mu^V_{MEDIUM})^{w_V}, (\mu^{WBB}_{LOW})^{w_{WBB}}),$$
$$m^C_1(Sell) = \min((\mu^C_{BIG_S})^{w_C}, (\mu^V_{BIG})^{w_V}, (\mu^{WBB}_{BIG})^{w_{WBB}}),$$
$$m^\varepsilon_1(Buy) = \min((\mu^\varepsilon_{BIG_B})^{w_C}, (\mu^V_{BIG})^{w_V}, (\mu^{WBB}_{MEDIUM})^{w_{WBB}}),$$
$$m^\varepsilon_1(Buy, Hold) = \min((\mu^\varepsilon_{BIG_B})^{w_C}, (\mu^V_{BIG})^{w_V}, (\mu^{WBB}_{LOW})^{w_{WBB}}),$$
$$m^\varepsilon_1(Sell) = \min((\mu^\varepsilon_{BIG_S})^{w_C}, (\mu^V_{BIG})^{w_V}, (\mu^{WBB}_{MEDIUM})^{w_{WBB}}),$$
$$m^\varepsilon_1(Sell, Hold) = \min((\mu^\varepsilon_{BIG_S})^{w_C}, (\mu^V_{BIG})^{w_V}, (\mu^{WBB}_{LOW})^{w_{WBB}}),$$

The power operator is used in the above rules as only such weighting procedure guarantees reasonable results when dealing with the *min* operator for aggregation [8, 31, 42]. Some comments are necessary. By $m^C_1(Buy)$ we denote here the mass of belief in *Buy* decision originated from the first source of evidence based on ΔC, the index 1 marks the first occurrence of this type of belief in the set of rules (there are n^C_{BUY} such masses of belief in the set of rules). Similarly, $m^\varepsilon_1(Sell, Hold)$ is the first occurrence of the mass of belief in the ambiguous $(Buy, Hold)$ decision originated from the second source of evidence based on ε (there are $n^\varepsilon_{(SELL,HOLD)}$ such masses of belief in the set of rules) and so on.Therefore, the general set of 108 rules consists of two subsets originated from two sources of evidence based on ΔC and ε, respectively:

$m^C_{i_B}(Buy),\ i_B = 1\ to\ n^C_{BUY};$
$m^C_{i_S}(Sell),\ i_S = 1\ to\ n^C_{SELL};$
$m^C_{i_H}(Hold),\ i_H = 1\ to\ n^C_{HOLD};$
$m^C_{i_{BH}}(Buy,Hold),\ i_{BH} = 1\ to\ n^C_{(BUY,HOLD)};$
$m^C_{i_{SH}}(Sell,Hold),\ i_{SH} = 1\ to\ n^C_{(Sell,HOLD)};$
$m^\varepsilon_{j_B}(Buy),\ j_B = 1\ to\ n^\varepsilon_{BUY};$
$m^\varepsilon_{j_S}(Sell),\ j_S = 1\ to\ n^\varepsilon_{SELL};$
$m^\varepsilon_{j_H}(Hold),\ j_H = 1\ to\ n^\varepsilon_{HOLD};$
$m^\varepsilon_{j_{BH}}(Buy,Hold),\ j_{BH} = 1\ to\ n^\varepsilon_{(BUY,HOLD)};$
$m^\varepsilon_{j_{SH}}(Sell,Hold),\ j_{SH} = 1\ to\ n^\varepsilon_{(Sell,HOLD)}.$

Summarizing the particular masses of belief in *Buy* decision based on the first source of evidence, we get $m^c_*(Buy) = \sum_{i_B=1}^{n^C_{BUY}} m^C_{i_B}$. Similarly, the other summarized masses of belief $m^C_*(Sell)$, $m^C_*(Hold)$, $m^C_*(Buy,Hold)$, $m^C_*(Sell,Hold)$, $m^\varepsilon_*(Buy)$, $m^\varepsilon_*(Sell)$, $m^\varepsilon_*(Hold)$, $m^\varepsilon_*(Buy,Hold)$, $m^\varepsilon_*(Sell,Hold)$ have been obtained. To get the normalized basic assignment functions for both sources of evidence, the above summarized masses of belief originated from these sources have been divided by the normalization factors

$S^C = m^C_*(Buy)+m^C_*(Sell)+m^C_*(Hold) + m^C_*(Buy,Hold)+ m^C_*(Sell,Hold)$ and
$S^\varepsilon = m^\varepsilon_*(Buy) + m^\varepsilon_*(Sell) = m^\varepsilon_*(Hold)+m^\varepsilon_*(Buy,Hold)+ m^\varepsilon_*(Sell,Hold)$, respectively.
As the result, the basic assignment functions fulfilling the normalization conditions
$m^C(Buy)+m^C(Sell)+m^C(Hold)+m^C(Buy,Holg)+m^C(Sell,Hold)=1$,
$m^\varepsilon(Buy)+m^\varepsilon(Sell)+m^\varepsilon(Hold)+m^\varepsilon(Buy,Holg)+m^\varepsilon(Sell,Hold)=1$ have been obtained.

Finally, using the Dempster's rule (3.81) (see Chapter 3), we get the combined basic assignment function

$$m^{C\varepsilon}(Buy), m^{C\varepsilon}(Sell), m^{C\varepsilon}(Hold), m^{C\varepsilon}(Buy,Hold), m^{C\varepsilon}(Sell,Hold). \tag{6.32}$$

Using this *bpa*, from (3.79) and (3.80) (see Chapter 3) the competing decisions *Buy*, *Sell* and *Hold* can be presented in the form of belief intervals $BI(Buy)= [Bel(Buy),Pl(Buy)]$, $BI(Sell)=[Bel(Sell),Pl(Sell)]$
and $BI(Hold)= [Bel(Hold),Pl(Hold)]$.

Finally, to choose the best trading decision among *Buy*, *Sell* and *Hold*, it is enough to compare these intervals using the method for interval comparison presented in Subsection 3.3.3 (see Chapter 3). Nevertheless, such approach leads, in practice, to the poor trading results, since the presented above rule-based evidential reasoning system may generate *Buy* or *Sell* signals almost in each *Bar* as the reaction on random fluctuations of prices and volumes. Therefore, the signals generated by the rule-base system were used as the most important part of the more general strategy.

6.2.3.2 Trading Strategy

To reduce the influence of the market "noise" to the decision making process, the moving averages of the interval $BI(Buy)$, $BI(Sell)$ and $BI(Hold)$ signals are used to indicate the direction of a trend. The short-term averages $BI^N(Buy)$, $BI^N(Sell)$ and $BI^N(Hold)$ of these signals have been calculated averaging the signals of the last N *Bars* and the long-term averages have been obtained as the averages of the last M *Bars*: $BI^M(Buy)$, $BI^M(Sell)$, $BI^M(Hold)$. Obviously, $M > N$.

Besides moving averages, the *StopLoss* and *TakeProfit* orders were used as they are probably the most important tools to minimize losses and protect profits on an open position. A *StopLoss* is an order to close a previously opened position at a price less profitable for the customer than the opening price. A *TakeProfit* order closes a position at a price more profitable than the opening price. The optimized values of *StopLoss* (*SL*) and *TakeProfit* (*TP*) are obtained in the stage of teaching the model using historical data.

In a nutshell, the trading strategy can be described as follows: if in the current ith *Bar* there are no open positions then:

if $BI_i^N(Hold) > \max(BI_i^N(Buy), BI_i^N(Sell))$ (in interval sense), then the decision is *Hold* and no positions should be opened in this *Bar*;
if $BI_i^N(Buy) > \max(BI_i^N(Hold), BI_i^N(Sell))$ then if $\Delta C_i > 0$ and
$BI_i^N(Buy) > BI_{i-1}^N(Buy) > 0$ then *Buy*, and *Hold* in all other cases.

If in the current ith *Bar* the long position (*Buy*) is open, then: if the current price is lower than the *StopLoss* level or higher than the *TakeProfit* level or the short-term moving average $BI_i^N(Buy)$ crosses above a longer-term moving average $BI_i^M(Buy)$ then the long position should be closed. In all other cases, the long position should be held. The strategy for opening and closing short positions (*Sell*) is a mirror reflection of the described above strategy for managing long positions.

Basing on this trading strategy, the model simulating the decision making process in the trading with the use of historical data has been developed and implemented using the well known specialized software *Wealth-Lab Developer* 3.0.
The main output of this model is the total return R, i.e., a profit gained by the model during a chosen time period. In the developed stock trading expert system, the total return R is the function of the parameters ΔC_{Big}, ε_{Big}, ΔV_{Big}, ΔWBB_{Big} defining the values of membership functions used in rule-base system, the weighs of attributes in antecedents: w_C, w_V, w_{WBB}, periods for calculation of the *Bollinger band* and short and long-term averages: n, N, M, the values of the *StopLoss* and *TakeProfit* levels: SL, TP.

The developed trading system has been optimized. Optimization pertains to the ability to determine the combination of the values of trading system's parameters which result in the most favorable performance for the trading system. These optimization parameters include a variety of technical indicator periods, periodicity, stops, targets, and more. So, in our case, the optimization problem has been formulated as follows:

$$\max R(\Delta C_{Big}, \varepsilon_{Big}, \Delta V_{Big}, \Delta WBB_{Big}, w_C, w_V, w_{WBB}, n, N, M, SL, TP),$$

$$s.t.\ w_C + w_V + w_{WBB} = 1,\ N < M,$$

where R is the total return gained by the system during a teaching time period. Of course, the duration of this period should be great enough to guarantee the statistical validity of simulated results. Finally, the quality of a trading system is estimated comparing the total return obtained in the teaching time period with that gained in the test period usually immediately following the teaching period. An effective trading system usually provides good results in the teaching period and at least satisfactory results in the testing period. It is well known that the results are usually deteriorating with widening the test period. Therefore, in practice, the reoptimization, i.e., the repetition of the optimization procedure is usually used after some period of real trading (test period) with the use of previously optimized system. Hence, it is necessary to optimize durations of teaching and test periods. We have tested the developed trading systems using the prices of FWIG20 futures contract (WIG20 is the main index of Warsaw Stock Exchange). All the results were obtained with the use of only one contract in trading without reinvestments and taking into account the transaction costs. The best results with high profit, smooth profit curve and high percentage of winning trades we have obtained using 1-hour time frame (1h-Bars). The optimal durations of teaching and test period (in 2007-2008 years) were found to be 22 and 4 weeks, respectively. The typical profit curve is shown in Fig. 6.10, where the teaching and test time periods are separated by vertical black line. The

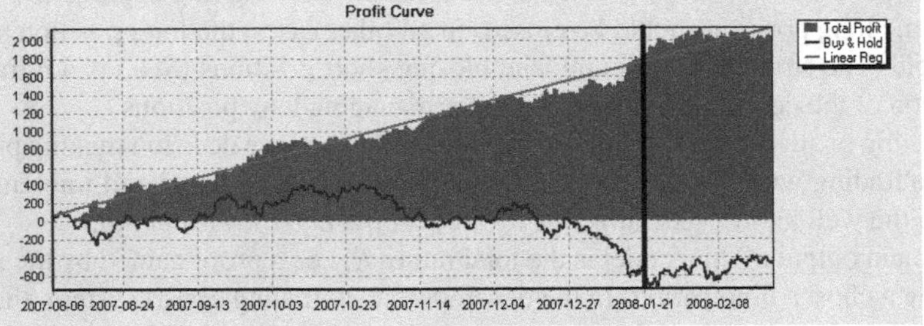

Fig. 6.10 Profit curie obtained trading FWIG20 futures contract: the teaching period: 06.08.2007 - 18.01.2008, the test period: 21.01.2008 - 22.02.2008

results obtained in the teaching and test periods are presented in Table 6.3 and Table 6.4. In the last column of Table 6.4, the total profit gained by the system during first six months of 2008 year is presented.

The average Initial Margin Deposit was \$1135 per one contract. To estimate the efficiency of the system, we propose to treat this Deposit as the average month investment. In this case, the average month profitability in the test periods in first six months of 2008 year is \$4882/1135/6=0.72. The positive high returns with relatively high percentage of winning trades in all test periods illustrate well the trading system

Table 6.3 Results in the teaching periods

	06.08.2007-18.01.2008	10.09.2007-22.02.2008	15.10.2007-28.03.2008	19.11.2007-02.05.2008	24.12.2007-06.06.2008
Profit,$	8181	5086	4178	3613	7110
Number of trades	144	95	86	112	109
Winning trades,%	57.64	56.84	59.3	54.64	62.38
Gross profit,$	14040	11645	10373	10890	10900
Gross loss,$	-5859	-6559	-6195	-7277	-3790

Table 6.4 Results in the test periods

	21.01.2008-22.02.2008	25.02.2008-28.03.2008	31.03.2008-02.05.2008	05.05.2008-06.06.2008	09.06.2008-11.07.2008	21.01.2008-11.07.2008
Profit,$	1160	1522	695	850	650	4882
Number of trades	34	24	18	21	26	123
Winning trades,%	55.88	54.17	55.55	57.14	61.54	56.91
Gross profit,$	3577	3122	1459	1391	1527	11077
Gross loss,$	-2417	-1600	-764	-541	-877	-6195

reliability and stability. Taking into account also the high value of the ratio Gross profit/ Gross loss=11077/6195=1.78 we can say the developed trading system may be successfully used in the practical trading.

6.3 Summary and Discussion

In this chapter, two approaches to building stock trading expert systems are presented.

The first of them is based on fuzzy logic representation of trading rules. In the framework of this approach, two stock trading expert systems are developed and compared. The first expert systems is built using the special adaptation of well-known Mamdani's approach. Another system is based on the recognition that Mamdani's approach had been developed for fuzzy logic controllers, not for solving decision making problem. Therefore, the so-called "logic-motivated fuzzy logic operators", based on the other mathematical representation of t-norm and Yager's implication rule, were used for the expert system engineering. The efficiency of such expert systems is naturally measured by comparing the system outputs versus the stock price movement. It is shown that "logic-motivated fuzzy logic operators" approach splits out into two different frameworks for expert systems building referred here as $Yager_{ave}$ and $Yager_{max}$, respectively. The results obtained using the real data of NYSE and Warsaw Stock Exchange allow us to say that the optimized expert system based on "logic-motivated fuzzy logic operators" framework provides substantially greater benefits and is more reliable. It is shown that $Yager_{ave}$ approach is the most adaptable, profitable and robust method for building fuzzy logic based stock trading systems which provides positive final profits even in the case of risky trading by long positions (buying) opposite to the downtrend.

The second considered approach to building expert and decision support systems is based on evidential reasoning. Unlike the other known methods, in the framework of this approach it is possible (in compliance with the key principle of the Dempster-Shafer theory) to assign a belief mass to a group of events, and aggregate different sources of evidence using the Dempster's rule of combination.

The basic principles and features of this approach are illustrated using numerical examples. In addition, to prove its practicability, the stock trading expert system based on evidential reasoning has been developed, optimized and tested on the real data from Warsaw Stock Exchange. It is shown that this system provides high returns (profits) with smooth profit curve and high percentage of winning trades when trading the futures contract on the main index of Warsaw Stock Exchange (WIG20).

References

1. Achelis, S.B.: Technical Analysis from A to Z: Covers Every Trading Tool-From the Absolute Breadth Index to the Zig Zag. Probus Publisher, Chicago (1995)
2. Baba, N., Kozaki, M.: An intelligent forecasting system of stock price using neural networks. In: Proceedings of IJCNN 1992, pp. 317–377 (1992)
3. Bayes, T.: An assay toward solving a problem in the doctrine of chances. Phil. Trans. Roy. Soc. (London) 53, 370–418 (1763)
4. Binaghi, E., Madella, P.: Fuzzy Dempster- Shafer reasoning for rule-based classifiers. Intelligent Syst. 14, 559–583 (1999)
5. Binaghi, E., Gallo, I., Madella, P.: A neural model for fuzzy Dempster-Shafer classifiers. International Journal of Approximate Reasoning 25, 89–121 (2000)
6. Brock, W., Lakonishok, J., LeBaron, B.: Simple technical trading rules and the stochastic properties of stock returns. Journal of Finance 47, 1731–1764 (1992)
7. Dacorogna, M.M., Muller, U.A., Jost, C., Pictet, O.V., Olsen, R.B., Ward, J.R.: Heterogeneous real-time trading strategies in the foreign exchange market. The European Journal of Finance 1, 383–403 (1995)
8. Dimova, L., Sevastianov, P., Sevastianov, D.: MCDM in a fuzzy setting: investment projects assessment application. International Journal of Production Economics 100, 10–29 (2006)
9. Dourra, H., Siy, P.: Investment using technical analysis and fuzzy logic. Fuzzy Sets and Systems 127, 221–240 (2002)
10. Dymova, L., Sevastianov, P., Bartosiewicz, P.: A new approach to the rule-base evidential reasoning: stock trading expert system application. Expert Systems with Applications 37, 5564–5576 (2010)
11. Fodor, J.C.: On fuzzy implications operators. Fuzzy Sets and Systems 42, 293–300 (1991)
12. Gorzalczany, M.B., Piasta, Z.: Neuro-fuzzy approach versus rough-set inspired methodology for intelligent decision support. Information Sciences 120, 45–68 (1999)
13. Haefke, C., Helmenstein, C.: Predicting stock market averages to enhance profitable trading strategies. In: Proceedings of the Third International Conference on Neural Networks in the Capital Markets, London, pp. 378–389 (2000)

14. Hodges, J., Bridges, S., Sparrow, C., Wooley, B., Tang, B., Jun, C.: The development of an expert system for the characterization of containers of contaminated waste. Expert Systems with Applications 17, 167–181 (1999)
15. Ishizuka, M., Fu, K.S., Yao, J.: Inference procedure and uncertainty for the problem reduction method. Inform. Sci. 28, 179–206 (1982)
16. Kaufmann, A., Gupta, M.: Introduction to Fuzzy Arithmetic-Theory and Applications. Van Nostrand Reinhold, New York (1985)
17. Kendall, S.M., Ord, K.: Time series, 3rd edn. Oxford University Press, New York (1990)
18. Kim, K.J., Han, I.: Genetic algorithms approach to feature discretization in artificial neural networks for the prediction of stock price index. Expert Systems with Applications 19, 125–132 (2000)
19. Kuo, R.J., Chen, C.H., Hwang, Y.C.: An intelligent stock trading decision support system through integration of genetic algorithm based fuzzy neural network and artificial neural network. Fuzzy Sets and Systems 118, 21–45 (2001)
20. Mahfoud, S., Mani, G.: Financial forecasting using genetic algorithms. Applications of Artificial Intelligence 10, 543–566 (1996)
21. Mamdani, E., Assilian, S.: An experiment in linguistic synthesis with a fuzzy logic controller. Int. J. Mach. Studies 1, 1–13 (1975)
22. Mehta, K., Bhattacharyy, S.: Adequacy of training data for evolutionary mining of trading rules. Decision Support Systems 37, 461–474 (2004)
23. Parson, S.: Current approaches to handling imperfect information in data and knowledge bases. IEEE Transactions on Knowledge and Date Engineering 8, 353–372 (1996)
24. Pawlak, Z.: Rough Sets. International Journal of Information and Computer Science 11, 145–172 (1982)
25. Pictet, O.V., Docorogna, M.M., Chopard, B., Shirru, M., Tomassini, M.: Using genetic algorithms for robust optimization in financial applications. Neural Network World 5, 573–587 (1995)
26. Rutkowska, D.: Neuro-fuzzy architectures and hybrid learning. Physica-Verlag, Heidelberg (2002)
27. Rutkowski, L.: Flexible Neuro-Fuzzy Systems: Structures, Learning and Performance Evaluation. Kluwer, Boston (2004)
28. Rutkowski, L., Cpałka, K.: Designing and Learning of Adjustable Quasi-Triangular Norms With Applications to Neuro-Fuzzy Systems. IEEE Transaction on Fuzzy Systems 13, 140–151 (2005)
29. Santiprabhob, P., Nguyen, H.T., Pedrycz, W., Kreinovich, V.: Logic-Motivated Choice of Fuzzy Logic Operators. In: FUZZ-IEEE, pp. 646–649 (2001)
30. Sevastianov, P., Dymova, L.: Synthesis of fuzzy logic and Dempster-Shafer Theory for the simulation of the decision-making process in stock trading systems. Mathematics and Computers in Simulation 80, 506–521 (2009)
31. Sevastjanov, P., Figat, P.: Aggregation of aggregating modes in MCDM, Synthesis of Type 2 and Level 2 fuzzy sets. Omega 35, 505–523 (2007)
32. Shen, L., Loh, H.T.: Applying rough sets to market timing decisions. Decision Support Systems 37, 583–597 (2004)
33. Straszecka, E.: Combining uncertainty and imprecision in models of medical diagnosis. Information Sciences 176, 3026–3059 (2006)
34. Sun, R.: Robust reasoning: integrating rule-based and similarity based reasoning. Artificial Intelligence 75, 241–295 (1995)
35. Torn, A., Zilinskas, A.: Global optimization. Springer, Berlin (1989)
36. Tsumoto, S.: Automated extraction of hierarchical decision rules from clinical databases using rough set model. Expert Systems with Applications 24, 189–197 (2003)

37. Turksen, I.B., Kreinovich, V., Yager, R.R.: A new class of fuzzy implications (axioms of fuzzy implication revisited). Fuzzy Sets and Systems 100, 267–272 (1998)
38. Wadman, D., Schneider, M., Schnaider, E.: On the use of interval mathematics in fuzzy expert system. International Journal of Intelligent Systems 9, 241–259 (1994)
39. Wang, Y.-F.: Mining stock price using fuzzy rough set system. Expert Systems with Applications 24, 13–23 (2003)
40. Warren, A.W.: Data Filtering For Trend Channel Analysis. Stocks and Commodities 11, 103–111 (1993)
41. Xu, D.-L., Liu, J., Yang, J.-B., Liu, G.-P., Wang, J., Jenkinson, I., Ren, J.: Inference and learning methodology of belief-rule-based expert system for pipeline leak detection. Expert Systems with Applications 32, 103–113 (2007)
42. Yager, R.R.: Multiple objective decision-making using fuzzy sets. International Journal of Man-Machine Studies 9, 375–382 (1979)
43. Yager, R.R.: Generalized probabilities of fuzzy events from belief structures. Inform. Sci. 28, 45–62 (1982)
44. Yager, R.R., Filev, D.P.: Including probabilistic uncertainty in fuzzy logic controller modeling using Dempster-Shafer theory. IEEE Trans. Syst., Man Cybernet. 25, 1221–1230 (1995)
45. Yang, J.B.: Rule and utility based evidential reasoning approach for multi-attribute decision analysis under uncertainties. European Journal of Operational Research 131, 31–61 (2001)
46. Yang, J.B., Liu, J., Wang, J., Sii, H.S., Wang, H.: Belief rule-base inference methodology using the evidential reasoning approach - RIMER. IEEE Transactions on Systems Man and Cybernetics Part A-Systems and Humans 36, 266–285 (2006)
47. Yang, J.B., Liu, J., Xu, D.L., Wang, J., Wang, H.: Optimization Models for Training Belief-Rule-Based Systems. IEEE Transactions on Systems, Man and Cybernetics, Part A-Systems and Humans 37, 569–585 (2007)
48. Yang, J.B., Xu, D.L.: On the evidential reasoning algorithm for multiple attribute decision analysis under uncertainty. IEEE Transactions on Systems, Man, and Cybernetics - Part A: Systems and Humans 32, 289–304 (2002)
49. Yang, J.B., Xu, D.L.: Nonlinear information aggregation via evidential reasoning in multi-attribute decision analysis under uncertainty. IEEE Transactions on Systems, Man, and Cybernetics - Part A: Systems and Humans 32, 376–393 (2002)
50. Yen, J.: Generalizing the Dempster-Shafer theory to fuzzy sets. IEEE Trans. Syst., Man Cybernet. 20, 559–570 (1990)
51. Zadeh, L.A.: Fuzzy Sets. Information and Control 8, 338–353 (1965)
52. Zimmerman, H.J., Zysno, P.: Latent connectives in human decision making. Fuzzy Sets and Systems 4, 37–51 (1980)

Chapter 7
Application of Interval and Fuzzy Analysis in Economic Modeling

In this chapter, a new approach to solving interval and fuzzy equations based on the generalized procedure of interval extension called "interval extended zero" method is proposed. The central for the proposed approach is the treatment of "interval zero" as an interval centered around 0. It is shown that such proposition is not of heuristic nature, but is a direct consequence of interval subtraction operation. Some methodological problems concerned with this definition of interval zero are discussed. It is shown that the resulting solution of interval linear equations based on the developed method may be naturally treated as a fuzzy number. An important advantage of a new method is that it substantially decreases the excess width effect. On the other hand, we show that it can be used as the reliable practical tool for solving linear interval and fuzzy equations as well as the systems of them. The fundamentals of the proposed approach we have published in [67, 68]. In this chapter, we present the generalization of the obtained and some new results.

The applications of the proposed approach are performed by the solution of well known Leontief's input-output problem in the interval setting and the solution of the problem of fuzzy Internal Rate of Return in budgeting.

7.1 Basics of "Interval Zero Extension" Method

This section deals with the problem of solving interval and fuzzy linear equations. The generalized procedure of interval extension called "interval extended zero" method is proposed, which leads to the solution of linear interval equations in the form of fuzzy number.

Therefore, some interval representations of such fuzzy solution are proposed. It is shown that they may be naturally treated as the modified operations of interval division.

The methodological problems concerned with this new approach to interval extension are discussed.

L. Dymowa: Soft Computing in Economics and Finance, ISRL 6, pp. 241–291.
springerlink.com

It is shown that a new method substantially decreases the so-called excess width effect, i.e., undesirable and sometimes drastic rising of the width of resulting intervals as the consequence of interval computations.

7.1.1 The Problem Formulation

The problems of interval and fuzzy equations solution are of perennial interest, because of their direct relevance to practical modeling and optimization of real world processes including finance [14, 21], economy [15, 20, 54, 74], mechanics [26].

Nevertheless, the problem of interval or fuzzy equations solution is not a trivial one even for the linear equations such as

$$A \cdot X = B, \tag{7.1}$$

where A and B are intervals or fuzzy values. As it is stated in [12], "...for certain values of A and B, Eq. (7.1) has no solution for X. That is, for some triangular fuzzy numbers A and B there is no fuzzy set X so that, using regular fuzzy arithmetic, $A \cdot X$ is exactly equal to B".

The same problem takes place in the case of interval form of Eq. (7.1). In other words, the classical solution too often fails to exist [17, 19]. Therefore, in the framework of modern approaches to the solution of fuzzy and interval forms of Eq. (7.1) as well as of linear fuzzy and interval systems, an equality of the left and right hand sides of Eq. (7.1) is not obligatory requirement [60].

The different numerical methods for solving a specific fuzzy linear system $A \cdot X = B$, where the entries of matrix A are real values and the entries of vector B are fuzzy numbers were proposed in [3, 5, 6, 7, 30, 77]. Ferreira et al. [34] proposed a numerical method for solving systems of interval polynomial equations. This method provides only a real valued solution. Abbasbandy and Otadi [4] obtained the real valued roots of fuzzy polynomials using fuzzy neural networks. Shieh [72] considered the equation $B = A + X$ where A, B are known fuzzy numbers and the fuzzy number X has to be calculated. It is shown in [72] that this problem can be formulated in the form of a fuzzy relation.

Although many different numerical methods were proposed for solving interval and fuzzy equations, including such complicated ones as neural net solutions [16, 18] and fuzzy extension of the Newton's method [1, 2], only particular solutions valid in the specific conditions were obtained.

Currently, the dominant approaches to the solution of linear fuzzy and interval systems are based on the treating of Eq. (7.1) as a set of real valued equations whose parameters belong to the corresponding intervals or fuzzy values A and B [49]. In this framework, the important ideas are the concepts of the united solution set (USS), its subsets called the tolerable solution set (TSS) and the controllable solution set [49]. Basing on these ideas, Buckley and Qu [19] and Buckley et al. [17] proposed three solutions of a fuzzy linear system: joint or vector solution X_J, marginal solution X_E and the solution X_1. It was shown that $X_J \subseteq X_E \subseteq X_1$, that means that the vector solution X_J is the more important one. Muzzioli and Reynaerts [60] proposed a generalization of the vector solution to the fuzzy linear system $A_1x+b_1=A_2x+b_2$. We note here that the vector solution plays an important role in our further analysis since it serves as an external constraint for the solution proposed in this section. Although the significant advances were achieved using these approaches, two problems are still open. The first of them is of methodological nature: how to interpret a solution of fuzzy or interval linear equation if it is not a solution of an equation in the classical sense? The second problem is the so-called "excess width effect" or inadmissible wide intervals or fuzzy values representing final solutions. In this section, we propose an approach alleviating these problems.

Since there is a certain pluralism in choosing an appropriate method for solving fuzzy or interval linear systems, we propose to turn back to the classical approach, but looking at the problem from the other point of view.

In our opinion, the root of the problem is that the equations

$$F(X)-B=0,\ F(X)=B,$$

where B is an interval or fuzzy value, $F(X)$ is some interval or fuzzy function, are not equivalent ones. Moreover, the main problem is that the conventional interval or fuzzy extension of usual equation, which leads to the interval or fuzzy equation such as $F(X)-B=0$, is not a correct procedure. Really, in the left hand side of this extended equation we have an interval, whereas in the right hand side we have the real valued zero. Since an interval can not be equal to the real value we shall call here this observation "interval equation's right hand side problem".

Less problems we meet when dealing with interval or fuzzy equation in the form of $F(X)=B$, but in many cases its roots are inverted intervals, i.e., such that $\overline{x}<\underline{x}$. This fact deserves a more detailed analysis and we study it thoroughly in this section. We prefer to treat fuzzy values as the sets of α-cuts since in this framework we have no problem of membership function's shape. Therefore, all fuzzy arithmetic problems are reduced to the interval ones. The key to the proposed approach is the observation that "zero" value in interval analysis does not mean "nothing" and a more natural is the treating of "interval zero" as an interval centered around zero. This allows us to avoid formally what we call here "interval equation's right hand side problem" and to use the classical approach to solving linear interval equations. As the result, an underdetermined equation with two variables representing the bounds of an interval root we are looking for is inferred. Of course, such equations normally have no solutions, but if there are known interval constraints on the variables, the solution can be obtained in the interval form with the use of constraint satisfaction method *CSM* [29]. We shall show that such constraints could be obtained from the basis definitions of interval analysis and that they completely coincide with the vector solution X_J [17, 19] reduced to the case of single interval equation.

The technique is based on the fuzzy extension principle [76]. The values of uncertain parameters in an equation are substituted for corresponding intervals or fuzzy values and all arithmetic operations are substituted for relevant interval/fuzzy operations.

Let us recall some basic principles of the fuzzy arithmetic [76] needed for further analysis. In general, for an arbitrary form of membership function the technique of fuzzy-interval calculations is based on the representation of initial fuzzy values in the form of the so-called α-cuts which are the intervals associated with the corresponding degrees of membership. All calculations are made with those α-cuts according to the well known interval arithmetic rules and the resulting fuzzy intervals are obtained as the set of corresponding final α-cuts. Thus, if A is a fuzzy number, then $A = \bigcup_{\alpha} \alpha A_{\alpha}$, where A_{α} is a crisp interval $\{x : \mu_A(x) \geq \alpha\}$, αA_{α} is a fuzzy value with the support A_{α}.

Therefore, if A, B, Z are fuzzy values and @ is an operation from $\{+, -, *, /\}$, then

$$Z = A@B = \bigcup_{\alpha}(A@B)_{\alpha} = \bigcup_{\alpha} A_{\alpha}@B_{\alpha}. \tag{7.2}$$

Since in the case of α-cut presentation, the fuzzy arithmetic is based on the interval arithmetic rules, the basic definitions of the applied interval analysis should also be presented. One of the most inconvenient features of interval arithmetic is the fast increasing of width of intervals obtained as the results of interval calculations. To reduce this undesirable effect, several different modifications of interval arithmetic were proposed. The most known are: non- standard interval arithmetic [56] based on the special form of interval subtraction and division, generalized interval arithmetic [36], segment interval analysis [66], centralized interval arithmetic [59], MV-form [22]. All these approaches provide good results only in the specific conditions. On the other hand, in practice the so-called "naive" form proposed by Moore [59] is proved to be the best one [39]. According to it, if $[x] = [\underline{x}, \overline{x}]$ and $[y] = [\underline{y}, \overline{y}]$ are crisp intervals and @ $\in \{+, -, *, /\}$, then

$$[x]@[y] = \{x@y, \forall x \in [x], \forall y \in [y]\}. \tag{7.3}$$

As the direct outcome of the basic definition (7.3), the following expressions were obtained:

$$[x] + [y] = [\underline{x} + \underline{y}, \overline{x} + \overline{y}], \tag{7.4}$$

$$[x] - [y] = [\underline{x} - \overline{y}, \overline{x} - \underline{y}], \tag{7.5}$$

$$[x] * [y] = [\min(\underline{x}\underline{y}, \overline{x}\overline{y}, \underline{x}\overline{y}, \overline{x}\underline{y}), \max(\underline{x}\underline{y}, \overline{x}\overline{y}, \underline{x}\overline{y}, \overline{x}\underline{y})], \tag{7.6}$$

$$[x]/[y] = [\underline{x}, \overline{x}] * [1/\overline{y}, 1/\underline{y}], (0 \notin [y]). \tag{7.7}$$

There are many inherent problems within the applied interval analysis, e.g., a division by zero-containing interval, but in general, it can be considered as a reliable mathematical tool for modeling under conditions of uncertainty. It is proved that a solution of an equation with fuzzy parameters can be obtained using the α-cuts representation of these parameters [47]. As the result we obtain a system of interval equations (see Chapter 3 for more detail).

Let us look at this problem from more general point of view. An important methodological problem of interval equations solution which is not widely discussed in the literature is what we call "interval equation's right hand side problem". Suppose there exists some basic non-interval equation $f(x) = 0$. Its natural interval extension can be obtained using replacement of its variables by interval ones and all arithmetic operations by relevant interval operations. As a result we get an interval equation $[f]([x]) = 0$. Observe that this equation is senseless since its left hand part represents an interval value, whereas the right hand part is non-interval degenerated zero. Obviously, if $[f](x) = [\underline{f}, \overline{f}]$, then equation $[f]([x]) = 0$ is true only when $\underline{f} = \overline{f} = 0$. It is easy to show that, in general, the equation $[f]([x]) = 0$ can be verified only for the inverted interval $[x]$, i.e., when $\overline{x} < \underline{x}$. Inverted intervals are analyzed in the framework of modal interval arithmetic [35], but it is very hard and perhaps even impossible to meet a real-life situation when the notation $\overline{x} < \underline{x}$ is meaningful.
It is known [59] that if an expression can be presented in the different, but algebraically equivalent forms, they provide different interval results after interval extension. The same is true for the equations.

Let us consider interval extensions of the simplest linear equation

$$ax = b \tag{7.8}$$

and its algebraically equivalent forms

$$x = \frac{b}{a}, \tag{7.9}$$

$$ax - b = 0, \tag{7.10}$$

for a, b being intervals ($0 \notin a$).

Since there are no strong rules in interval analysis for choosing the best form of equation among its algebraically equivalent representations to be extended, it is natural to compare the results we get from interval extensions of Eq. (7.8) - Eq. (7.10). Let $[a] = [\underline{a}, \overline{a}]$ and $[b] = [\underline{b}, \overline{b}]$ be intervals. For the sake of simplicity, let us first consider the case of $[a] > 0, [b] > 0$, i.e., $\underline{a}, \overline{a} > 0$ and $\underline{b}, \overline{b} > 0$. Then interval extension of Eq. (7.8) is $[\underline{a}, \overline{a}][\underline{x}, \overline{x}] = [\underline{b}, \overline{b}]$. Using conventional interval arithmetic rule (7.6) from this equation we obtain $[\underline{ax}, \overline{ax}] = [\underline{b}, \overline{b}]$. It is clear that equality of the right and left hand sides of this equation is possible only if $\underline{ax} = \underline{b}$ and $\overline{ax} = \overline{b}$ and finally we have

$$\underline{x} = \frac{\underline{b}}{\underline{a}}, \overline{x} = \frac{\overline{b}}{\overline{a}}. \tag{7.11}$$

As a consequence of the rule (7.7), interval extension of Eq. (7.9) results in the expressions

$$\underline{x} = \frac{\underline{b}}{\overline{a}}, \overline{x} = \frac{\overline{b}}{\underline{a}}. \tag{7.12}$$

It is important that the solution (7.12) is equivalent to what is called as joint or vector solution X_J (see [17, 19]) in the simplest case of single equation. In this case, the definition of X_J is automatically reduced to $[x] = \{x' \in R : \exists a \in [a], \exists b \in [b], ax = b\}$.

It is easy to see that this definition (based on the extension principle) is equivalent to the conventional rule of interval division (7.7). This is not surprising as the conventional interval arithmetic rules (7.4)-(7.7) are based on the "natural extension principle" [59]. The solution (7.12) plays an important role in our analysis. For the simplicity, here we shall call it "conventional interval solution".

Consider some examples.

Example 7.1. Let $a = [3,4]$, $[b] = [1,2]$. Then from Eqs. (7.11) we get $\underline{x} = 0.333, \overline{x} = 0.5$ and from Eqs. (7.12) $\underline{x} = 0.25, \overline{x} = 0.666$.

Example 7.2. Let $a = [1,2]$, $[b] = [3,4]$. Then from Eqs. (7.11) we get $\underline{x} = 3, \overline{x} = 2$ and from Eqs. (7.12) $\underline{x} = 1.5, \overline{x} = 4$.

Example 7.3. . Let $a = [0.1, 0.3]$, $[b] = [1,1]$ (i.e., b is a real number). Then from Eqs. (7.11) we obtain $\underline{x} = 10, \overline{x} = 3.333$ and from Eqs. (7.12)) $\underline{x} = 3.333, \overline{x} = 10$.

We can see that interval extension of Eq. (7.8) may result in inverted intervals $[x]$, i.e., such that $\overline{x} < \underline{x}$ (see Examples 2 and 3), whereas extension of Eq. (7.9) gives us correct intervals ($\underline{x} < \overline{x}$). It is worth noting that interval extension of Eq. (7.9) will always provide the correct resulting intervals because Eqs. (7.12) are inferred directly from the basic definition (7.7). For our purposes it is quite enough to state that interval extension of Eq. (7.9) guarantees the resulting intervals be correct ones in all cases, whereas interval extension of Eq. (7.8) may result in practically senseless inverted intervals. It is worth noting that in Example 3, the formal interval extension of Eq. (7.8) leads to contradictory interval equation since in the right hand side of extended Eq. (7.8) we have a degenerated interval b (real value), whereas the left hand side is an interval. In all such cases the solution of interval extension of Eq. (7.8) is an inverted interval. This, at first glance, strange result is easy to explain from common methodological positions. Really, the rules of interval mathematics are constructed in such a way that any arithmetical operation on intervals results in an interval as well. These rules conform to the well known common viewpoint that any arithmetical operation with uncertainties should increase total uncertainty (and entropy) of a system. Therefore, placing the degenerated intervals in right hand side of Eq. (7.8) would be equivalent to the requirement to reduce an uncertainty of the left hand side down to zero. This is possible only in the case of inverse character of the interval $[x]$ that in turn can be interpreted as a request to introduce negative entropy into a system.

The standard interval extension of Eq. (7.10) is $[\underline{ax}, \overline{ax}] - [\underline{b}, \overline{b}] = 0$. With the use of interval arithmetic rules (7.4) and (7.5), from this equation we obtain $[\underline{ax} - \overline{b}, \overline{ax} - \underline{b}] = 0$ and finally

$$\underline{x} = \frac{\overline{b}}{\underline{a}}, \overline{x} = \frac{\underline{b}}{\overline{a}}.$$

It is easy to see that in any case $\underline{x} > \overline{x}$, i.e., we obtain an inverted interval. Obviously, such solution may be considered only as absurd one. We can say this fact is the direct consequence of that conventional interval extension of Eq. (7.10) is in contradiction with the basic assumptions of interval analysis since the right hand side of this equation always is the degenerated zero, whereas the left hand side is represented by interval.

Summarizing, we can say that only Eq. (7.9) can be considered as the reasonable base for interval extension. On the other hand, from this base we obtain Eqs .(7.12) which often result in a drastic extension of the output interval in comparison with input intervals (see Example 3, where the width of the resulting interval $[x]$ is almost 35 times greater than the width of the initial interval $[a]$).

It is worthy to note here that we can use Eq. (7.9) only in the simplest cases of linear equations, whereas the key methodological problem we deal with is to find an adequate interval extension of linear and nonlinear equations in their most general form $f(x) = 0$.

It is easy to see that there is no way to improve the interval solutions we obtain from interval extensions of Eq. (7.8) and Eq. (7.9). In general, an effect of interval's width increasing can not be eliminated at all since it reflects the reality and is consistent with the indeterminacy (entropy) increase principle. Nevertheless, it does not mean that we could not attempt to improve interval arithmetic rules to reduce the width of resulting intervals to the maximum possible extent.
That is why, let us turn to the consideration of Eq. (7.10) and look at it from another point of view.

Strictly speaking, in the framework of conventional interval analysis any interval extension of Eq. (7.10) is not a correct operation since we obtain an interval mathematical expression only in the left hand side of the equation, whereas in its right hand side the usual zero integer is not changed. In our opinion, the root of the problem is that the conventional approach to the interval extension do not involve an operation we call "interval zero extension". Formally, when extending equation Eq. (7.10), one obtains not only interval in its left hand side, but interval zero in the right hand side, and in general, this interval zero cannot be degenerated interval $[0,0]$.

In other words, we propose an operation called "interval zero extension" to obtain an "interval zero" in the right side of extended Eq. (7.10). Since "interval zero" is not a degenerated interval, such approach makes it possible to solve the problem of the correct interval extension of Eq. (7.10).

First of all, what is "interval zero"? In conventional interval analysis [39], it is assumed that any interval containing zero may be considered as the "interval zero". This is a satisfactory definition to suppress the division by zero in the conventional interval arithmetic, but for our purposes a more restrictive definition is needed. Let us look at this problem from another point of view. Without a loss of generality, we can define the degenerated (usual) zero as the result of the operation $a - a$, where a is any real valued number or variable. Hence, in a similar way we can define an "interval zero" as the result of operation $[a] - [a]$, where $[a]$ is an interval. It is easy to see that for any interval $[a]$ from basic definitions (7.3) and (7.5) we get $[\underline{a},\overline{a}] - [\underline{a},\overline{a}] = [\underline{a} - \overline{a}, \overline{a} - \underline{a}] = [-(\overline{a} - \underline{a}), \overline{a} - \underline{a}]$. Therefore, in any case the result of interval subtraction $[a] - [a]$ is an interval centered around 0. Another approach to the interval subtraction $[a] - [a]$ and division $[a]/[a]$ is the so-called "dependence" hypothesis. It is based on the assumption that any $x \in [a]$ is dependent on corresponding x that belongs to the other sample of $[a]$. Hence, $[a] - [a] = 0$ and $[a]/[a] = 1$. The "dependence" hypothesis is well known and in some particular cases provides quite good results [38]. Nevertheless, in our case we prefer to use the other reasoning. It is clear that generally the result of interval calculations in the left hand side of Eq. (7.10) is not obligatory presented by the subtraction of dependent intervals. Therefore, to define an "interval zero", the subtraction of independent intervals should be considered. Let $[a]$ and $[b]$ be independent intervals. Then from basic definition (7.5) we get $[a] - [b]$=$[\underline{a} - \overline{b}, \overline{a} - \underline{b}]$ and in the asymptotic case when $\underline{a} \to \underline{b}$ and $\overline{a} \to \overline{b}$ we obtain $([a] - [b]) \to ([a] - [a])$=$[-(\overline{a} - \underline{a}), \overline{a} - \underline{a}]$. Obviously, there is no need for "dependence" hypothesis in such reasoning. Therefore, in all cases when we write anything like $[a] - [a]$ we threat such expression only in interval arithmetic sense according to the general definition (7.5).

Thus, if we want to treat a result of subtraction of two identical intervals as "interval zero", then the most general definition of such "zero" should be "interval zero is an interval centered around 0". It must be emphasized that introduced definition says nothing about the length of "interval zero". Really, when extending equation such (7.10) with previously unknown values of variables in the left hand side,the only thing we can say about the left hand side is that it should be an interval centered around with a not defined length. Hence, in general case, as the result of interval extension of Eq. (7.10) we get

$$[\underline{a}, \overline{a}][\underline{x}, \overline{x}] - [\underline{b}, \overline{b}] = [-y, y], \tag{7.13}$$

where y is an undefined parameter. The method for solving Eq. (7.13) has been developed. The right hand side of Eq. (7.13) is an interval centered around zero which can be treated as an interval extension of the right hand side of Eq. (7.10), in other words, as an interval extension of 0. This is the reason for us to call our approach "interval extended zero" method. Of course, the value of y in Eq. (7.13) is not yet defined and this seems to be quite natural since the values of $\underline{x}, \overline{x}$ are also not defined.

At first, consider the case of positive interval numbers $[a]$ and $[b]$, i.e., $\underline{a}, \overline{a}, \underline{b}, \overline{b} > 0$. Then from Eq. (7.13) we get

$$\begin{cases} \underline{ax} - \overline{b} = -y, \\ \overline{ax} - \underline{b} = y. \end{cases} \tag{7.14}$$

Summing the left hand sides and the right ones in Eqs. (7.14) we obtain only one linear equation with two unknown variables $\underline{x}$ and $\overline{x}$:

$$\underline{ax} + \overline{ax} - \underline{b} - \overline{b} = 0. \tag{7.15}$$

It is impossible to get a single real valued solution of Eq. (7.15)) as it is an underdetermined equation. On the other hand, if there are some constraints on the values of unknown variables, then Eq. (7.15) with these constraints may be considered as the so-called constraint satisfaction problem [29] and its interval solution may be obtained.

In our case of continuous variables the appropriate definition of the constraint satisfaction problem (*CSP*) can be presented as follows: given a vector $(x_1,...,x_n)$ of unknowns, a (C,x) constraint system is defined by a set of constraints $C = \{c_1,...,c_p\}$ and a bounded domain $x = x_1 \times ... \times x_n$ as follows:

$$C_i : f_i(x_1,...,x_n) = 0,\ i = 1,...,n,$$

$$C_j : f_j(x_1,...,x_n) \leq 0,\ j = m+1,...,p,$$

$$x_k \in [\underline{x}_k, \overline{x}_k],\ k = 1,...,n.$$

The constraints are presented by linear or nonlinear analytic expressions which do not need to be differentiable. Each variable lies in a closed interval. The Cartesian product of variable domains x is called a box. The solution set is the set of tuples in x that satisfy all the constraints from C. The purpose of interval-based algorithms is to generate a set of n-dimensional boxes whose union encloses the solution set. These boxes can be generated using, e.g., a branch-and-prune algorithm.

Nevertheless, in our case there is no need for such complicated numerical method as the analytic solution can be obtained. In our case, $x = \underline{x} \times \overline{x}$. So the set of constraints should be defined. As the first of them, we can treat the Eq. (7.15) itself. The other constraint on the variables $\underline{x}$ and $\overline{x}$ is the solution of Eq. (7.15) in assumption that $\underline{x} = \overline{x}$. In this degenerated case we get the solution of Eq. (7.15) as $x_m = \frac{\underline{b}+\overline{b}}{\underline{a}+\overline{a}}$. It is easy to see that x_m is the upper bound for $\underline{x}$ and the lower bound for $\overline{x}$: if $\underline{x} > x_m$ or $\overline{x} < x_m$ we get the degenerated solution of Eq. (7.15), i.e., $\underline{x} > \overline{x}$. The lower bound for $\underline{x}$ and the upper bond for $\overline{x}$ should be defined too. It is known that there are no strong rules for definition of the constraints in the framework of *CSP*. Generally, they can be introduced as "external" ones with respect to the considered problem. However, in our case there is no need for "external" constraints. We can use the conventional interval solution (7.12) to define the constraints as it provides the widest interval solution and may be obtained from the basic definition of interval arithmetic.

Therefore, we define the natural lower bound for $\underline{x}$ and the upper bond for $\overline{x}$ as follows: $\underline{x} = \frac{\underline{b}}{\overline{a}}, \overline{x} = \frac{\overline{b}}{\underline{a}}$.

Thus, we have $[\underline{x}] = [\frac{\underline{b}}{\overline{a}}, x_m]$ and $[\overline{x}] = [x_m, \frac{\overline{b}}{\underline{a}}]$. These intervals can be narrowed taking into account Eq. (7.15) which in the spirit of *CSP* is treated as a constraint. It is clear that the right bound of $\underline{x}$ and the left bound of $\overline{x}$, i.e., x_m, can not be changed as they present the degenerated (real valued) solution of Eq. (7.15). So let us focus of the left bound of $\underline{x}$ and the right bound of $\overline{x}$. From (7.15) we have

$$\underline{x} = \frac{\underline{b}+\overline{b}-\overline{a}\overline{x}}{\underline{a}}, \overline{x} \in [x_m, \frac{\overline{b}}{\underline{a}}],\ \overline{x} = \frac{\underline{b}+\overline{b}-\underline{a}\underline{x}}{\overline{a}}, \underline{x} \in [\frac{\underline{b}}{\overline{a}}, x_m]. \tag{7.16}$$

Obviously, when $\overline{x}$ is maximal in the interval $[x_m, \frac{\overline{b}}{\underline{a}}]$, i.e., $\overline{x} = \frac{\overline{b}}{\underline{a}}$, we get the minimal value of $\underline{x}$, i.e., $\underline{x}_{\min} = \frac{\underline{b}+\overline{b}}{\underline{a}} - \frac{\overline{a}\overline{b}}{\underline{a}^2}$. Similarly, from (7.16) we get the maximal in the interval $[\frac{\underline{b}}{\overline{a}}, x_m]$ value of $\overline{x}$, i.e., $\overline{x}_{\max} = \frac{\underline{b}+\overline{b}}{\overline{a}} - \frac{\underline{a}\underline{b}}{\overline{a}^2}$ when $\underline{x} = \frac{\underline{b}}{\overline{a}}$.

Generally, it is possible that $\underline{x}_{\min} < \frac{\underline{b}}{\overline{a}}$ and $\overline{x}_{\max} > \frac{\overline{b}}{\underline{a}}$. Therefore, the maximal lower bound of $\underline{x}$ and the minimal upper bound of $\overline{x}$ can be presented by the expressions $\underline{x}^L_{\max} = \max\left(\frac{\underline{b}}{\overline{a}}, \frac{\underline{b}+\overline{b}}{\underline{a}} - \frac{\overline{a}\overline{b}}{\underline{a}^2}\right)$, $\overline{x}^U_{\min} = \min\left(\frac{\overline{b}}{\underline{a}}, \frac{\underline{b}+\overline{b}}{\overline{a}} - \frac{\underline{a}\underline{b}}{\overline{a}^2}\right)$, respectively. Therefore, we get the following interval solution of Eq. (7.15):

$$[\underline{x}] = \left[\underline{x}^L_{\max}, \frac{\underline{b}+\overline{b}}{\underline{a}+\overline{a}}\right],\ [\overline{x}] = \left[\frac{\underline{b}+\overline{b}}{\underline{a}+\overline{a}}, \overline{x}^U_{\min}\right]. \tag{7.17}$$

It is important that in the framework of *CSP*, the following relations between $\underline{x}^L_{\max}$ and $\overline{x}^U_{\min}$ are fulfilled in calculations: if $\underline{x}^L_{\max} = \frac{\underline{b}}{\overline{a}}$, then $\overline{x}^U_{\min} = \frac{\underline{b}+\overline{b}}{\overline{a}} - \frac{\underline{a}\underline{b}}{\overline{a}^2}$; if $\overline{x}^U_{\min} = \frac{\overline{b}}{\underline{a}}$, then $\underline{x}^L_{\max} = \frac{\underline{b}+\overline{b}}{\underline{a}} - \frac{\overline{a}\overline{b}}{\underline{a}^2}$. It is worth noting that in our simplest case, the method based on *CSP*. i.e., *CSM* leads to the solution, which have one of the bounds (upper or lower, see (7.17)) the same of corresponding bound of conventional interval solution (in the considered case of positive $[a]$ and $[b]$, the solution is as follows: $\underline{x}^L_{\max} = \frac{\underline{b}}{\overline{a}}$, $\overline{x}^U_{\min} = \frac{\underline{b}+\overline{b}}{\overline{a}} - \frac{\underline{a}\underline{b}}{\overline{a}^2}$).

Taking into account our remark concerned with equivalence of the conventional interval division rule and the so-called vector solution X_J [19, 17], we can say X_J serves as an external constrain or envelopment of our solution.

It is seen that Eqs. (7.17) define all possible solutions of Eq. (7.15). The values $\underline{x}^L_{min}, \overline{x}^U_{max}$ constitute the interval which produces the widest interval zero after of its substitution in Eqs. (7.14). In other words, the maximum interval solution's width $w_{\max} = \overline{x}^U_{\min} - \underline{x}^L_{\max}$ corresponds to the maximum value of y: $y_{\max} = \frac{\overline{a}\overline{b}}{\underline{a}} - \underline{b}$. Substitution of the degenerated solution $\underline{x} = \overline{x} = x_m$ in Eqs. (7.14) produces the minimum value of y: $y_{min} = \frac{\overline{a}\cdot\overline{b} - \underline{a}\cdot\underline{b}}{\underline{a}+\overline{a}}$.

Indeed, using the conventional *CSM* we get an approximate interval solution of underdetermined real valued equation which has no solution in usual sense. The similar problem we meet in the case of interval and fuzzy equations which often have no solution in usual (classical) sense too. Therefore, an approximate solution of interval equation obtained using the *CSM* and "interval extended zero" method may be treated as the natural extension of *CSM*.

It is clear that for any permissible solution $\underline{x}' > \underline{x}^L_{max}$ we have corresponding $\overline{x}' < \overline{x}^U_{min}$, for each $\underline{x}'' > \underline{x}'$ inequalities $\overline{x}'' < \overline{x}'$ and $y'' < y'$ take place. Thus, the formal interval solution (7.17) factually represents the continuous set of nested interval solutions of Eq. (7.15). Hereinafter, we show that this set of interval solutions can be in a natural way interpreted as a fuzzy number.

We can see that the values of y characterize the closeness of the right hand side of Eq. (7.13) to degenerated zero and the minimum value y_{min} is defined exclusively by interval parameters $[a]$ and $[b]$. Hence, the values of y may be considered, in a certain sense, as a measure of interval solution's uncertainty caused by initial uncertainty of Eq. (7.13). Therefore we can introduce

$$\alpha = 1 - \frac{y - y_{min}}{y_{max} - y_{min}}, \tag{7.18}$$

which may be associated with each permissible solution of Eq. (7.13), $[\underline{x},\overline{x}]$. We can see that α rises from 0 to 1 with a decrease in the interval's length from maximum value to 0, i.e., with an increase of the solution's certainty. Consequently, the values of α may be treated as labels of α-cuts representing some fuzzy solution to interval Eq. (7.13). Finally, we obtain the solution in the form of triangular fuzzy number (see Fig. 7.1):

$$\tilde{x} = \left\{ \underline{x}^L_{\max}, \frac{\underline{b}+\overline{b}}{\underline{a}+\overline{a}}, \overline{x}^U_{\min} \right\} \tag{7.19}$$

This result needs some comments. Using the proposed approach to the solution

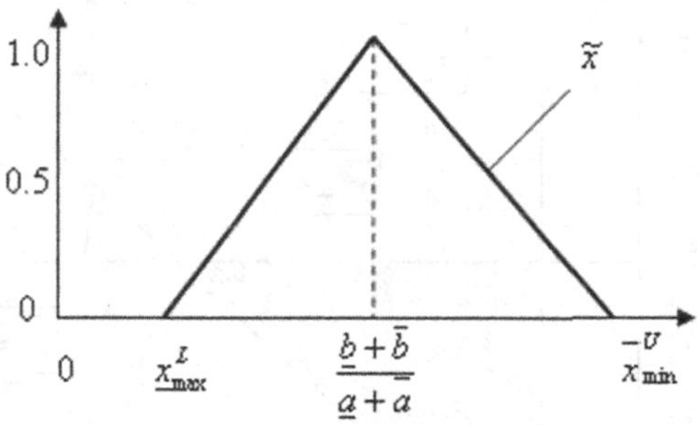

Fig. 7.1 Fuzzy root of fuzzy Eq. (7.13)

of interval linear equation based on the Eq. (7.13), we obtain a triangular fuzzy number with a support which in all cases is included in the interval obtainable from conventional interval extension of Eq. (7.9).

An important characteristic of the fuzzy solution $\tilde{x}$ is its mode $x_m = \frac{\underline{b}+\overline{b}}{\underline{a}+\overline{a}}$ (see Fig. 7.1). It was shown above that x_m is a degenerated solution of Eq. (7.15), i.e., $\underline{x} = \overline{x} = x_m$. On the other hand, x_m is an asymptotic solution when intervals $[a]$ and $[b]$ contract to the points. Thus, x_m plays in a fuzzy solution the role similar to a middle point of usual interval.

At first glance, it seems somewhat surprising to have a fuzzy solution of interval equation. On the other hand, the proposed "interval extended zero" method is based on some restricting assumptions. The main assumption is the treating of "interval zero" as an interval centered around 0. As a consequence, we obtain a solution in the form of triangular fuzzy number, which is more certain result than the interval representing its support since such fuzzy number inherits more information about possible real valued solutions.

Another explanation of these results can be provided using the following inexact, but intuitively obvious reasoning. When we have interval uncertainty in the parameters of an equation, then we usually produce the interval of possible values of the solution. Intuitively, however, some values from this interval are more plausible (probable) and some are less plausible (probable). For example, if we know that $x = a + b$ and we know that a and b are in the intervals $[-1,1]$, then strictly speaking, we can only guarantee that x belongs to the interval $[-2,2]$. However, intuitively, the value 0 is much more plausible (probable) than say 2 because the only way to get 2 is to have the worst-case situation of $a = b = 1$, while 0 can come from many different values of a and b. It is therefore desirable to supplement the widest interval solution with the fuzzy number describing which values from this interval are more plausible and which are less plausible.

The developed method can be illustrated graphically. Consider the example $[a] = [1,3]$, $[b] = [3,5]$. Evidently, the shaded area in Fig. 7.2 is formed by the natural constrains (7.12) and $\underline{x} \leq \overline{x}$. Since in addition for each pair $\underline{x}, \overline{x}$, Eq. (7.15) should

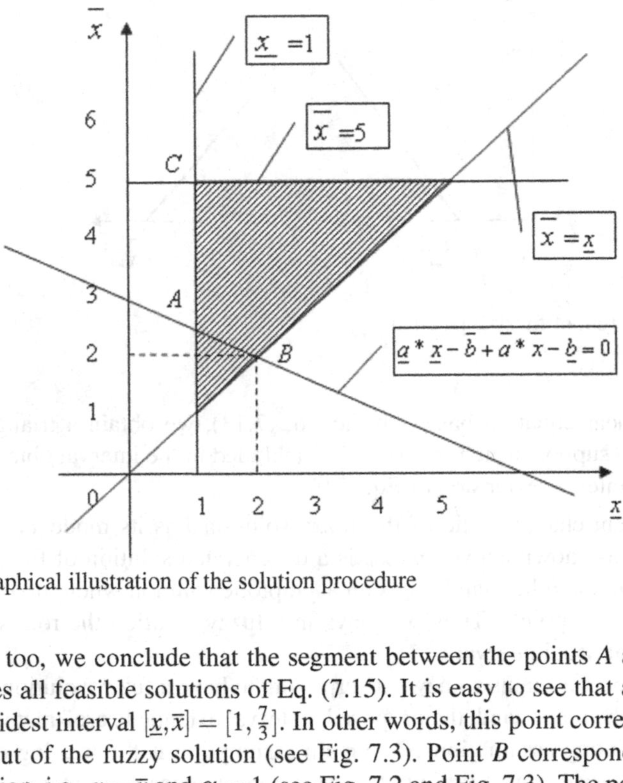

Fig. 7.2 Graphical illustration of the solution procedure

be verified too, we conclude that the segment between the points A and B in Fig. 7.2 includes all feasible solutions of Eq. (7.15). It is easy to see that at point A we have the widest interval $[\underline{x}, \overline{x}] = [1, \frac{7}{3}]$. In other words, this point corresponds to the lowest α-cut of the fuzzy solution (see Fig. 7.3). Point B corresponds to the real value solution, i.e., $\underline{x} = \overline{x}$ and $\alpha = 1$ (see Fig. 7.2 and Fig. 7.3). The points between A and B correspond to the intermediate α-cuts $(0 \leq \alpha \leq 1)$.

It is important that introduced "interval zero" $[-y, y]$ in the right hand side of Eq. (7.13) may be naturally treated as an "error" of an approximate fuzzy solution. Obviously, substituting the widest interval solution $[\underline{x}, \overline{x}] = [1, \frac{7}{3}]$ obtained in the considered example in Eq. (7.13) we get the maximal "error" of the approximate solution $[-4,4]$. On the other hand, substituting the conventional interval solution Eq. (7.12) (which is the same as joint or vector solution X_J [17, 19]) into Eq. (7.13) or Eqs. (7.14), we get in the right hand side of these equations the interval "error"

$[-4, 12]$. It is seen that the proposed "interval extended zero" method provides the substantially smaller "error" than the conventional or joint solution X_J can produce. It is important also that using the "interval extended zero" method we get an "error" in the form of the centered around zero interval which can be naturally treated as an "interval zero". It is shown in the Fig. 7.3 that the width of the fuzzy solution obtained in the considered example is substantially smaller than that of the conventional interval solution.

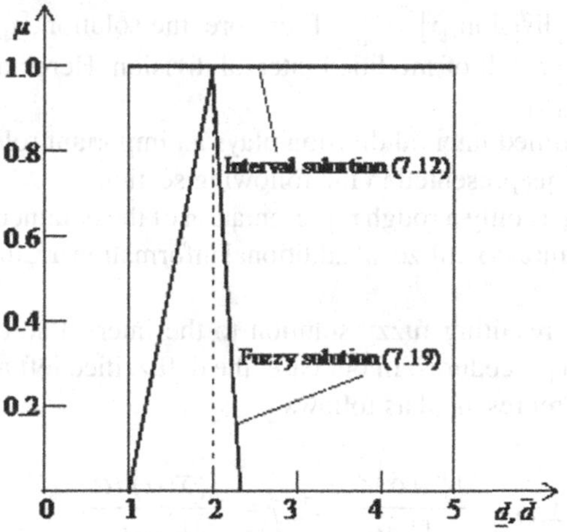

Fig. 7.3 Fuzzy and interval solutions of Eq. (7.13) for $[a] = [1, 3]$, $[b] = [3, 5]$

In the similar way, the fuzzy solutions of Eq. (7.13) have been obtained for other placements of intervals $[a]$ and $[b]$. In these cases, interval extensions of (7.10) are usually resulted in the expressions different from (7.12) and Eq. (7.15) takes another forms too.

In the case of $[a] < 0$, $[b] > 0$, i.e., $\underline{a}, \overline{a} < 0$, $\underline{b}, \overline{b} > 0$ we get $\underline{ax} - \overline{b} + \overline{ax} - \overline{b} = 0$, $\tilde{x} = \left\{\underline{x}^L_{\max}, \frac{\underline{b}+\overline{b}}{\underline{a}+\overline{a}}, \overline{x}^U_{\min}\right\}$, $\underline{x}^L_{\max} = \max(\frac{\overline{b}}{\overline{a}}, \frac{\underline{b}+\overline{b}}{\underline{a}} - \frac{\overline{a}\underline{b}}{\underline{a}^2})$, $\overline{x}^U_{\min} = \min(\frac{\underline{b}}{\underline{a}}, \frac{\underline{b}+\overline{b}}{\overline{a}} - \frac{\underline{a}\overline{b}}{\overline{a}^2})$.

In the case of $[a] > 0$, $[b] < 0$, i.e., $\underline{a}, \overline{a} > 0$, $\underline{b}, \overline{b} < 0$ we get $\overline{a}\underline{x} - \overline{b} + \underline{a}\overline{x} - \underline{b} = 0$, $\tilde{x} = \left\{\underline{x}^L_{\max}, \frac{\underline{b}+\overline{b}}{\underline{a}+\overline{a}}, \overline{x}^U_{\min}\right\}$, $\underline{x}^L_{\max} = \max\left(\frac{\underline{b}}{\underline{a}}, \frac{\underline{b}+\overline{b}}{\overline{a}} - \frac{\underline{a}\overline{b}}{\overline{a}^2}\right)$, $\overline{x}^U_{\min} = \min\left(\frac{\overline{b}}{\overline{a}}, \frac{\underline{b}+\overline{b}}{\underline{a}} - \frac{\overline{a}\underline{b}}{\underline{u}^2}\right)$.

In the case of $[a] < 0$, $[b] < 0$, i.e., $\underline{a}, \overline{a} < 0$, $\underline{b}, \overline{b} < 0$ we get $\underline{a}\overline{x} - \overline{b} + \overline{a}\underline{x} - \underline{b} = 0$, $\tilde{x} = \left\{\underline{x}^L_{\max}, \frac{\underline{b}+\overline{b}}{\underline{a}+\overline{a}}, \overline{x}^U_{\min}\right\}$, $\underline{x}^L_{\max} = \max\left(\frac{\overline{b}}{\underline{a}}, \frac{\underline{b}+\overline{b}}{\overline{a}} - \frac{\underline{ab}}{\overline{a}^2}\right)$, $\overline{x}^U_{\min} = \min\left(\frac{\underline{b}}{\overline{a}}, \frac{\underline{b}+\overline{b}}{\underline{a}} - \frac{\overline{a}\overline{b}}{\underline{a}^2}\right)$.

In the case of $[a] > 0$, $0 \in [b]$, we get $\overline{a}\underline{x} - \overline{b} + \overline{ax} - \underline{b} = 0$, $\tilde{x} = \left\{\underline{x}^L_{\max}, \frac{\underline{b}+\overline{b}}{2\overline{a}}, \overline{x}^U_{\min}\right\}$, $\underline{x}^L_{\max} = \max\left(\frac{\underline{b}}{\underline{a}}, \frac{\underline{b}+\overline{b}}{\overline{a}} - \frac{\overline{b}}{\underline{a}}\right)$, $\overline{x}^U_{\min} = \min\left(\frac{\overline{b}}{\underline{a}}, \frac{\underline{b}+\overline{b}}{\overline{a}} - \frac{\underline{b}}{\underline{a}}\right)$.

In the case of $[a] < 0$, $0 \in [b]$, we get $\underline{a}\overline{x} - \overline{b} + \underline{ax} - \underline{b} = 0$, $\tilde{x} = \left\{\underline{x}^L_{\max}, \frac{\underline{b}+\overline{b}}{2\overline{a}}, \overline{x}^U_{\min}\right\}$, $\underline{x}^L_{\max} = \max\left(\frac{\overline{b}}{\overline{a}}, \frac{\underline{b}+\overline{b}}{\underline{a}} - \frac{\underline{b}}{\overline{a}}\right)$, $\overline{x}^U_{\min} = \min\left(\frac{\underline{b}}{\overline{a}} \frac{\underline{b}+\overline{b}}{\underline{a}} - \frac{\overline{b}}{\overline{a}}\right)$.

Obviously, we can assume the support of obtained fuzzy number to be a solution of the analyzed problem. Such a solution may be treated as a "pessimistic" one since it corresponds to the lowest α-cut of the resulting fuzzy value. We use here the word "pessimistic" to emphasize that this solution is charged with the largest imprecision as it is obtained in the most uncertain conditions possible on the set of considered α -cuts.

On the other hand, we can treat it as an approximate solution of the initial interval equation (7.10), which in turn can be formally presented in the algebraically equivalent form of interval division $[x] = \frac{[b]}{[a]}$. Therefore, the solution $[\underline{x}_{\max}, \overline{x}_{\min}]$ can be formally treated as the result of modified interval division. Hereinafter, such result will be denoted as $[x]_{\text{mod}}$.

The concept of modified interval division plays an important role in the solution of linear interval systems presented in the following section.

The solution $[x]_{\text{mod}}$ is only a rough representation of the obtained fuzzy solution. Therefore, it seems natural to utilize all additional information available in the fuzzy solution.

We can reduce the resulting fuzzy solution to the interval solution using well known defuzzification procedures. In our case, the defuzzified left and right bounds of the solution can be represented as follows:

$$\underline{x}_{def} = \frac{\int_0^1 \underline{x}(\alpha)d\alpha}{\int_0^1 d\alpha}, \overline{x}_{def} = \frac{\int_0^1 \overline{x}(\alpha)d\alpha}{\int_0^1 d\alpha}. \tag{7.20}$$

Expressions (7.20) present the simplest form of type reduction and the result is an interval. They emphasize that contribution of the α-cut to an overall estimation rises with increase of α. Hence, we can say that the use of Esp. (7.20) provides the less type reduction "error" than using the support of the fuzzy solution, as these expressions utilize more information of this solution. Of course, the set of complementary parameterized weighted functions of α can be used in these integrals. Nevertheless, if there are no additional reasons to introduce such functions, the simplest form of the type reduction (like (7.20)) should be used. For example, in the case of $[a],[b] > 0$, from (7.14) and (7.18) we get the expressions for $\underline{x}(\alpha)$ and $\overline{x}(\alpha)$. Substituting them into Esp. (7.20) we have

$$\underline{x}_{def} = \frac{\overline{b}}{\underline{a}} - \frac{y_{max} + y_{min}}{2\underline{a}}, \overline{x}_{def} = \frac{\underline{b}}{\overline{a}} + \frac{y_{max} + y_{min}}{2\overline{a}}. \tag{7.21}$$

Hereinafter, such interval solutions which can be treated also as the results of modified interval division will be denoted for all placements of $[a]$ and $[b]$ as $[x]_{\text{mod}\,def}$.

It is easy to prove that the obtained interval $[\underline{x}_{def}, \overline{x}_{def}]$ is always included into support interval of the initial fuzzy solution, i.e., $[\underline{x}_{def}, \overline{x}_{def}] \subset [\underline{x}^L_{max}, \overline{x}^U_{min}]$. To illustrate, let us consider the example: $[a] = [14,60], [b] = [25,99]$. In the conventional interval arithmetic, Eqs. (7.12) provide $[\underline{x},\overline{x}] = [0.42,7.07]$. Using the proposed method we get $[\underline{x}^L_{max},\overline{x}^U_{min}] = [0.42,1.97]$ and $[\underline{x}_{def},\overline{x}_{def}] = [1.04,1.82]$. It is easy to see that $[\underline{x}_{def},\overline{x}_{def}] \subset [\underline{x}^L_{max},\overline{x}^U_{min}] \subset [\underline{x},\overline{x}]$. Moreover, the length of $[\underline{x},\overline{x}]$ is 4.3 times greater than that of $[\underline{x}^L_{max},\overline{x}^U_{min}]$ and 8.5 times greater than that of $[\underline{x}_{def},\overline{x}_{def}]$. Thus, the proposed method provides a considerable reduce in the resulting interval's length in comparison with that obtained using conventional interval arithmetic rules.

The developed approach to solving interval linear equations was used to develop a new interval Gauss elimination algorithm allowing us to solve the interval extended Leontiefs input-output problem without drastic increasing of resulting intervals (see Subsection 7.2.2). The method presented here may be considered as a general framework for solving not only linear interval equations and sets them, but interval and fuzzy nonlinear equations as well.

7.1.2 Solution Linear Fuzzy Equations

Let us consider a fuzzy linear equation

$$A \cdot X = C, \tag{7.22}$$

where A , C are fuzzy values.

Using the α -cut representation of fuzzy values

$$A = \bigcup_{\alpha} [a]_{\alpha}, C = \bigcup_{\alpha} [c]_{\alpha},$$

where $[a]_{\alpha}$,$[c]_{\alpha}$ are intervals on the α -cuts, and the "interval extended zero" method described in the previous subsection, the problem (7.22) can be reformulated as the set of crisp interval equations on the corresponding α-cuts.

Then taking into account that the "interval extended zero" method leads to the fuzzy solution of an interval equation, the problem (7.22) finally takes the following form:

$$\bigcup_{\alpha} ([a]_{\alpha} X_{\alpha} - [c]_{\alpha}) = [-y,y]_{\alpha}, \tag{7.23}$$

where X_{α} is a fuzzy value to be found as a solution of interval equation on the corresponding α -cut. The role of the undefined parameter y has been clarified in the Subsection 7.1.1. As the solution of (7.23) provides the set of X_{α}, the generalized solution of (7.22) in the spirit of α -cuts based approach can be presented as $X = \bigcup_{\alpha} X_{\alpha}$.

Let us consider two simple examples. In the first of them, the values A and C are the symmetrical trapezoidal fuzzy numbers $A = [14,30,44,60]$, $C = [25,50,74,99]$ (see Fig. 7.4).

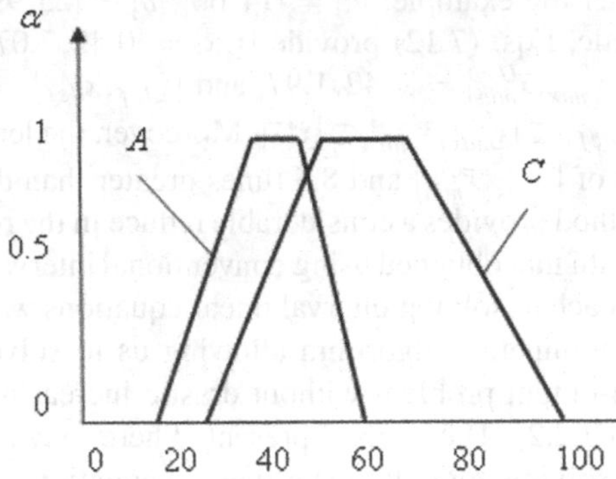

Fig. 7.4 Symmetrical fuzzy parameters of Eq. (7.22)

The resulting solution in the form of the set of X_α is presented in Fig. 7.5. It is seen that X_α have a common peak.

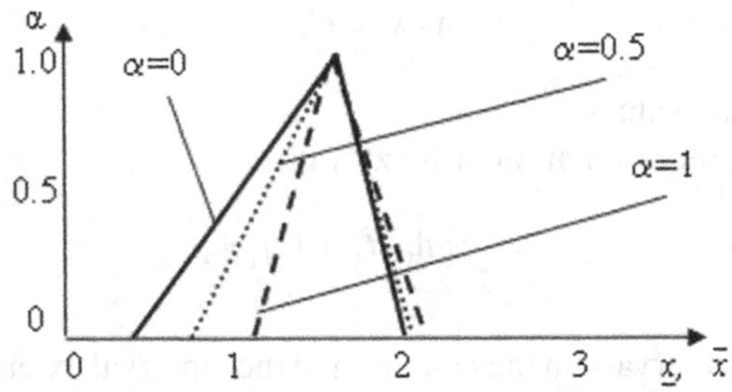

Fig. 7.5 Solution of Eqs. (7.23) in the case of symmetrical fuzzy parameters

Nevertheless, if A and C are asymmetrical fuzzy number, then the resulting X_α have no a common peak. For example, if $A = [14,26,40,60]$, $C = [25,60,80,99]$ (see Fig. 7.6), we get the result presented in Fig. 7.7.

The fuzzy solutions on the α -cuts in the case of symmetrical A and C have a common peak which in this case corresponds to real valued solutions $\frac{\underline{b}_\alpha+\overline{b}_\alpha}{\underline{a}_\alpha+\overline{a}_\alpha}$ (see Fig. 7.1) of interval equations on α-cuts. It is easy to see that these real valued solutions are the same for the symmetrical A and B, whereas they are different ones for the asymmetrical A and C. Therefore, in the last case there is no common peak. It is seen that fuzzy solutions on α-cuts are not nested. This is the cost of the use of *CSM* which is not a linear procedure (see (7.17)). A natural question arises: how to use such set of not nested fuzzy solutions on the α -cuts? We use the standard disjunction procedure, i.e., the final fuzzy solution is constructed as $X = \bigcup\limits_{\alpha} X_\alpha$. In practice, this solution can be simplified as it is usually enough to take into account only the lower and upper α-cuts: $X = X_0 \cup X_1$. Using this approach, the set of X_α

shown in Fig. 7.7 can be finally transformed to the fuzzy solution presented in Fig. 7.8. Therefore, the lack of nesting of the local solutions we obtain on the α-cuts is not a practical problem.

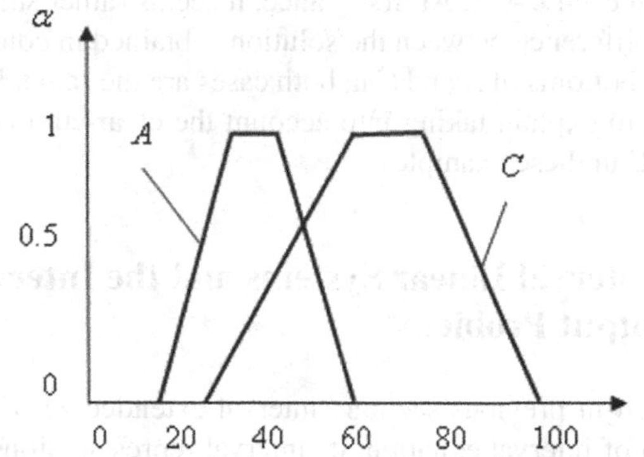

Fig. 7.6 Asymmetrical fuzzy parameters of Eq. (7.22)

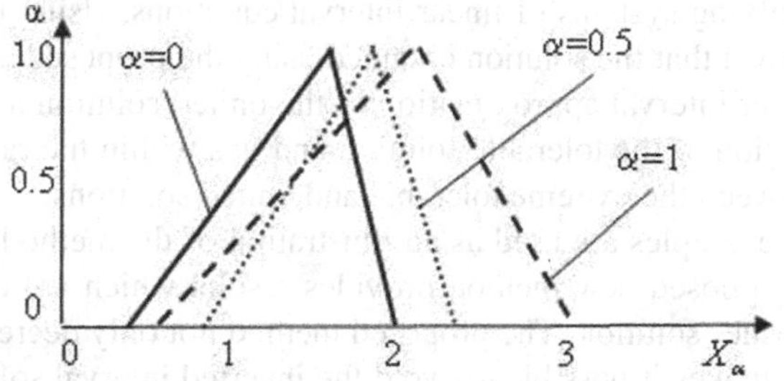

Fig. 7.7 Solution of Eqs.(7.23) in the case of asymmetrical fuzzy parameters

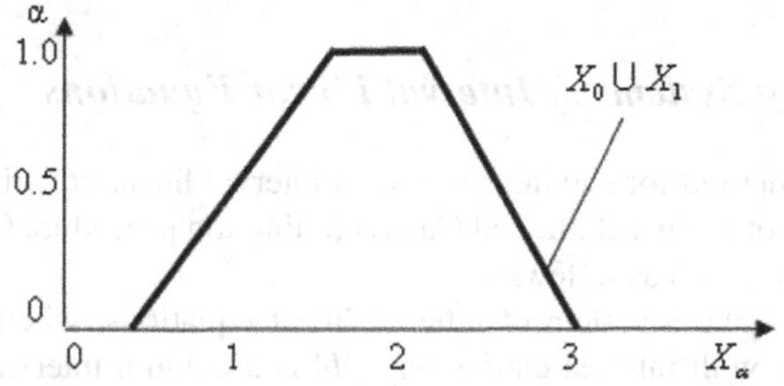

Fig. 7.8 Aggregated solution of Eq. (7.23) in the case of asymmetrical fuzzy parameters

It is important that in the both considered examples, the maximal length of generalized solution, i.e., the length of bottom of the trapezoidal fuzzy number $X = \bigcup_{\alpha} X_\alpha$, is substantially smaller than that obtained using the conventional approach, i.e., expressions (7.12) on each α-cut. At first glance, it seems rather surprising that there is an appreciable difference between the solutions obtained in considered examples in spite of that the bottoms of A and C in both cases are the same. Nevertheless, this difference is easy to explain taking into account the clear-cut distinction between the tops of A and C in these examples.

7.2 Solving Interval Linear Systems and the Interval Leontiev's Input-Output Problem

Since the presented in previous section "interval extended zero" method leads to the fuzzy solution of interval equation, its interval representations are proposed. It is shown that they may be naturally treated as the modified operations of interval division. These operations of interval division are used for the modified interval extensions of known numerical methods for solving systems of usual linear equations and finally for solving systems of linear interval equations. Using the well known example, it is shown that the solution obtained using the proposed approach can be treated as the inner interval approximation of the united solution and the outer interval approximation of the tolerable solution and lies within the range of possible AE-solutions between the extreme tolerable and united solutions.

Seven known examples are used as an illustration of the method's efficacy. It is shown that the proposed new method provides results which are close to the so-called maximal inner solution. The proposed method not only decreases the excess width effect, but makes it possible to avoid the inverted interval solutions too. The influence of system's size and zero entries on the resulting excess width effect is analyzed using the Leontief's input-output model of economics as an application of the proposed method.

7.2.1 Solving Systems of Interval Linear Equations

The developed method for solving systems of interval linear equations is based on the modification of the usual interval Gauss elimination procedure ($UIGEP$) which briefly can be presented as follows.

Let $[A][x] = [b]$ be a system of interval linear equations, where $[A]$ is an $n \times n$ interval matrix with interval entries $[a_{ij}]$, $[b]$ is a column interval vector with n entries $[b_i]$, $[x]$ is a interval solution vector.

In the forward elimination stage, the system is reduced to the triangular form using elementary row operations:

$$[a_{ij}]^{(k+1)} = [a_{ij}]^{(k)} - \frac{[a_{ik}]^{(k)}[a_{kj}]^{(k)}}{[a_{kk}]^{(k)}},$$

$$[b_i]^{(k+1)} = [b_i]^{(k)} - \frac{[a_{ik}]^{(k)}[b_k]^{(k)}}{[a_{kk}]^{(k)}},$$

$$[a_{kk}]^{(k)} \neq 0, \qquad (k = 1,2,...,n-1; i,j = k+1,k+2,...,n), \tag{7.24}$$

where k is the row number.

In the backward elimination stage, the interval solution vector is obtained as follows:

$$[x_n] = \frac{[b_n]^{(n)}}{[a_{nn}]^{(n)}},$$

$$[x_i] = \frac{[b_i]^{(i)} - \sum\limits_{j=i+1}^{n} [a_{ij}]^{(i)}[x_j]}{[a_{ii}]^{(i)}},$$

$$(i = n-1, n-2, ..., 2, 1). \tag{7.25}$$

The usual interval arithmetic rules are applied in *UIGEP*. To improve the numerical stability of the above algorithm, partial pivoting, based on the means of interval entries is employed in the forward elimination stage.

To develop the modified interval Gauss elimination procedure (*MIGEP*), the usual operation of interval division in (7.24) and (7.25) is substituted for the modified interval divisions presented in previous Section.

First of all we compare the results obtained using our method with the so-called formal, united, controllable and tolerable solution sets which play an important role in the theory of interval analysis [71]. The key concept in the solution of interval equations is the so-called formal solution of the interval equation sometimes referred to as the algebraic solution. By definition, an interval solution (interval vector, matrix, etc.) is called a formal solution of the interval equation (system of equations, inequalities, etc.) if substituting this interval into the equation and executing all interval arithmetic operations results in a valid equality. There are also the following three solution sets which have been the subject of (more or less) active research in modern interval analysis:

- United solution set

$\Xi_{uni}([A],[b]) = \{x \in \Re^n | (\exists A \in [A])(\exists b \in [b])(Ax = b)\}$ formed by the solutions of real valued system $Ax = b$ with $A \in [A]$ and $b \in [b]$. It is undoubtedly the most popular solution sets due to historical origination of interval analysis from sensitivity problems. $\Xi_{uni}([A],[b])$ is sometimes called the simply solution set. Its analogue for dynamical systems is the well-known attainability set (see [45, 52]).

- Tolerable solution set

$\Xi_{tol}([A],[b]) = \{x \in \Re^n | (\forall A \in [A])(\exists b \in [b])(Ax = b)\}$ formed by all real valued vectors x such that $Ax = b$ for any $A \in [A]$ (see, e.g., [31, 61, 69]) It was actually the first of the solution sets whose definition involves different logical quantifiers.

- Controllable solution set

$\Xi_{ctr}([A],[b]) = \{x \in \Re^n | (\forall b \in [b])(\exists A \in [A])(Ax = b)\}$ formed by all real valued vectors $x \in \Re^n$, such that for any desired $\forall b \in [b]$ we can find an appropriate $\exists A \in [A]$ satisfying $Ax = b$ (see [70]).

Since obtaining united, tolerable and controllable solution sets is an *NP*-hard problem, these solutions are usually illustrated using examples of 2×2 linear interval systems. Therefore, we consider the popular interval linear system from [9] repeatedly considered by many authors [71] as an example:

$$A = \begin{bmatrix} [2,4] & [-2,1] \\ [-1,2] & [2,4] \end{bmatrix} \; B = \begin{bmatrix} [-2,2] \\ [-2,2] \end{bmatrix}$$

In this case, the controllable solution set is empty [71]. The tolerable, united and formal solutions of this system are presented in Fig. 7.9 together with the solution obtained using the usual interval Gauss elimination procedure ($[x] = ([-5,5],[-4,4])$) and the solution obtained using the modified interval Gauss elimination ($[x]_{mod} = ([-0.96,0.96],[-0.92,0.92])$). We can see that the usual interval Gauss elimination procedure produces the outer interval estimate (enclosure) of the united solution set and the formal solution (x=$([-\frac{1}{3},\frac{1}{3}],[-\frac{1}{3},\frac{1}{3}])$) really provides a good inner approximation of the tolerable solution set. It is seen that in this example, our solution obtained using the modified interval Gauss elimination procedure can be treated as an inner interval approximation of the united solution and an outer interval approximation of the tolerable solution. However, the united solution set (often called simply solution set), tolerable solution set and controllable solution set, are only extreme points of a large family of all possible *AE*-solution sets. The definition of *AE*-solution set seems to be much more complicated than those of the united, tolerable and controllable solution sets, but is similar to them. The complete definition and detailed description of *AE*-solution sets are presented in [65, 71]. It is known that in the case of $n \times n$ linear interval system there are $2^{n(n+1)}$ *AE*-solution sets (see [71] p.347). For example, we can consider 61 generalized *AE*-solution sets for an interval linear 2×2 -system. Therefore, in Fig. 7.10 we present a comparison of our solution ($[x]_{mod} = ([-0.96,0.96],[-0.92,0.92])$) obtained using the modified interval Gauss elimination procedure with some of the *AE*-solutions obtained in [71] (p. 349) for the considered example of a 2×2 system. It is easy to see that our solution lies within the range of possible *AE*-solutions between the extreme tolerable and united solutions. To show the features and the advantages of our approach, we shall use some examples which are characterized by some specific features and can be considered as critical tests. Two versions of *MIGEP* were examined: the first is based on the widest interval result of modified division $[x]_{\text{mod}}$, the second - on the defuzzified result of modified division $[x]_{\text{mod}\,def}$ (see previous section).

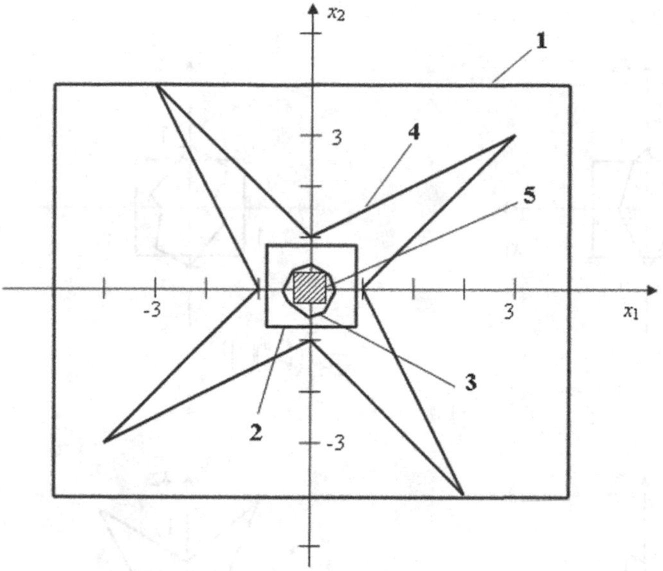

Fig. 7.9 Solutions of the example 2×2 system of linear interval equations: 1-the solution obtained using the usual interval Gauss elimination procedure, 2- the solution obtained using the modified interval Gauss elimination, 3-tolerable solution, 4-united solution, 5-formal solution

Therefore, the solutions obtained using *UIGEP* and *MIGEP* will be denoted as $[x_i]$, $[x_i]_{\text{mod}}$, $[x_i]_{\text{mod}\, def}$, $i = 1, \ldots, n$, respectively. Seven well known in the literature examples were used to compare the methods:

Example 1 from [37]:

$$A = \begin{bmatrix} [0.7, 1.3] & [-0.3, 0.3] & [-0.3, 0.3] \\ [-0.3, 0.3] & [0.7, 1.3] & [-0.3, 0.3] \\ [-0.3, 0.3] & [-0.3, 0.3] & [0.7, 1.3] \end{bmatrix} \quad B = \begin{bmatrix} [-14, 7] \\ [9, 12] \\ [3, 3] \end{bmatrix}$$

Example 2 from [63]:

$$A = \begin{bmatrix} [3.7, 4.3] & [-1.5, -0.5] & [0.0, 0.0] \\ [-1.5, -0.5] & [3.7, 4.3] & [-1.5, -0.5] \\ [0.0, 0.0] & [-1.5, -0.5] & [3.7, 4.3] \end{bmatrix} \quad B = \begin{bmatrix} [-14, 14] \\ [9, 9] \\ [-3, 3] \end{bmatrix}$$

Example 3 from [63]:

$$A = \begin{bmatrix} [3.7, 4.3] & [-1.5, -0.5] & [0.0, 0.0] \\ [-1.5, -0.5] & [3.7, 4.3] & [-1.5, -0.5] \\ [0.0, 0.0] & [-1.5, -0.5] & [3.7, 4.3] \end{bmatrix} \quad B = \begin{bmatrix} [-14, 0] \\ [-9, 0] \\ [-3, 0] \end{bmatrix}$$

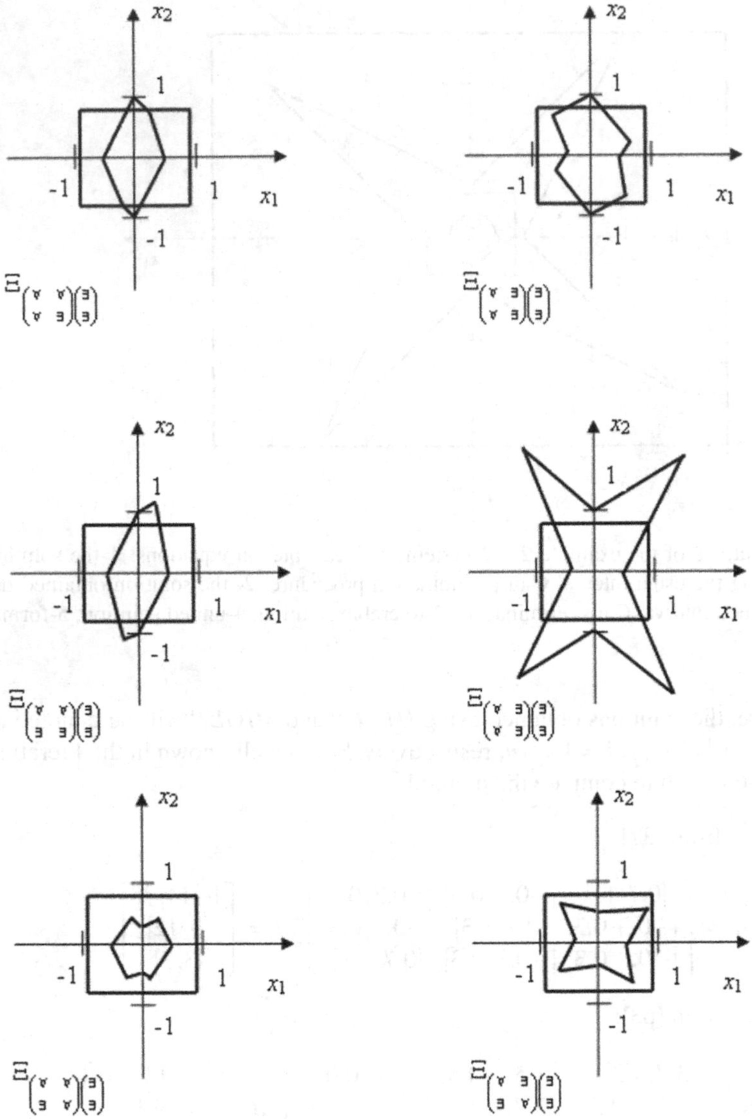

Fig. 7.10 A comparison of the solution ($[x]_{mod} = ([-0.96, 0.96], [-0.92, 0.92])$) obtained using the modified interval Gauss elimination procedure with some of *AE*-solutions from [71] (p. 349): the original notation of *AE*-solutions ($\Xi()$) used in [71] is retained

Example 4 from [63]:

$$A = \begin{bmatrix} [3.7,4.3] & [-1.5,-0.5] & [0.0,0.0] \\ [-1.5,-0.5] & [3.7,4.3] & [-1.5,-0.5] \\ [0.0,0.0] & [-1.5,-0.5] & [3.7,4.3] \end{bmatrix} \quad B = \begin{bmatrix} [0,14] \\ [0,9] \\ [0,3] \end{bmatrix}$$

Example 5 from [63]:

$$A = \begin{bmatrix} [3.7,4.3] & [-1.5,-0.5] & [0.0,0.0] \\ [-1.5,-0.5] & [3.7,4.3] & [-1.5,-0.5] \\ [0.0,0.0] & [-1.5,-0.5] & [3.7,4.3] \end{bmatrix} \quad B = \begin{bmatrix} [2,14] \\ [-9,-3] \\ [-3,1] \end{bmatrix}$$

Example 6 from [63]:

$$A = \begin{bmatrix} [3.7,4.3] & [-1.5,-0.5] & [0.0,0.0] \\ [-1.5,-0.5] & [3.7,4.3] & [-1.5,-0.5] \\ [0.0,0.0] & [-1.5,-0.5] & [3.7,4.3] \end{bmatrix} \quad B = \begin{bmatrix} [2,14] \\ [3,9] \\ [-3,1] \end{bmatrix}$$

Example 7 from [37]:

$$A = \begin{bmatrix} [2,3] & [0,1] \\ [1,2] & [2,3] \end{bmatrix} \quad B = \begin{bmatrix} [0,120] \\ [60,240] \end{bmatrix}$$

The numbered examples correspond to the numbers in Table 7.1. These examples are characterized by some specific features and can be considered as critical tests. In Table 1, the results obtained with the use of the usual interval Gauss elimination procedure ($[x_i]$) and the modified interval Gauss elimination procedure ($[x_i]_{\text{mod}}$ and $[x_i]_{\text{mod}\,def}$) are compared with those obtained by Markov [57, 58] using the Jacobi type iteration method since Markov's results can be treated as the maximal inner solution [51].

It is seen that only in the examples 2, 5 and 7, does Markov's method provides non-inverted interval solutions which can be treated as inner interval estimates of the solution set since it was proved by Shary [71] that "If a proper (non inverted) interval vector $[x]$ is a formal solution to the equation $[A][x] = [b]$ then $[x]$ is an inner interval estimate of the solution set." Hence, the inverted solutions obtained by Markov in the other examples are not maximal inner solutions. Of course, within the frameworks of directed interval arithmetic, "modal" arithmetic or the extended interval arithmetic developed by Kaucher [46], inverted interval solutions make a sense from a purely mathematical point of view. But it is very difficult to use inverted intervals in economical or mechanical applications. We can see that in examples 2, 5 and 7 the results obtained using our methods are close enough to Markov's solutions (an exception is the example 7 where the Markov's method provides the considerable narrower solution than our method). But the most important is the fact that in all the considered examples, our methods provide non-inverted interval solutions which are considerable narrower that those obtained by the usual interval Gauss elimination procedure.

Table 7.1 A comparison of obtained solutions

	Markov's method	$[x_i]$ mod def	$[x_i]$ mod	$[x_i]$
1.	[-9.13, -13.05]	[-14.19, 8.80]	[-37.29, 31.91]	[-101, 91]
	[16.77, 7.16]	[3.45, 11.60]	[-10.22, 25.21]	[-62.25, 99]
	[11, -2.68]	[-8.59, 12.43]	[-22.12, 25.92]	[-66, 90]
2.	[-2.93, -2.93]	[-2.43, 2.43]	[-6.30, 6.30]	[-6.38, 6.38]
	[-0.94, -0.94]	[-2.67, 2.67]	[-6.21, 6.21]	[-6.40, 6.40]
	[-0.37, -0.37]	[-1.50, 1.50]	[-3.15, 3.15]	[-3.40, 3.40]
3.	[-3.46, -0.94]	[-3.40, -1.28]	[-4.73, 0.00]	[-6.38, 0.00]
	[-2.31, -1.77]	[-3.16, -1.19]	[-4.23, 0.00]	[-6.40, 0.00]
	[-0.90, -0.94]	[-1.62, -0.54]	[-2.25, 0.00]	[-3.40, 0.00]
4.	[0.94, 3.46]	[1.28, 3.40]	[0.00, 4.73]	[0.00, 6.38]
	[1.77, 2.31]	[1.19, 3.16]	[0.00, 4.23]	[0.00, 6.40]
	[0.94, 0.90]	[0.54, 1.62]	[0.00, 2.25]	[0.00, 3.40]
5.	[0.39, 2.87]	[0.67, 2.41]	[-0.69, 3.76]	[-1.09, 4.29]
	[-1.11, -1.09]	[-1.82, -0.07]	[-3.03, 1.17]	[-4.02, 1.24]
	[-0.82, -0.18]	[-1.17, 0.05]	[-1.84, 0.71]	[-2.44, 0.78]
6.	[1.46, 3.54]	[1.69, 3.62]	[0.52, 4.82]	[0.52, 6.25]
	[2.46, 2.28]	[1.66, 3.32]	[0.46, 4.27]	[0.45, 6.07]
	[0.11, 0.52]	[-0.15, 1.23]	[-0.87, 2.02]	[-0.88, 2.73]
7.	[0, 17.14]	[-12.4, 30.33]	[-53.22, 71.74]	[-120, 90]
	[30, 68.57]	[-1.33, 67.55]	[-42.00, 106.44]	[-60, 240]

To estimate the quality of obtained results, we employ the special relative index of uncertainty: $RIU = \left(\frac{\max((x_m - \underline{x}), (\overline{x} - x_m))}{x_m}\right) \cdot 100\%$, where $x_m = (\underline{x} + \overline{x})/2$. This index may serve as a quantitative measure of the excess width effect: the more RIU, the more distinctly the excess width effect becomes apparent. In our analysis, we shall use only maximal values of RIU obtained on the entries of interval matrices and vectors.

It is easy to see that the proposed approach to the solution of linear interval systems based on the "interval extended zero" method may be easily used for the interval extension of other methods for the solution of linear systems. To illustrate the capability of our approach we have extended the Gauss-Jordan and LU decomposition methods as well as the iterative Gauss-Jacobi and Gauss-Seidel methods. As a basis of comparison we have taken the known Leontief's input-output model of economics.

7.2.2 Application to the Interval Leontief'S Input-Output Model of Economics

The general representation of the Leontief's input-output model IOM [53] is

$$(I - A) \times x = f, \tag{7.26}$$

where I is the identity matrix, A is a main interval technological coefficients matrix, f is an interval vector of final outputs (usually sales and inventory), x is a total output products vector.

First of all, we shall compare the interval extensions of the Gauss-Jordan and LU decomposition methods as well as the iterative Gauss-Jacobi and Gauss-Seidel methods to find the best of them using the known example of interval IOM taken from the paper [54], where the interval IOM of the national economics was presented by six basic sectors: agriculture (sector 1), industry (sector 2), construction (sector 3), transportation-post (sector 4), business and catering trade (sector 5), and other service departments (sector 6). In [54], the following interval vector of final outputs was used:

f=$\{[5000,5200],[91000,92000],[5200,5500],[1100,1300],[3400,3600],$
$[5900,6180]\}$.

The main interval technological coefficients matrix A of the problem analyzed in [54] is shown in Fig. 7.11.

$$A=\begin{bmatrix} [0.1389,0.1396] & [0.0804,0.0806] & [0.0033,0.0036] & [0.0001,0.0001] & [0.0321,0.0327] & [0.0052,0.0054] \\ [0.1565,0.1571] & [0.5043,0.5047] & [0.5634,0.5643] & [0.3401,0.3421] & [0.2405,0.2411] & [0.2642,0.2654] \\ [0.0001,0.0002] & [0.0004,0.0005] & [0.0067,0.0069] & [0.0013,0.0015] & [0.0090,0.0090] & [0.0145,0.0149] \\ [0.0110,0.0113] & [0.0178,0.0178] & [0.0296,0.0298] & [0.0140,0.0146] & [0.1103,0.1110] & [0.0413,0.0417] \\ [0.0214,0.0216] & [0.0749,0.0750] & [0.0917,0.0917] & [0.0490,0.0490] & [0.0400,0.0402] & [0.0454,0.0458] \\ [0.0268,0.0271] & [0.0284,0.0407] & [0.0142,0.0144] & [0.0358,0.0363] & [0.1086,0.1090] & [0.0981,0.0987] \end{bmatrix}$$

Fig. 7.11 Main interval technological coefficients matrix A [54]

Using the classical interval Gauss elimination procedure we obtain
$[x]=\{[27940,28913],[226211,232376],[5815,6224],[9079,9735],[23674,24686],$
$[17797,21969]\}$.

The use of the modified interval Gauss elimination procedure based on the "interval extended zero" method provides fuzzy vector solutions to interval linear systems. Therefore, to make such results comparable with those obtained using the usual interval Gauss elimination procedure, only the supports ($[x]_{mod}$) of the triangular fuzzy solutions will be used in further analysis. Using our extensions of the usual Gauss method, Gauss-Jordan, LU decomposition, Gauss-Jacobi and Gauss-Seidel methods based on the "interval extended zero" approach, for the considered example of the Leontief's IOM we obtained the results presented in Table 7.2. In all these extensions, we have used $[x]_{mod}$ as the result of interval division. It is seen that the modified interval Gauss method provides results with the minimal excess width effect ($RIU = 2.68\%$) and all the studied modified methods provide solutions which are substantially narrower than that of the classical interval Gauss elimination procedure which gives $RIU = 10.2\%$.

I light of this, in our further analysis we shall examine only the modified interval Gauss method in comparison with the usual interval Gauss method.

Table 7.2 Results obtained using different interval extended methods

Modified Gauss method	Gauss-Jordan method	LU decomposition	Gauss-Jacobi method	Gauss-Seidel method
$[28294, 28555]$	$[27957, 28896]$	$[28286, 28564]$	$[28291, 28558]$	$[28291, 28558]$
$[228367, 230179]$	$[226315, 232261]$	$[228227, 230325]$	$[228307, 230246]$	$[228307, 230246]$
$[5933, 6106]$	$[5817, 6223]$	$[5926, 6113]$	$[5927, 6112]$	$[5927, 6112]$
$[9304, 9507]$	$[9087, 9727]$	$[9288, 9524]$	$[9308, 9504]$	$[9308, 9504]$
$[24009, 24345]$	$[23689, 24670]$	$[23978, 24378]$	$[24047, 24308]$	$[24047, 24308]$
$[19198, 20256]$	$[17850, 21631]$	$[18766, 20693]$	$[18830, 20631]$	$[18830, 20631]$
$RIU = 2.68\%$	$RIU = 9.57\%$	$RIU = 4.88\%$	$RIU = 4.56\%$	$RIU = 4.56\%$

Since the defuzzified modified division $[x]_{\text{mod}\,def}$ provides substantially narrower results than the modified division $[x]_{mod}$ (see Table 7.1), in the following we shall use only $[x]_{\text{mod}\,def}$ in the modified interval Gauss elimination procedure.

To be sure of the method efficacy, we have tested it on the examples of matrix A with greater sizes. To do this, two interval technological coefficients matrices 10×10 and 1000×1000 were generated using the following procedure. At a first step, the real valued means a_{ij} of interval entries were randomly generated. Since in the Leontief's method the sums of the entries in the rows of A should be less or equal to 1, an additional normalization procedure was used. At a second step, the interval entries were obtained as $[a_{ij} - \Delta \cdot a_{ij}, a_{ij} + \Delta \cdot a_{ij}]$.

To examine the influence of Δ on the results we chose Δ=0.05, 0.1 and 0.15.

Therefore, in the considered examples, we have the following initial relative indexes of uncertainty: RIU_{in}=5%, RIU_{in}=10% and RIU_{in} =15% (see Table 7.3).

The interval entries of the vector f have been generated in a similar way as $[f_i - \Delta \cdot f_i, f_i + \Delta \cdot f_i]$. The values of f_i were randomly varied from 0 to 2500000. In Table 7.3, the modified interval Gauss method and the usual interval Gauss method are denoted as mod.Gauss and usual Gauss. $AvgW$, $MaxW$ and $MinW$ are the average, maximal and minimal widths of resulting interval solutions, respectively.

It can be seen that the results obtained using the modified interval Gauss method are not charged by the appreciable excess width effect (only a small difference between the values of RIU_{in} and RIU is observed) and are substantially (2-3 times) narrower that those obtained by the usual interval Gauss Elimination procedure.

In practice, the main technological coefficients matrix A often has a considerable number of zero entries, i.e., it is a sparse matrix. This problem is typical also for the finite element method used for modeling mechanical systems.

Since such zero entries may affect the results, we studied the influence of the percentage of zero entries in A on the resulting interval solution.

To generate the matrix A, the two-step approach described above was used and then the randomly chosen entries were filled with zeros.

In Table 7.4, we present the results obtained for the percentages of zero entries 5%, 30% and 70%. In all cases, matrix A was generated in such a way that RIU_{in} =5%.

It can be seen that there is no considerable influence of zero entries on the final excess width effect.

Table 7.3 Influence of matrix's size and RIU_{in} on the results

Matrix	RIU_{in}	Method	*RIU*	*AvgW*	*MaxW*	*MinW*
10×10	5%	*usual Gauss*	12.33%	44115140	55768471	31647227
		mod. Gauss	5.49%	18616304	24029404	14159352
10×10	10%	*usual Gauss*	25.92%	73256293	102820082	50264174
		mod. Gauss	11.34%	31262290	42765608	20188133
10×10	15%	*usual Gauss*	40.07%	68147565	113734175	42096300
		mod. Gauss	19.24%	28358578	47740859	16719267
1000×1000	5%	*usual Gauss*	14.87%	48784126	62546499	35461871
		mod. Gauss	7.31%	20224392	29853675	11388946
1000×1000	10%	*usual Gauss*	29.13%	96271977	125150270	70118707
		mod. Gauss	14.22%	39796769	60694099	22058638
1000×1000	15%	*usual Gauss*	42.57%	152979563	193242858	112521620
		mod. Gauss	20.00%	63118358	94800262	35317744

Table 7.4 Influence of zero entries

Matrix	Zero entries	Method	*RIU*	*AvgW*	*MaxW*	*MinW*
10×10	5%	*usual Gauss*	12.51%	35849343	60405911	21129885
		mod. Gauss	5.29%	15208019	23861280	9155307
10×10	30%	*usual Gauss*	10.22%	30103368	56240358	12246081
		mod. Gauss	4.89%	13428105	25360970	5738751
10×10	70%	*usual Gauss*	9.51%	16904647	24958373	5499126
		mod. Gauss	3.65%	8158243	12342386	2106151
1000×1000	5%	*usual Gauss*	13.26%	34671500	47923356	21220081
		mod. Gauss	6.53%	15142112	22884583	7014559
1000×1000	30%	*usual Gauss*	12.54%	27795873	41735433	14433180
		mod. Gauss	6.16%	12429883	20387022	4980193
1000×1000	70%	*usual Gauss*	10.81%	16383036	29803632	4161731
		mod. Gauss	5.20%	7858973	14772038	1748289

7.3 Solving Nonlinear Interval and Fuzzy Equations

In this section, the generalized procedure of interval extension called "interval extended zero" presented in Section 7.1 is extended to the case of nonlinear interval and fuzzy equations. The known "test" example of quadratic fuzzy equation is used to perform the advantages of a new method. In this example, only the positive solution can be obtained using known methods, whereas generally a negative fuzzy root can not be excepted. The sources of this problem are clarified. It is shown that opposite to the known methods, a new approach makes it possible to get both the positive and negative solutions of quadratic fuzzy equations. Generally, the developed method can be applied for solving a wide range of nonlinear interval and fuzzy equations if some initial constraints on solution's bounds are known.

Although the problem of solving nonlinear interval and fuzzy equations is of perennial interest [1, 2, 14, 16, 18, 23, 32, 48, 62], currently there are no universal methods for solving such equations proposed in the literature. Therefore, this problem is now open.

There are many different numerical methods proposed in the literature for solving interval and fuzzy equations including such complicated as neural net solutions [16, 18] and fuzzy extension of Newton's method in [1, 2], but only particular solutions valid in specific conditions were obtained. For example, only a positive root of the quadratic fuzzy equation have been obtained in [2, 19], although a negative solution can exist too. In our opinion, the root of such problems is the interpretation of interval and fuzzy extensions. It is known that the equations $F(X) - B = 0, F(X) = B$, where B is an interval or fuzzy value, $F(X)$ is some interval or fuzzy function, are not equivalent ones. Moreover, the main problem is that the conventional interval and fuzzy extensions of usual equation which lead to the interval or fuzzy equations such as $F(X) - B = 0$ are not correct procedures. Less problems we meet when dealing with the interval or fuzzy equation of the form $F(X) = B$, but in many cases its roots are inverted interval or fuzzy values, i.e., such that $\overline{x} < \underline{x}$.

To alleviate these problems in the case of linear interval and fuzzy equations, in Section 7.1, we proposed a new "interval extended zero" method.

In the current section, we show that "interval extended zero" method may be successfully used for solving nonlinear interval and fuzzy equations. Using the same example as in [2, 19], we get not only a positive fuzzy solution, but a negative too.

The general approach described in Section 7.1 can be adapted for solving nonlinear equations. The method we present in this section can be applied for solving a wide range of nonlinear interval and fuzzy equations if some initial constraints on the solution's bounds are known.

Nevertheless, to present our method more transparent, we consider the well known example of quadratic fuzzy equation [2, 19] that factually can be treated as the test task:

$$ax^2 + bx = c, \tag{7.27}$$

where $a = (3,3,4,5), b = (1,2,3), c = (1,1,2,3)$ are trapezoidal and triangular fuzzy numbers (see Fig. 7.12). Although it is stated in [2, 19] that Eq. (7.27) have no a negative fuzzy root, we obtain such root. Moreover, using the results of our analysis in Section 7.1, we clarify the origins of the problem the authors of [2, 19] faced with.

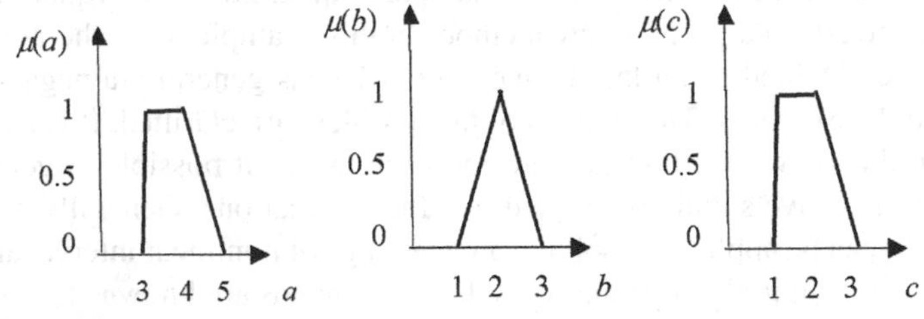

Fig. 7.12 Fuzzy parameters of Eq. (7.27)

As we prefer to use the α-cut representation of fuzzy numbers, fuzzy Eq. (7.27) is decomposed to the set of interval equations on the corresponding α-cuts. Obviously, when dealing with Eq. (7.27), on the lowest α-cut, i.e., for $\alpha = 0$, we get

$$[3,5]x^2 + [1,3]x = [1,3]. \tag{7.28}$$

Consider the case of $x > 0$, i.e., $\underline{x}, \overline{x} > 0$. Then from Eq. (7.28) we obtain

$$3\underline{x}^2 + \underline{x} = 1, \; 5\overline{x}^2 + 3\overline{x} = 3$$

and finally $\underline{x} = 0.4343$, $\overline{x} = 0.5307$. Nevertheless, in the assumption of negative $x < 0$, i.e., $\underline{x}, \overline{x} < 0$, from Eq. (7.28) we get

$$3\underline{x}^2 + 3\underline{x} = 1, \; 5\overline{x}^2 + \overline{x} = 3$$

and "...$\underline{x} \cong -0.629, \overline{x} \cong -0.98$, hence $\underline{x} > \overline{x}$ and therefore a negative root does not exist" [2].

To clarify the origins of this problem, let as consider the simplest interval linear equation $ax = b$, where a, b are intervals. Using conventional interval arithmetic rules [59], from this equation we get $[\underline{ax}, \overline{ax}] = [\underline{b}, \overline{b}]$ and finally: $\underline{x} = \frac{\underline{b}}{\underline{a}}, \overline{x} = \frac{\overline{b}}{\overline{a}}$.

Consider some examples.

For $a = [3,4]$, $[b] = [1,2]$ from $\underline{x} = \frac{\underline{b}}{\underline{a}}$, $\overline{x} = \frac{\overline{b}}{\overline{a}}$ we get $\underline{x} = 0.333, \overline{x} = 0.5$, for $a = [1,2]$, $[b] = [3,4]$ we get $\underline{x} = 3, \overline{x} = 2$, for $a = [3,4]$, $[b] = [0.7, 0.8]$ we get $\underline{x} = 0.23, \overline{x} = 0.2$

It is seen that interval equation $ax = b$ often have only inverted exact interval solution, i.e., such that $\underline{x} > \overline{x}$.

Obviously, in the case of the degenerated b, i.e., $\underline{b} = \overline{b}$ only inverted interval solutions can be obtained. It is seen that exact correct (non inverted) solutions of interval equation $ax = b$ exist only in some special conditions.

On the other hand, the united solution set (often called simply solution set), tolerable solution set and controllable solution set [71] can be analyzed as the approximate solutions, but this is out of scope of this book.

Only what we can say is that the interval equation in the form of $ax = b$ is not a reliable representation of the interval equation problem. We can see that Eq. (7.28) has the structure similar to that of $ax = b$ which is an unreliable representation of the interval equation problem. As with lowering the width of the right hand side of $ax = b$ this equation can provide inverted interval roots, we can expect such senseless results from nonlinear Eq. (7.28) as well. For example, changing $c = (1,1,2,3)$ by the more narrow fuzzy value $c^1 = (1, 1.5, 2)$, instead of (7.28) we get $[3,5]x^2 + [1,3]x = [1,2]$ and finally $3\underline{x}^2 + \underline{x} = 1$, $5\overline{x}^2 + 3\overline{x} = 2$. The positive roots of these equations are $\underline{x} = 0.4343, \overline{x} = 0.4$. So we have inverted interval solution $\underline{x} > \overline{x}$.

Indeed, when the right hand side of Eq. (7.28) is contracted to a point, i.e., it is a real value, Eq. (7.28) becomes completely senseless, since its interval left hand side can not be equal to any real value. Nevertheless, in considered particular case, the authors of [2, 19] obtained the positive fuzzy solution of Eq. (7.27) (see Fig. 7.13). To avoid above problems, in the spirit of "interval extended zero" method we

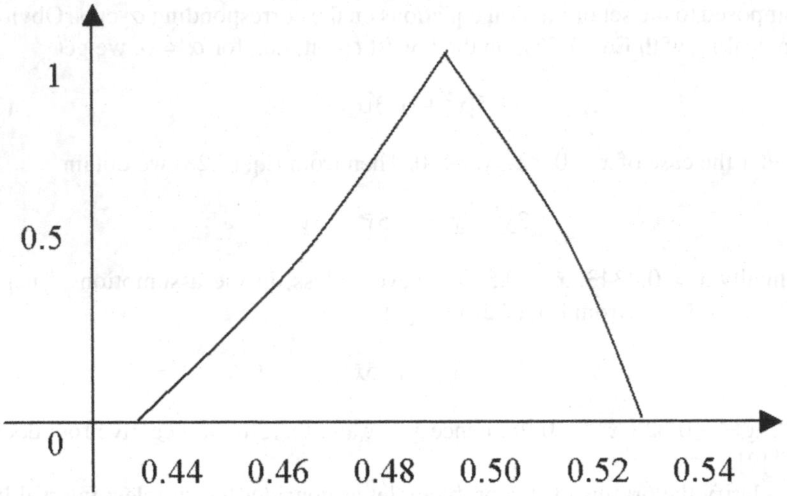

Fig. 7.13 Fuzzy root of Eq. (7.27) obtained in [1,6]

represent Eq. (7.27) on each α-cut in the following form:

$$[\underline{a},\overline{a}][\underline{x},\overline{x}]^2+[\underline{b},\overline{b}][\underline{x},\overline{x}]-[\underline{c},\overline{c}]=[-y,y], \tag{7.29}$$

where y is the undefined parameter (see Section 7.1) and index α is omitted for the simplicity. Using conventional interval arithmetic rules, from Eq.(7.29) we get

$$[\underline{ax}+\underline{b},\overline{ax}+\overline{b}][\underline{x},\overline{x}]-[\underline{c},\overline{c}]=[-y,y]. \tag{7.30}$$

Firstly, consider the case of the positive interval root of Eq. (7.30), i.e., $\underline{x},\overline{x}>0$. Then from (7.30) we obtain

$$\underline{ax}^2+\underline{bx}-\overline{c}=-y,\ \overline{ax}^2+\overline{bx}-\underline{c}=y. \tag{7.31}$$

The sum of Eqs. (7.31) results in the following equation

$$\underline{ax}^2+\underline{bx}-\overline{c}+\overline{ax}^2+\overline{bx}-\underline{c}=0. \tag{7.32}$$

Since in the case of real valued a,b,c, the positive root of (7.27) is presented by the expression $x=\frac{-b+\sqrt{b^2+4ac}}{2a}$, the "natural constraints" on the positive interval solution of (7.31) can be represented as follows:

$$x_{min}=\frac{-\overline{b}+\sqrt{\underline{b}^2+4\underline{ac}}}{2\overline{a}},\ x_{max}=\frac{-\underline{b}+\sqrt{\overline{b}^2+4\overline{ac}}}{2\underline{a}}. \tag{7.33}$$

Similar to the case of linear interval equation (see Section 7.1) we consider the real valued (degenerated) solution of Eq. (7.32), x_m, as the natural top bound for positive $\underline{x}$, i.e., $\underline{x} \leq x_m$ and bottom bound for positive $\overline{x}$, i.e., $x_m \leq \overline{x}$. For the case of $\underline{x} = \overline{x} = x_m$ from (7.32) we get

$$x_m = \frac{-(\underline{b}+\overline{b})+\sqrt{(\underline{b}+\overline{b})^2+4(\underline{a}+\overline{a})(\underline{c}+\overline{c})}}{2(\underline{a}+\overline{a})}. \quad (7.34)$$

Eq. (7.32) with described above constraints $x_{\min} \leq \underline{x} \leq x_m, x_m \leq \overline{x} \leq x_{\max}$ is a typical constraint satisfaction problem [29] and its interval solution can be obtained. From Eq. (7.32) we get the expressions $\underline{x} = \underline{f}(\overline{x}) = \frac{-\underline{b}+\sqrt{\underline{b}^2+4\underline{a}(\underline{c}+\overline{c}-\overline{a}\overline{x}^2-\overline{b}\overline{x})}}{2\underline{a}}$, $\overline{x} = \overline{f}(\underline{x}) = \frac{-\overline{b}+\sqrt{\overline{b}^2+4\overline{a}(\underline{c}+\overline{c}-\underline{a}\underline{x}^2-\underline{b}\underline{x})}}{2\overline{a}}$.

Generally, the interval solution of the above constraint satisfaction problem can be represented as follows:

$$[\underline{x}] = [x_{\min}, x_m] \cap [\underline{x}_1^*, \underline{x}_2^*], \; [\overline{x}] = [x_m, x_{\max}] \cap [\overline{x}_1^*, \overline{x}_2^*], \quad (7.35)$$

where
$\underline{x}_1^* = \min \underline{f}(\overline{x})$, $\underline{x}_2^* = \max \underline{f}(\overline{x})$ $(x_m \leq \overline{x} \leq x_{\max})$;
$\overline{x}_1^* = \min \overline{f}(\underline{x})$, $\overline{x}_2^* = \max \overline{f}(\underline{x})$ $(x_{\min} \leq \underline{x} \leq x_m)$.

It is easy to see that in our case

$$\underline{x}_1^* = \frac{-\underline{b}+\sqrt{\underline{b}^2+4\underline{a}(\underline{c}+\overline{c}-\overline{a}x_{\max}^2-\overline{b}x_{\max})}}{2\underline{a}}, \; \underline{x}_2^* = \frac{-\underline{b}+\sqrt{\underline{b}^2+4\underline{a}(\underline{c}+\overline{c}-\overline{a}x_m^2-\overline{b}x_m)}}{2\underline{a}},$$
$$\overline{x}_1^* = \frac{-\overline{b}+\sqrt{\overline{b}^2+4\overline{a}(\underline{c}+\overline{c}-\underline{a}x_m^2-\underline{b}x_m)}}{2\overline{a}}, \; \overline{x}_2^* = \frac{-\overline{b}+\sqrt{\overline{b}^2+4\overline{a}(\underline{c}+\overline{c}-\underline{a}x_{\min}^2-\underline{b}x_{\min})}}{2\overline{a}}.$$

It is clear that Exp. (7.35) leads to the interval solution

$$[\underline{x}] = [\underline{x}_{\min}, \underline{x}_{\max}], \; [\overline{x}] = [\overline{x}_{\min}, \overline{x}_{\max}], \quad (7.36)$$

where $\underline{x}_{\min} = \max(x_{\min}, \underline{x}_1^*)$, $\underline{x}_{\max} = \min(x_m, \underline{x}_2^*)$, $\overline{x}_{\min} = \max(x_m, \overline{x}_1^*)$, $\overline{x}_{\max} = \min(x_{\max}, \overline{x}_2^*)$.

As in the linear case (see Section 7.1), substituting the widest possible interval solution $[\underline{x}_{\min}, \overline{x}_{\max}]$ into Eqs. (7.31) we get the maximal value of y, i.e., $y_{\max}$, and substituting in this equation the shortness possible solution $[\underline{x}_{\max}, \overline{x}_{\min}] = [x_m, x_m]$ we obtain $y_{\min}$. As in the linear case, the formal interval solution (7.36) factually represents the continuous set of nested interval solutions of Eqs. (7.31) and we can use the expression $\eta = 1 - \frac{y-y_{min}}{y_{max}-y_{min}}$ (similar to (7.18)) to calculate the values of y on η-cuts (these η-cuts are introduced to represent the fuzzy solution on the α-cuts). For η rising from 0 to 1 using the last expression we get the values of y and substituting them into (7.31) we obtain the set of interval solutions $[\underline{x}, \overline{x}]_\eta$ on corresponding η-cuts. In Fig. 7.14, the positive fuzzy solution for the lowest α-cut (a=[3,5], b=[1.3], c=[1,3]) is presented.

Using the proposed method, the negative root ($\underline{x},\overline{x}<0$) of fuzzy Eq.(7.30) can be obtained as well. For this case we get the following set of expressions:

$$\underline{a}\overline{x}^2+\overline{b}\underline{x}-\overline{c}=-y,\ \overline{a}\underline{x}^2+\underline{b}\overline{x}-\underline{c}=y. \tag{7.37}$$

$$\underline{a}\overline{x}^2+\overline{b}\underline{x}-\overline{c}+\overline{a}\underline{x}^2+\underline{b}\overline{x}-\underline{c}=0. \tag{7.38}$$

$$x_m=\frac{-(\underline{b}+\overline{b})-\sqrt{(\underline{b}+\overline{b})^2+4(\underline{a}+\overline{a})(\underline{c}+\overline{c})}}{2(\underline{a}+\overline{a})}. \tag{7.39}$$

$$x_{min}=\frac{-\overline{b}-\sqrt{\overline{b}^2+4\overline{a}\overline{c}}}{2\underline{a}},x_{max}=\frac{-\underline{b}-\sqrt{\underline{b}^2+4\underline{a}\underline{c}}}{2\overline{a}}. \tag{7.40}$$

$$\underline{x}=\underline{f}(\overline{x})=\frac{-\overline{b}-\sqrt{\overline{b}^2+4\overline{a}(\underline{c}+\overline{c}-\underline{a}\overline{x}^2-\underline{b}\overline{x})}}{2\overline{a}},$$

$$\overline{x}=\overline{f}(\underline{x})=\frac{-\underline{b}-\sqrt{\underline{b}^2+4\underline{a}(\underline{c}+\overline{c}-\overline{a}\underline{x}^2-\overline{b}\underline{x})}}{2\overline{a}}. \tag{7.41}$$

$$[\underline{x}]=[x_{\min},x_m]\cap[\underline{x}_1^*,\underline{x}_2^*],\ [\overline{x}]=[x_m,x_{\max}]\cap[\overline{x}_1^*,\overline{x}_2^*], \tag{7.42}$$

where $\underline{x}_1^*=\min\underline{f}(\overline{x})$, $\underline{x}_2^*=\max\underline{f}(\overline{x})$ $(x_m\leq\overline{x}\leq x_{\max})$; $\overline{x}_1^*=\min\overline{f}(\underline{x})$, $\overline{x}_2^*=\max\overline{f}(\underline{x})$ $(x_{\min}\leq\underline{x}\leq x_m)$.

The numerical algorithm we used to obtain the negative root is similar to that we presented above for the positive root. The negative fuzzy solution for the lowest α-cut (a=[3,5], b=[1.3], c=[1,3]) is presented in Fig. 7.14.

It is seen that our positive fuzzy solution in the considered example is wider than the interval solution obtained in [2, 19] (see Fig. 7.13). Nevertheless, it does not mean that the results from [2, 19] are more "true" since the methods proposed in [2, 19] except obtaining the negative roots. Besides, our results may be substantially shortened using the reduction of fuzzy solution to an interval one with a help of defuzzification procedure (7.20).

The fuzzy solution presented in Fig. 7.14 was obtained for the lowest α-cut (α=0). To get the complete fuzzy solution of (7.27), the fuzzy solutions for other α-cuts ($0<\alpha\leq 1$) should be obtained using the algorithm described above. The positive solutions obtained for $\alpha=0$, $\alpha=0.5$ and $\alpha=1$ are presented in Fig. 7.15.

For different α -cuts we have fuzzy solutions with different supports and peaks. As a fuzzy value can be represented by the disjunction of its α-cuts, we treat the shaded area in Fig.7.15 as the final fuzzy solution. It is interesting that opposite to the result of [2, 19] (see Fig.7.13) it has a trapezoidal form and this seems quite natural since some parameters of the considered fuzzy equation (a and c in Eq. (7.27)) are trapezoidal fuzzy values too.

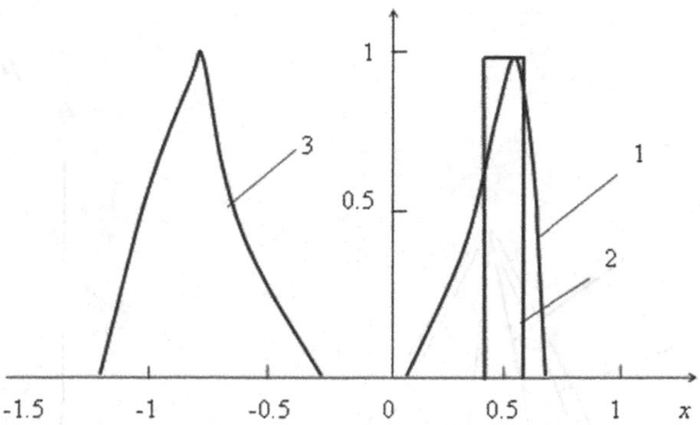

Fig. 7.14 Fuzzy roots of quadratic interval equation: 1,3-fuzzy solution obtained with use of "interval extended zero" method, 2-interval solution from [2, 19]

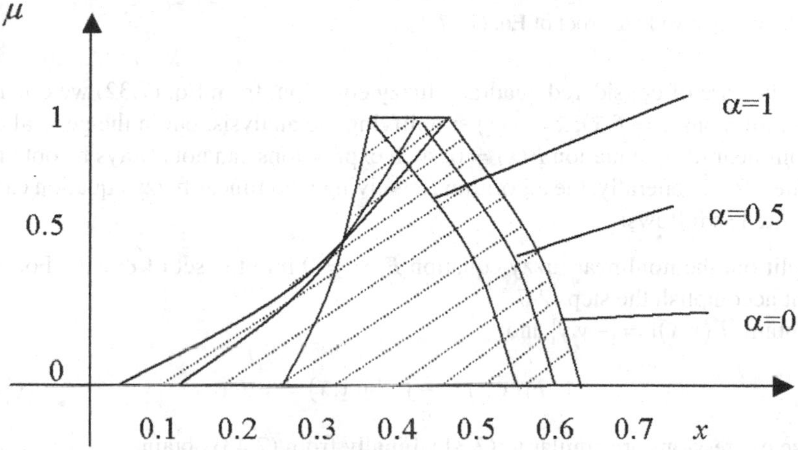

Fig. 7.15 Positive fuzzy root of Eq. (7.27)

The numerical algorithm we have used to obtain the negative root is similar to that we have presented above for the positive root. The result is presented in Fig. 7.16.

The resulting negative fuzzy root has a triangular form, whereas the positive root (see Fig. 7.15) is of trapezoidal type. This fact is a consequence of the special form of trapezoidal fuzzy parameters a and c (see Fig. 7.12), which have no fuzzy left parts.

It can be seen that the proposed method allows us to get positive and negative fuzzy solutions of quadratic interval and fuzzy equations, whereas the known approaches do not provide negative solutions.

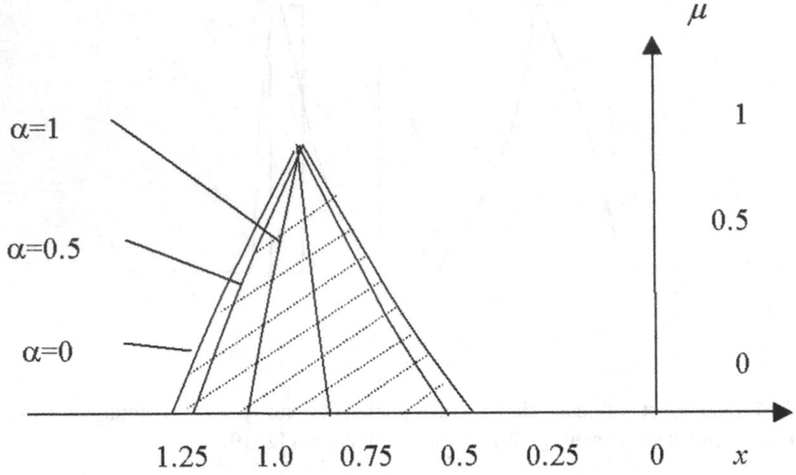

Fig. 7.16 Negative fuzzy root of Eq. (7.27)

In the case of considered quadratic fuzzy equation, from Eq. (7.32) we obtained the expressions $\underline{x} = \underline{f}(\overline{x})$, $\overline{x} = \overline{f}(\underline{x})$ simplifying the analysis, but in the general case of nonlinear fuzzy equation $F(x) = 0$, such expressions can not always be obtained.

Therefore, generally, the algorithm for solving a nonlinear fuzzy equation can be presented as follows:

1. Split out the nonlinear fuzzy equation $F(x) = 0$ into the set of α-cuts. For each α-cut accomplish the steps 2-7.
2. Obtain $[F(\underline{x},\overline{x})] = [-y, y]$ and

$$F_1(\underline{x},\overline{x}) = -y,\ F_2(\underline{x},\overline{x}) = y. \tag{7.43}$$

These expressions are similar to (7.31). Finally from (7.43) obtain

$$g(\underline{x},\overline{x}) = F_1(\underline{x},\overline{x}) + F_2(\underline{x},\overline{x}) = 0$$

(the last equation is the analog of Eq. (7.32)).
3. Obtain x_m as the numerical solution of equation $g(\underline{x},\overline{x}) = 0$.
4. Define $x_{\min}$ and $x_{\max}$ as the natural constraints (like in the case of quadratic equation) or as the external constraints originated from the mechanical or economical features of the considered problem.
5. Let $\underline{x}(\overline{x})$ be a numerical solution of $g(\underline{x},\overline{x}) = 0$ for given $\overline{x}$ and $\overline{x}(\underline{x})$ be a numerical solution of $g(\underline{x},\overline{x}) = 0$ for given $\underline{x}$. Then obtain
$\underline{x}_1^* = \min \underline{x}(\overline{x})$, $(\overline{x} \in [x_m, x_{\max}])$, $\underline{x}_2^* = \max \underline{x}(\overline{x})$, $(\overline{x} \in [x_m, x_{\max}])$,
$\overline{x}_1^* = \min \overline{x}(\underline{x})$, $(\underline{x} \in [x_{\min}, x_m])$, $\overline{x}_2^* = \max \overline{x}(\underline{x})$, $(\underline{x} \in [x_{\min}, x_m])$,
$[\underline{x}] = [x_{\min}, x_m] \cap [\underline{x}_1^*, \underline{x}_2^*]$, $[\overline{x}] = [x_m, x_{\max}] \cap [\overline{x}_1^*, \overline{x}_2^*]$,
$[\underline{x}] = [\underline{x}_{\min}, \underline{x}_{\max}]$, $[\overline{x}] = [\overline{x}_{\min}, \overline{x}_{\max}]$,

where $\underline{x}_{\min} = \max(x_{\min}, \underline{x}_1^*)$, $\underline{x}_{\max} = \min(x_m, \underline{x}_2^*)$, $\overline{x}_{\min} = \max(x_m, \overline{x}_1^*)$, $\overline{x}_{\max} = \min(x_{\max}, \overline{x}_2^*)$.

6. Substituting the widest possible interval solution $[\underline{x}_{\min}, \overline{x}_{\max}]$ into (7.43) obtain $y_{\max}$ and substituting in this equation the shortness solution $[\underline{x}_{\max}, \overline{x}_{\min}]$ obtain $y_{\min}$ (usually $\overline{x}_{\max} = \overline{x}_{\min} = x_m$).

7. Introduce the set of η-cuts as follows:

$$\eta = 1 - \frac{y - y_{\min}}{y_{\max} - y_{\min}}. \tag{7.44}$$

For each η-cut $(0 \leq \eta \leq 1)$ from (7.44) obtain y, substitute it in (7.43) and obtain the numerical solution of nonlinear system on η -cut : $[\underline{x}, \overline{x}]_\eta$.

To obtain the complete solution of initial nonlinear fuzzy equation $F(x) = 0$, the steps 2-7 should be repeated for all α -cuts and solutions obtained on α-cuts should be disjointed into the final solution.

In the following section, we shall show that the developed method can be successfully used for solving more complicated nonlinear problems.

7.4 Fuzzy Internal Rate of Return in Budgeting

In the capital budgeting, we usually deal with projects taking a long time — as a rule some years — for their realization. In such cases, the description of uncertainty within the framework of traditional probability methods usually is impossible due to the absence of objective information about the probabilities of future events.

This is the reason for the growing interest in the application of interval and fuzzy methods in budgeting, which has been observing during the last two decades. There are many financial parameters proposed in the literature for the project quality assessment, but the two primary among them — Net Present Value (*NPV*) and Internal Rate of Return (*IRR*) — are necessarily used in the financial analysis.

Whereas the problem of *NPV* fuzzy estimation is now well studied and many authors have contributed to its solution, obtaining the fuzzy *IRR* seems to be rather an open problem.

This problem is a consequence of inherent properties of fuzzy and interval mathematic, but it seems unnatural to have only a real valued *IRR* in the fuzzy environment when all other financial parameters are fuzzy. In this section, the problem of *IRR* estimation in the fuzzy setting is considered in the framework of more general problem of solving nonlinear fuzzy equations. Finally, the concept of restricted fuzzy *IRR* as the solution of the corresponding nonlinear fuzzy equation is proposed and analyzed.

7.4.1 The Problem Formulation

There are a lot of financial parameters proposed in literature [10, 13, 24, 55] for budgeting. The main ones being: Net Present Value (*NPV*), Internal Rate of Return (*IRR*), Payback period (*PB*), Profitability Index (*PI*). It is shown in [11] that the most important parameters are *NPV* and *IRR*. A good review of other useful financial parameters may be found in [8].

Net Present Value is usually calculated as follows:

$$NPV = \sum_{t=1}^{T} \frac{P_t}{(1+d)^t} - KV, \tag{7.45}$$

where d is a discount rate, KV is a starting capital investment, P_t is a total income (cash flow) in a year t, T is a duration of an investment project in years. Usually the discount rate is equal to the average bank interest rate in a country of an investment or other value corresponding to a profit rate of alternative capital investments. The value of *IRR* is a solution of non-linear equation with respect to d:

$$\sum_{t=1}^{T} \frac{P_t}{(1+d)^t} - KV = 0. \tag{7.46}$$

If P_t, KV (or at least one of them) are fuzzy numbers with the use of fuzzy extension of Eq. (7.46), i.e., by replacement of its parameters and variable with fuzzy ones and all arithmetic operations with relevant fuzzy operations, Eq. (7.46) can be transformed to the fuzzy equation. The problem is to find a fuzzy solution of such fuzzy equation, i.e., to obtain fuzzy *IRR*.

An economic nature of *IRR* can be explained as follows. If an actual bank discount rate or return of any other alternative investment under consideration is less than *IRR* of considered project, then the investment in this project is more preferable.

Only the cases when Eq. (7.46) has single root will be analyzed. The reasons behind this was presented above in Subsection 4.5.1 (see Chapter 4).

Currently, traditional approaches to the evaluation of *NPV*, *IRR* and other financial parameters are subjected to quite deserved criticism, since future incomes P_t and rates d are rather uncertain parameters. Uncertainties which one meets in capital budgeting cannot be adequately described in terms of probability. Really, in budgeting we usually deal with a business-plan that takes a long time for its realization. In such cases, the description of uncertainty via probability representation of P_t, KV and d usually is impossible due to a lack of information about probabilities of future events. Thus, what we really have in such cases are some subjective expert's judgments. In real-world situations, investors or experts involved are able to estimate only intervals of possible values P_t and d and the expected (more probable) values inside these intervals. That is why during the last two decades the growing interest to applications of the interval arithmetic [59] and fuzzy sets theory methods [76] in budgeting has been observing. After pioneer works by Ward [73] and

Buckley [14], some other authors contributed to the development of fuzzy capital budgeting theory [21, 25, 27, 28, 33, 40, 41, 42, 43, 44, 50, 64]. It is safe to say now that almost all problems of the fuzzy *NPV* estimation are solved. An unresolved problem is the fuzzy estimation of the *IRR*. Ward [73] considered Eq. (7.46) and stated that this expression cannot be applied to the fuzzy case because the left hand side of Eq. (7.46) is fuzzy, 0 in the right hand side is a crisp value and an equality is impossible. Hence, Eq. (7.46) is senseless from the fuzzy viewpoint. Kuchta [50] proposed a method for fuzzy *IRR* estimation where the α-cut representation of fuzzy numbers [47] was used. The method is based on an assumption (see [50, p. 380]) that a set of equations for *IRR* determination on each α-cut may be presented (in our notation) as follows:

$$(KVV^{\alpha})_1 + \sum_{t=1}^{T} \frac{(P_t^{\alpha})_1}{(1+IRR_1^{\alpha})^t} = 0, \quad (KVV^{\alpha})_2 + \sum_{t=1}^{T} \frac{(P_t^{\alpha})_2}{(1+IRR_2^{\alpha})^t} = 0, \qquad (7.47)$$

where $KVV = -KV$, indexes "1", "2" stand for the left and right bounds of corresponding intervals, respectively, $P_t^{\alpha} = [(P_t^{\alpha})_1, (P_t^{\alpha})_2]$ are crisp interval representations of fuzzy cash flows at time t on the α -cuts.

Of course, from the equations (7.47) all crisp intervals $d^{\alpha} = [d_1^{\alpha}, d_2^{\alpha}]$, expressing the fuzzy valued *IRR* may be obtained. On the other hand, Eqs. (7.47) are not a direct consequence of conventional fuzzy and interval arithmetics rules.

Eqs. (7.47) were obtained in [50] using fuzzy extension of (7.46) assuming that P_t, KV (or at least one of them) are fuzzy numbers and representing them by the sets of α-cuts. Since Eqs. (7.47) should be verified on each α -cut, it is quite enough to consider only crisp interval extension of (7.46), which is the particular case of more general equation

$$F(d) - B = 0, \qquad (7.48)$$

where B is an interval ($B = KV$ in our case) and $F(d)$ is an interval valued function of interval argument d (in our case $F(d) = \sum_{t=1}^{T} \frac{P_t}{(1+d)^t}$).

Using regular interval arithmetic [39], this equation can be transformed to $[F_1(d) - B_2, F_2(d) - B_1] = 0$, and finally we get two equations $F_1(d) - B_2 = 0, F_2(d) - B_1 = 0$. Of course, if we deal with a linear interval function $F(d) = A \cdot d$ (A is an interval) , then $F_1(d) = A_1 \cdot d_1$ and $F_2(d) = A_2 \cdot d_2$, but if $F(d) = \frac{A}{d}$ we have $F_1(d) = \frac{A_1}{d_2}$, $F_2(d) = \frac{A_2}{d_1}$ since F_1 is the left bound (*min* value in interval) and F_2 is the right bound (*max* value in interval) of interval value $F(d)$.

Hence, the use of regular interval arithmetic rules leads to the following equations:

$$(KVV^{\alpha})_1 + \sum_{t=1}^{T} \frac{(P_t^{\alpha})_1}{(1+IRR_2^{\alpha})^t} = 0, \quad (KVV^{\alpha})_2 + \sum_{t=1}^{T} \frac{(P_t^{\alpha})_2}{(1+IRR_1^{\alpha})^t} = 0. \qquad (7.49)$$

There is no way to get a correct not inverted interval solution of (7.49). Only inverted intervals *IRR*, i.e., such that $IRR_1^{\alpha} > IRR_2^{\alpha}$ can be obtained. Since it is hard or even impossible to interpret reasonable such results, they can not be used in practice.

It was shown in Subsection 4.5.1 (see Chapter 4) that only approximate real valued *IRR* (represented by usual non interval numbers) may be obtained from (7.49). In Section 7.1, we shown that the main problem is that the conventional interval extension (and the fuzzy as well) of usual equation, which leads to the interval or fuzzy equation such (7.49) is not a correct procedure. Less problems we meet when dealing with interval or fuzzy equation in the form

$$F(d) = B. \tag{7.50}$$

An important feature of interval and fuzzy mathematics is that Eq. (7.50) is not equivalent to Eq. (7.48). This fact deserves more detailed analysis and we have studied it thoroughly in Section 7.1.

Summarizing, we can say that the problem of *IRR* estimation in the fuzzy setting should be considered in the framework of more general problem of solving fuzzy equations.

7.4.2 Fuzzy Internal Rate of Return for Crisp Interval Cash Flows. Basics.

To make the main idea more transparent, let us consider a simplified example of a one-year project, when a real valued investment KV is made at the beginning of the first year and production starts right away. An interval profit $P = [\underline{P}, \overline{P}]$ is earned at the end of the first year and then the project is finished. In this case Eq. (7.46) transforms into the following form:

$$\frac{P}{1+d} - KV = 0. \tag{7.51}$$

Rewriting Eq. (7.51) as:

$$d = \frac{P}{KV} - 1 \tag{7.52}$$

we get the simplest, but the widest (see Section 7.1) interval solution for d:

$$[d] = [\overline{d}, \underline{d}] = \left[\frac{\underline{P}}{KV} - 1, \frac{\overline{P}}{KV} - 1\right]. \tag{7.53}$$

Using the "interval extended zero" method described in Section 7.1, we get the interval extension of Eq. (7.51) as follows:

$$\frac{[\underline{P}, \overline{P}]}{1 + [\underline{d}, \overline{d}]} - KV = [-y, y] \tag{7.54}$$

With a help of interval analysis rules (7.4)-(7.7), the interval equation (7.54) may be represented in the form

$$\left[\frac{\underline{P}}{1+\overline{d}} - KV, \frac{\overline{P}}{1+\underline{d}} - KV\right] = [-y, y] \tag{7.55}$$

and finally as

$$\begin{cases} \frac{P}{1+\underline{d}} - KV = -y, \\ \frac{\overline{P}}{1+\underline{d}} - KV = y. \end{cases} \tag{7.56}$$

From equations (7.56) we obtain the explicit dependence between bounds of interval $[d]$:

$$\overline{d} = \frac{\underline{P}}{2 \cdot KV - \frac{\overline{P}}{1+\underline{d}}} - 1. \tag{7.57}$$

On the other hand, the original Eq. (7.51) can be rewritten in algebraically equivalent form

$$P - KV(1+d) = 0. \tag{7.58}$$

With a help of "interval extended zero" method we get from Eq. (7.58):

$$[\underline{P},\overline{P}] - KV(1 + [\underline{d},\overline{d}]) = [-y,y] \tag{7.59}$$

and after some simple transformations

$$\overline{d} = \frac{\underline{P}+\overline{P}}{KV} - 2 - \underline{d}. \tag{7.60}$$

It is easy to see that in both cases when using representations (7.54) or (7.59) for the degenerated solution we have $\underline{d} = \overline{d} = \frac{\underline{P}+\overline{P}}{2 \cdot KV} - 1$. Also taking into account the widest possible interval solution (7.53) we conclude that in both cases the following restrictions are verifying:

$$\underline{d} \in \left[\frac{\underline{P}}{KV} - 1, \frac{\underline{P}+\overline{P}}{2 \cdot KV} - 1\right], \overline{d} \in \left[\frac{\underline{P}+\overline{P}}{2 \cdot KV} - 1, \frac{\overline{P}}{KV} - 1\right]. \tag{7.61}$$

Taking into account the natural restriction $0 \leq \underline{d} \leq \overline{d}$ (usually the values of IRR are positive, but generally this restriction is not obligatory) we can solve Eq. (7.54) in interval the form for $\underline{d}$ and $\overline{d}$ in the framework of constraint satisfaction problem [29] as follows

$$\begin{aligned} [\underline{d}] &= \left(\frac{\overline{P}}{2 \cdot KV - \frac{\underline{P}}{1+\overline{d}}} - 1\right) \cap \left[\max(0, \frac{\underline{P}}{KV} - 1), \frac{\underline{P}+\overline{P}}{2 \cdot KV} - 1\right] \mid \\ &\quad \mid \overline{d} \in \left[\frac{\underline{P}+\overline{P}}{2 \cdot KV} - 1, \frac{\overline{P}}{KV} - 1\right] \\ [\overline{d}] &= \left(\frac{\underline{P}}{2 \cdot KV - \frac{\overline{P}}{1+\underline{d}}} - 1\right) \cap \left[\frac{\underline{P}+\overline{P}}{2 \cdot KV} - 1, \frac{\overline{P}}{KV} - 1\right] \mid \\ &\quad \mid \underline{d} \in \left[\max(0, \frac{\underline{P}}{KV} - 1), \frac{\underline{P}+\overline{P}}{2 \cdot KV} - 1\right]. \end{aligned} \tag{7.62}$$

As a result we get from (7.62) the following interval solutions:

$$[\underline{d}] = \left[\frac{\overline{P}}{KV\cdot\left(2-\frac{\underline{P}}{\overline{P}}\right)} - 1, \frac{\underline{P}+\overline{P}}{2\cdot KV} - 1\right], [\overline{d}] = \left[\frac{\underline{P}+\overline{P}}{2\cdot KV} - 1, \frac{\overline{P}}{KV} - 1\right]. \quad (7.63)$$

Finally, using the approach described in Section 7.1 we get a fuzzy solution of Eq. (7.54) based on solution (7.63), i.e., the fuzzy *IRR* as

$$\tilde{d} = \left\{\frac{\overline{P}}{KV\cdot\left(2-\frac{\underline{P}}{\overline{P}}\right)} - 1, \frac{\underline{P}+\overline{P}}{2\cdot KV} - 1, \frac{\overline{P}}{KV} - 1\right\}. \quad (7.64)$$

In a similar way from equations (7.59)-(7.61) and restriction $0 \le \underline{d} \le \overline{d}$ we obtain

$$\begin{aligned} [\underline{d}] &= \left[\max\left\{0, \tfrac{\underline{P}}{KV} - 1\right\}, \tfrac{\underline{P}+\overline{P}}{2\cdot KV} - 1\right], \\ [\overline{d}] &= \left[\tfrac{\underline{P}+\overline{P}}{2\cdot KV} - 1, \min\left\{\tfrac{\underline{P}+\overline{P}}{KV} - 2, \tfrac{\overline{P}}{KV} - 1\right\}\right], \end{aligned} \quad (7.65)$$

$$\tilde{d} = \left\{\max\left\{0, \frac{\underline{P}}{KV} - 1\right\}, \frac{\underline{P}+\overline{P}}{2\cdot KV} - 1, \min\left\{\frac{\underline{P}+\overline{P}}{KV} - 2, \frac{\overline{P}}{KV} - 1\right\}\right\}. \quad (7.66)$$

Some illustrative examples are presented in Fig. 7.17.

Thus, the different but algebraically equivalent representations (7.51) and (7.58) of the equation for *IRR* provide after the interval extension the different fuzzy solutions (7.64) and (7.66). Basically, this result is not new, since it is well known [59] that different representations of some original algebraic expression often derive different numerical results after interval extension (see Section 3 for more detail).

Currently, this circumstance is not considered as the drastic drawback, but rather as a specific feature inherent in applied interval analysis [39]. That is why it seems quite natural in situations when we have some different interval or fuzzy solutions of some original problem, to use an intersection of obtained solutions as the most reliable solution of considered problem. Such intersections are performed in Fig. 7.17 by the shaded regions.

7.4.3 *Numerical Solution of the Nonlinear Fuzzy Problem of Internal Rate of Return Calculation*

As it was shown in previous subsection, the fuzzy *IRR* problem can be formulated with use of fuzzy extension of two different, but algebraically equivalent representations of initial non-fuzzy Eq. (7.46):

$$\sum_{t=1}^{T} \frac{P_t}{(1+d)^t} - KV = 0, \quad (7.67)$$

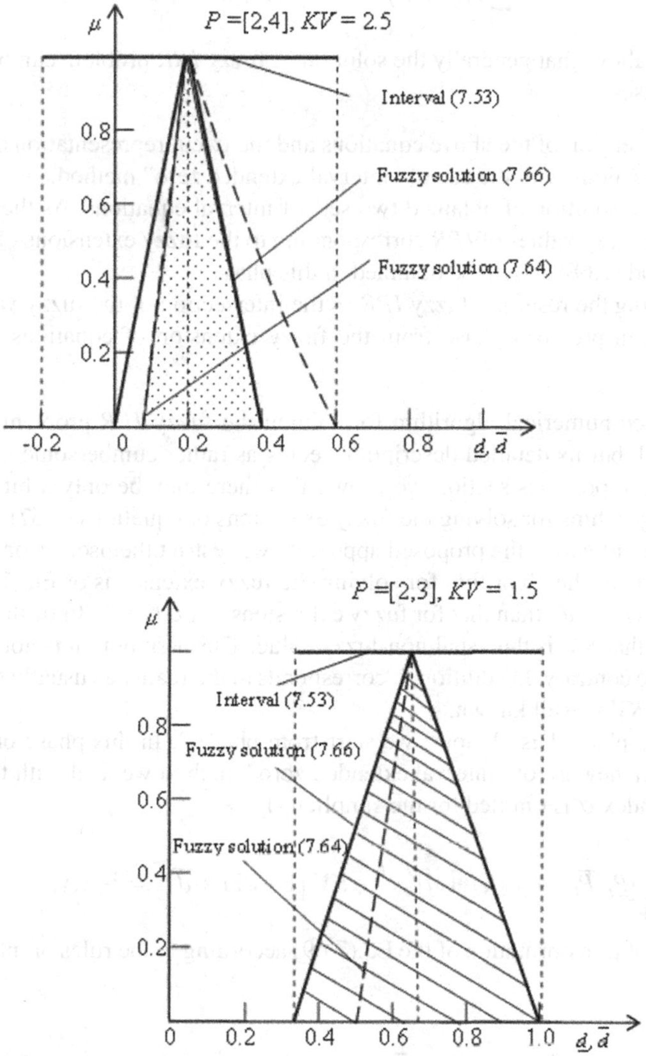

Fig. 7.17 Examples of fuzzy solutions of interval equations (7.54) and (7.59): μ is the membership function of fuzzy value

$$\sum_{t=1}^{T} P_t(1+d)^{T-t} - KV(1+d)^T = 0. \tag{7.68}$$

It is shown above that generally the solution of fuzzy *IRR* problem can be obtained in three phases:

- Fuzzy extension of the above equations and the α-cut representation of obtained fuzzy equations with the use of "interval extended zero" method.
- Numerical solution of obtained two sets of interval equations. As the result, the different fuzzy values of *IRR* corresponding to the fuzzy extensions of equations (7.67) and (7.68) should be obtained in this phase.
- Calculating the resulting fuzzy *IRR* as the intersection of the fuzzy valued *IRR*s obtained in previous phase from the fuzzy extension of equations (7.67) and (7.68).

The proposed numerical algorithm for solving the fuzzy *IRR* problem is not too complicated, but its detailed description seems as rather cumbersome one. On the other hand, in previous section we shown that there may be only a bit difference between algorithms for solving the fuzzy extensions of equations (7.67) and (7.68). Therefore, to illustrate the proposed approach, we restrict themselves only with the consideration of the algorithm for solving the fuzzy extensions of Eq.(7.68), as its description is shorter than that for fuzzy extensions of Eq.(7.67). In further analysis, we assume that *KV* is the usual non-fuzzy value. This assumption is not a simplification: quite contrary. In addition, it corresponds to the reality as usually the starting investment *KV* is well known.

Since the phase 1 is obvious, we start from phase 2. In this phase on each α - cut in the framework of "interval extended zero" method we deal with the interval equation (index α is omitted for the simplicity)

$$\sum_{t=1}^{T} [\underline{P}_t, \overline{P}_t] \left[1+\underline{d}, 1+\overline{d}\right]^{T-t} - KV \left[1+\underline{d}, 1+\overline{d}\right]^T = [-y, y]. \tag{7.69}$$

As a result of transformation of the Eq.(7.69) according to the rules of interval analysis we have

$$\begin{aligned} \left[\Sigma_{t=1}^{T} \underline{P}_t(1+\underline{d})^{T-t} - KV(1+\overline{d})^T,\ \Sigma_{t=1}^{T} \overline{P}_t(1+\overline{d})^{T-t} - KV(1+\underline{d})^T\right] \\ = [-y, y] \end{aligned} \tag{7.70}$$

and

$$\sum_{t=1}^{T} \underline{P}_t(1+\underline{d})^{T-t} - KV(1+\overline{d})^T = -y, \sum_{t=1}^{T} \overline{P}_t(1+\overline{d})^{T-t} - KV(1+\underline{d})^T = y. \tag{7.71}$$

The sum of the equations (7.71) provides us the expression connecting the unknown left $\underline{d}$ and right $\overline{d}$ bounds of *IRR*:

$$\sum_{t=1}^{T}\underline{P}_t(1+\underline{d})^{T-t}-KV(1+\overline{d})^{T}+\sum_{t=1}^{T}\overline{P}_t(1+\overline{d})^{T-t}-KV(1+\underline{d})^{T}=0. \quad (7.72)$$

To obtain the fuzzy solution of Eq. (7.72), according to the results of Section 7.3, the natural restriction $0 \leq \underline{d} \leq \overline{d}$ should be used. This restriction allows us to get a simple nonlinear equation with respect to $\overline{d}$ for each fixed value of $\underline{d}$. This equation can be solved using well-known numerical methods.

Finally, as in the linear case (see Section 7.1), the resulting fuzzy interval representation of *IRR* with corresponding membership function is obtained.

When dealing with fuzzy cash flows, the Eq. (7.68) is represented bt the set of α-cuts. For each α-cut the interval equation in the form of the Eq. (7.72) is obtained. Since the set of α-cuts is only an approximate representation of the fuzzy number (as we deal with the method based on the discretization), the precision of final result depends on the number of α -cuts. In practice, this number is usually choosing when analyzing the shapes of used fuzzy values: if we deal with complicated shapes, the number of α-cuts rises.

Thus, the fuzzy problem is reduced to the crisp interval one. The algorithm for its solving may be presented as follows:

1. By partition into n α-cuts, the fuzzy interval problem (Eq. (7.68)) is transformed into the set of crisp interval equations.
2. On each α-cut the crisp intervals $[\underline{P}_1,\overline{P}_1]_\alpha,[\underline{P}_2,\overline{P}_2]_\alpha,[\underline{P}_3,\overline{P}_3]_\alpha,\ldots,[\underline{P}_T,\overline{P}_T]_\alpha$ are calculated.
3. For each α-cut the maximum value $\underline{d}_\alpha$ is calculated by solving the non-linear equation (7.72) in the assumption of $\underline{d}_\alpha=\overline{d}_\alpha$. Substituting obtained values of $\underline{d}_\alpha=\overline{d}_\alpha$ into one of the expressions (7.71) on observed α-cut the corresponding $y_{\alpha\min}$ is obtained.
4. Substituting the $\underline{d}_{\alpha min}=0$ in (7.72) the corresponding $\overline{d}_{\alpha max}$ is obtained. Substituting the $\underline{d}_{\alpha min}$, $\overline{d}_{\alpha max}$ in (7.71) we get the values of $y_{\alpha\max}$.
5. For each α-cut the interval of possible values $\underline{d}_\alpha=[0,\overline{d}_\alpha]$ is divided into m equal parts and $\underline{d}_{\alpha i}, i=1,\ldots,m \quad (d_{1\alpha 0}=0, \quad \underline{d}_{\alpha m}=\overline{d}_\alpha)$ are calculated. Using $\underline{d}_{\alpha i}$ in (7.72) corresponding values $\overline{d}_{\alpha i}$ are obtained. Then using $\underline{d}_{\alpha i}$ and $\overline{d}_{\alpha i}$ in (7.71) the values $y_{\alpha i}$ are obtained. Finally, using the expression analogous to (7.18) (see Section 7.3) the values $\eta_{\alpha i}$ characterizing the degrees of membership of the crisp intervals $[\underline{d}_{\alpha i},\overline{d}_{\alpha i}]$ in the resulting fuzzy solution of (7.68) on the α-cuts are calculated.

To illustrate, let us consider the project with the crisp initial investment $KV=1$ (one million of some monetary units) and trapezoidal fuzzy positive cash flows expected for the subsequent three years and presented by their bottom and top α-cuts (see Table 7.5). The bounds of intermediate α-cuts are obtained using simple

Table 7.5 Sample project

Year	Bottom α-cuts of fuzzy cash flows	Top α-cuts of fuzzy Cash Flows
1	$[\underline{P}_1, \overline{P}_1] = [1, 2]$	$[\underline{P}_1, \overline{P}_1] = [1.1, 1.5]$
2	$[\underline{P}_2, \overline{P}_2] = [1.5, 3]$	$[\underline{P}_2, \overline{P}_2] = [1.6, 2]$
3	$[\underline{P}_3, \overline{P}_3] = [1, 2]$	$[\underline{P}_3, \overline{P}_3] = [1.2, 1.5]$

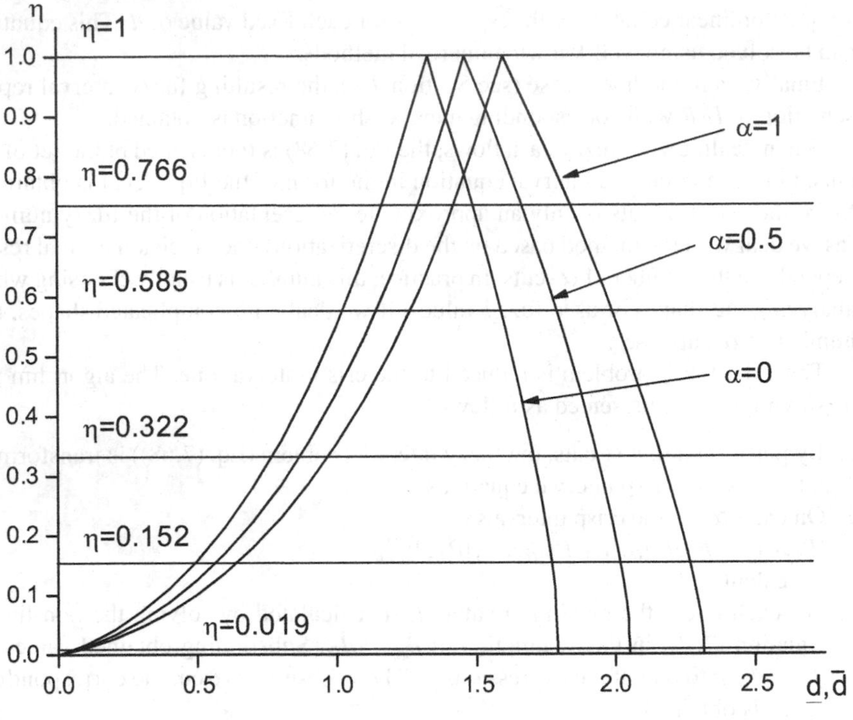

Fig. 7.18 Fuzzy *IRR* for three-year investment project

interpolation. The result obtained with the use of numerical algorithm for our example is presented in Fig.7.18.

Fig. 7.18 shows that on each $\eta_{\alpha i}$-cut we have $IRR = [\underline{d}_{\alpha i}, \overline{d}_{\alpha i}]$ being in turn the intervals on α-cuts. It seems to be natural for intervals $IRR = [\underline{d}_{\alpha i}, \overline{d}_{\alpha i}]$ with upper values of α to contribute more to the final fuzzy solution. Hence, for defuzzification of obtained results it is possible to use, for example, following expressions:

$$\underline{d}_i = \frac{\sum_{\alpha=0}^{n} \underline{d}_{\alpha i} \cdot \alpha}{\sum_{\alpha=0}^{n} \alpha}, \overline{d}_i = \frac{\sum_{\alpha=0}^{n} \overline{d}_{\alpha i} \cdot \alpha}{\sum_{\alpha=0}^{n} \alpha}. \tag{7.73}$$

The result of defuzzification of *IRR* using expressions (7.73) for considered example is shown in Fig. 7.19, where the result obtained for considered example with the use of fuzzy extension of Eq. (7.67) is presented as well. The final solution may be obtained as the intersection of these fuzzy solutions (the shaded area in Fig. 7.19). It can be seen that such intersection is a narrowed final solution of fuzzy *IRR* problem.

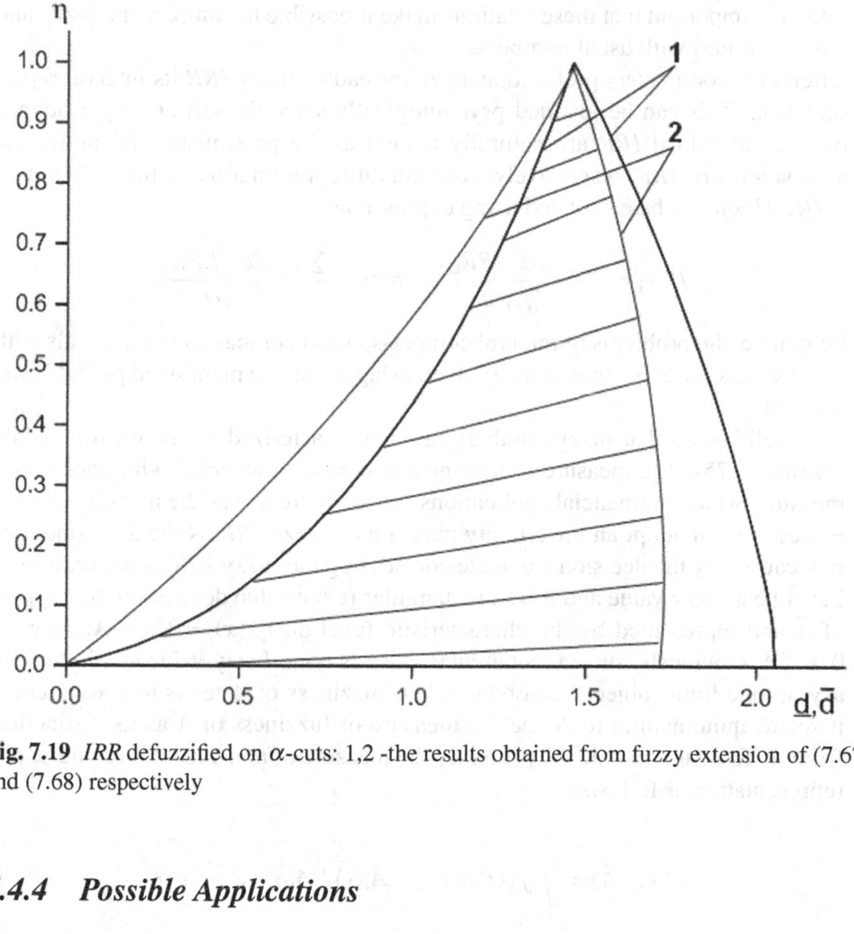

Fig. 7.19 *IRR* defuzzified on α-cuts: 1,2 -the results obtained from fuzzy extension of (7.67) and (7.68) respectively

7.4.4 Possible Applications

There may be different possible applications of fuzzy *IRR* in budgeting, but here we briefly describe only three more obvious ones.

- If several investment projects should be compared in regard to their fuzzy *IRR*s, the problem of fuzzy values comparison arises. There exist numerous definitions of ordering relation over fuzzy values (as well as crisp intervals) in the literature. The review of the problem of intervals and fuzzy values comparison was presented in Chapter 3, where a new approach to to the solution of this problem based on the probabilistic approach and the Dempster-Shafer theory of evidence is presented as well. So if IRR_1 and IRR_2 are fuzzy valued *IRR*s of two compared projects, the following probabilities may be calculated $P(IRR_1 > IRR_2)$, $P(IRR_1 = IRR_2)$, $P(IRR_1 < IRR_2)$. Since in practice, the routine task is the comparison of *IRR* (fuzzy value in our case) with known (non fuzzy) bank discount rate, it is important that these relations make it possible to compare the fuzzy and interval values with usual numbers.

- Often decision makers prefer to analyze instead of fuzzy *IRR* its interval representation. This can be justified psychologically since the left and right bounds of interval valued *IRR* are naturally treated as the pessimistic and optimistic assessments of *IRR*, respectively. An interval representation of fuzzy *IRR*, i.e., $[IRR_1, IRR_2]$ can be calculated using expressions

$$IRR_1 = \frac{\sum_{\alpha=0}^{n} \alpha \cdot IRR_{\alpha 1}}{\sum_{\alpha=0}^{n} \alpha}, \quad IRR_2 = \frac{\sum_{\alpha=0}^{n} \alpha \cdot IRR_{\alpha 2}}{\sum_{\alpha=0}^{n} \alpha}.$$

Of course, the problems of interval comparison and comparison of intervals with usual values arise, but they can be solved using the above mentioned probabilistic method.

- It is well known that fuzzy numbers can be characterized by the measure of its fuzziness [75]. The measure of fuzziness is directly connected with uncertainty measure, which in financial applications is usually treated as the measure of risk. Hence we can adopt an uncertainty measure of fuzzy *IRR* as the assessment of risk caused by the decision we made considering the fuzzy *IRR* as the criterion. Let $\tilde{A}$ be a fuzzy value and A be a rectangular fuzzy value defined on the support of $\tilde{A}$ and represented by the characteristic function $\eta_A(x) = 1, x \in A; \eta_A(x) = 0, x \notin A$. Obviously, such a rectangular value is not a fuzzy value at all, but it is asymptotic limit (object) we obtain when fuzziness of $\tilde{A}$ tends to zero. Hence, it seems quite natural to define the measure of fuzziness of $\tilde{A}$ as its distinction from A. To do this, we define primarily the measure of non-fuzziness using α-cut representation as follows:

$$MNF(\tilde{A}) = \int_0^1 f(\alpha)((A_{\alpha 2} - A_{\alpha 1})/(A_{02} - A_{01}))d\alpha,$$

where $f(\alpha)$ is some function of α , e.g., $f(\alpha) = 1$ or $f(\alpha) = \alpha$. Of course, the last expression makes sense only for the fuzzy or interval values, i.e., only for non zero width of support $A_{02} - A_{01}$. It is easy to see that if $f(\alpha) = 1$ and $\tilde{A} \to A$ then $MNF(\tilde{A}) \to 1$. Obviously, the measure of fuzziness can be defined as $MF(\tilde{A}) = 1 - MNF(\tilde{A})$.
We can say that a rectangular value A defined on the support of $\tilde{A}$ is a more uncertain object than $\tilde{A}$.
Really, only what we know about A is that all $x \in A$ belong to A with equal degrees, whereas the membership function $(0 \le \mu(x) \le 1)$ characterizing the fuzzy

value $\tilde{A}$ provides more information to the description and as a consequence, represents a more certain object. Therefore, we can treat the measure of non-fuzziness MNF as the uncertainty measure. Hence, if some decision is made concerning the fuzzy IRR , the uncertainty and, consequently, the risk of such decision can be calculated as $MNF(IRR)$. Since the fuzzy IRR and the corresponding risk assessment can play the role of local criteria, a multiple criteria problem arises. Generally, an investment evaluation is a multiple criteria task. The methods for its solving in the fuzzy setting are discussed in Chapter 4.

7.5 Summary and Discussion

In this chapter, we present a new approach to the solution of interval and fuzzy equations based on the generalized procedure of interval and fuzzy extension called "interval extended zero" method. The key idea is the treatment of the interval zero as an interval centered around zero. It is shown that such approach is the direct consequence of the interval subtraction operation. Some methodological problems concerned with this definition of the interval zero are discussed. The proposed method is based on the solution of an interval equation presented in the form of the nested interval solutions, which can be in a natural way treated as a fuzzy number. To solve a linear fuzzy equation, the α-cut representation of fuzzy parameters of an equation is used and interval equations on the corresponding α-cuts are obtained. The set of fuzzy solutions on these α -cuts is obtained using the "extended zero" method. It is important that these solutions, generally, are not nested. Finally, using the standard disjunction procedure, the resulting fuzzy solution is obtained as the aggregation of the fuzzy solutions on the α-cuts. It is shown that the proposed method can be used as the reliable practical tool for solving interval and fuzzy linear equations as well as systems of them. An important for practice advantage of the developed new method is that it provides substantially narrower solutions than conventional methods.

As the "interval extended zero" method generally provides the fuzzy solution of interval equation, its interval representations are proposed. It is shown that they may be naturally treated as the modified operations of interval division and used for interval extension of known numerical methods for solving systems of linear equations.

Using the well known example, we show that our solution obtained using the modified interval Gauss elimination procedure can be treated as the inner interval approximation of the united solution and the outer interval approximation of the tolerable solution. We show also that our solution lies within the range of possible AE-solutions between the extreme tolerable and united solutions. To illustrate the proposed method, we present the results obtained for known seven examples repeatedly used in the literature as the tests for numerical methods in the interval setting. Comparing our results with those obtained using Markov's Jacobi type iterative method and usual interval Gauss elimination procedure, we show that the proposed method not only allows us to decrease the excess width effect, but makes it possible to avoid inverted interval solutions too. It is important that our results are close to the so-called maximal inner solutions, i.e., approximate solutions with minimal excess width effect.

The influence of the system's size and zero entries on the resulting excess width effect is analyzed using the Leontief's input-output model of economics as an example.

It is shown that the "interval zero" method provides a fuzzy solution of nonlinear interval and fuzzy equations. It is important that opposite to the known approaches, the method makes it possible to get both the positive and negative fuzzy solutions of interval and fuzzy quadratic equation.

The proposed method may be used for the solution of more complicated fuzzy nonlinear equations and the corresponding general algorithm is presented as well.

In this chapter, we analyze the problem of calculation of *IRR* in the fuzzy setting and use the developed numerical method for the solution of nonlinear fuzzy equations to obtain the fuzzy *IRR*. The resulting fuzzy *IRR* can be used directly or may be defuzzified as well if the crisp final result is needed. It should be emphasized that for practical purposes the most useful approach is a reduction of fuzzy *IRR* to some crisp intervals by means of type reduction. The width of such interval can serve as a natural measure of risk connected with the use of *IRR* in financial analysis in the fuzzy setting.

References

1. Abbasbandy, S.: Extended Newton's method for a system of nonlinear equations by modified Adomian decomposition method. Applied Mathematics and Computation 170, 648–656 (2005)
2. Abbasbandy, S., Asady, B.: Newton's method for solving fuzzy nonlinear equations. Applied Mathematics and Computation 159, 349–356 (2004)
3. Abbasbandy, S., Jafarian, A.: Steepest descent method for system of fuzzy linear equations. Applied Mathematics and Computation 175, 823–833 (2006)
4. Abbasbandy, S., Otadi, M.: Numerical solution of fuzzy polynomials by fuzzy neural network. Applied Mathematics and Computation 181, 1084–1089 (2006)
5. Allahviranloo, T.: Successive over relaxation iterative method for fuzzy system of linear equations. Applied Mathematics and Computation 162, 189–196 (2005)
6. Allahviranloo, T.: The Adomian decomposition method for fuzzy system of linear equations. Applied Mathematics and Computation 163, 553–563 (2005)
7. Allahviranloo, T., Ahmady, E., Ahmady, N., Alketaby, K.S.: Block Jacobi two-stage method with Gauss-Sidel inner iterations for fuzzy system of linear equations. Applied Mathematics and Computation 175, 1217–1228 (2006)
8. Babusiaux, D., Pierru, A.: Capital budgeting, project valuation and financing mix: Methodological proposals. Europian Journal of Operational Research 135, 326–337 (2001)
9. Barth, W., Nuding, W.: Optimale Lösung von Intervallgleichungssystemen, vol. 12, pp. 117–125 (1974)
10. Belletante, B., Arnaud, H.: Choisir ses investissements. Chotar et Assosies Editeurs, Paris (1989)

11. Bogle, H.F., Jehenck, G.K.: Investment Analysis: US Oil and Gas Producers Score High in University Survey. In: Proceedings of Hydrocarbon Economics and Evaluation Symposium, Dallas, pp. 234–241 (1985)
12. Briggs, F.E.: On problems of estimation in Leontief models. Econometrica 25, 411–455 (1975)
13. Brigham, E.F.: Fundamentals of Financial Management. The Dryden Press, New York (1992)
14. Buckley, J.J.: The fuzzy mathematics of finance. Fuzzy Sets and Systems 21, 257–273 (1987)
15. Buckley, J.J.: Solving fuzzy equations in economics and finance. Fuzzy Sets and Systems 48, 289–296 (1992)
16. Buckley, J.J., Eslami, E.: Neural net solutions to fuzzy problems: The quadratic equation. Fuzzy Sets and Systems 86, 289–298 (1997)
17. Buckley, J.J., Eslami, E., Feuring, T.: Fuzzy Mathematics in Economics and Engineering. In: Studies in Fuzziness and Soft Computing. Physica Verlag, Heidelberg (2002)
18. Buckley, J.J., Eslami, E., Hayashi, Y.: Solving fuzzy equations using neural nets. Fuzzy Sets and Systems 86, 271–278 (1997)
19. Buckley, J.J., Qu, Y.: Solving linear and quadratic fuzzy equations. Fuzzy Sets and Systems 38, 43–59 (1990)
20. Buckley, J.J., Qu, Y.: Solving systems of linear fuzzy equations. Fuzzy Sets and Systems 43, 33–43 (1991)
21. Calzi, M.L.: Towards a general setting for the fuzzy mathematics of finance. Fuzzy Sets and Systems 35, 265–280 (1990)
22. Caprani, O., Madsen, K.: Mean value forms in interval analysis. Computing 25, 147–154 (1980)
23. Chang, J.-C., Chen, H., Shyu, S.-M., Lian, W.-C.: Fixed-Point Theorems in Fuzzy Real Line. Comput. Math. Appl. 47, 845–851 (2004)
24. Chansa-ngavej, C., Mount-Campbell, C.A.: Decision criteria in capital budgeting under uncertainties: implications for future research. Int. J. Prod. Economics 23, 25–35 (1991)
25. Chen, S.: An empirical examination of capital budgeting techniques: impact of investment types and firm characteristics. Eng. Economist. 40, 145–170 (1995)
26. Chen, S.-H., Yang, X.-W.: Interval finite element method for beam structures. Finite Elements in Analysis and Design 34, 75–88 (2000)
27. Chiu, C.Y., Park, C.S.: Fuzzy cash flow analysis using present worth criterion. Eng. Economist. 39, 113–138 (1994)
28. Choobineh, F., Behrens, A.: Use of intervals and possibility distributions in economic analysis. J. Oper. Res. Soc. 43, 907–918 (1992)
29. Cleary, J.C.: Logical Arithmetic. Future Computing Systems 2, 125–149 (1987)
30. Dehghan, M., Hashemi, B.: Iterative solution of fuzzy linear systems. Applied Mathematics and Computation 175, 645–674 (2006)
31. Deif, A.S.: Sensitivity Analysis in Linear Systems. Springer, Berlin (1986)
32. Dubois, D., Prade, H.: Operations on fuzzy numbers. J. Systems Sci. 9, 613–626 (1978)
33. Dimova, L., Sevastianov, D., Sevastianov, P.: Application of fuzzy sets theory, methods for the evaluation of investment efficiency parameters. Fuzzy Economic Review 5, 34–48 (2000)
34. Ferreira, J.A., Patricio, F., Oliveira, F.: On the computation of solutions of systems of interval polynomial equations. Journal of Computational and Applied Mathematics 173, 295–302 (2005)
35. Gardnes, E., Mielgo, H., Trepat, A.: Modal intervals: Reasons and ground semantics. In: Nickel, K. (ed.) Interval Mathematics 1985. LNCS, vol. 212, pp. 27–35. Springer, Heidelberg (1986)

36. Hansen, E.: A generalized interval arithmetic. In: Nickel, K. (ed.) Interval Mathematics. LNCS, vol. 29, pp. 7–18. Springer, Heidelberg (1975)
37. Hansen, E.: Bounding the solution of interval linear equations. SIAM J. Numer. Anal. 29, 1493–1503 (1992)
38. Hanss, M., Klimke, A.: On the reliability of the influence measure in the transformation method of fuzzy arithmetic. Fuzzy Sets and Systems 143, 371–390 (2004)
39. Jaulin, L., Kieffir, M., Didrit, O., Walter, E.: Applied Interval Analysis. Springer, London (2001)
40. Kahraman, C.: Fuzzy versus probabilistic benefit/cost ratio analysis for public work projects. Int. J. Appl. Math. Comp. Sci. 11, 705–718 (2001)
41. Kahraman, C., Ruan, D., Tolga, E.: Capital budgeting techniques using discounted fuzzy versus probabilistic cash flows. Information Sciences 142, 57–76 (2002)
42. Kahraman, C., Tolga, E., Ulukan, Z.: Justification of manufacturing technologies using fuzzy benefit/cost ratio analysis. Int. J. Product. Economy 66, 45–52 (2000)
43. Kahraman, C., Ulukan, Z.: Continous compounding in capital budgeting using fuzzy concept. In: Proceedings of 6th IEEE International Conference on Fuzzy Systems (FUZZ-IEEE 1997), Bellaterra, pp. 1451–1455 (1997)
44. Kahraman, C., Ulukan, Z.: Fuzzy cash flows under inflation. In: Proceedings of 7th IEEE International Fuzzy Systems Association World Congress (IFSA 1997), Univeristy of Economics, Prague CZech Republic Bellaterra, pp. 104–108 (1997)
45. Kalman, R.E., Falb, P.L., Arbib, M.A.: Topics in Mathematical System Theory. McGrow-Hill, New York (1969)
46. Kaucher, E.: Interval Analysis in the Extended Interval Space IR. Computing Supplement 2, 33–49 (1980)
47. Kaufmann, A., Gupta, M.: Introduction to fuzzy-arithmetic theory and applications. Van Nostrand Reinhold, New York (1985)
48. Kawaguchi, M.F., Da-Te, T.: A calculation method for solving fuzzy arithmetic equations with triangular norms. In: Proc. 2d IEEE Int. Conf. on Fuzzy Systems (FUZZ-IEEE), San Francisco, pp. 470–476 (1993)
49. Kearfott, B.: Rigorous Global Search: Continuous Problems. Kluwer Academic Publishers, The Netherlands (1996)
50. Kuchta, D.: Fuzzy capital budgeting. Fuzzy Sets and Systems 111, 367–385 (2000)
51. Kupriyanova, L.: Inner estimation of the united solution set to interval linear algebraic system. Reliable Computing 1, 15–31 (1995)
52. Lee, E.B., Markus, L.: Foundations of Optimal Control Theory. John Wiley, New York (1970)
53. Leontief, W.: Quantitative input-output relations in the economic system of the United States. Review of Economics and Statistics 18, 100–125 (1936)
54. Li, Q.-X., Liu, S.-F.: The foundation of the grey matrix and the grey input-output analysis. Applied Mathematical Modelling 32, 267–291 (2008)
55. Liang, P., Song, F.: Computer-aided risk evaluation system for capital investment. Omega 22, 391–400 (1994)
56. Markov, S.M.: A non-standard subtraction of intervals. Serdica 3, 359–370 (1977)
57. Markov, S.: An iterative method for algebraic solution to interval equations. Applied Numerical Mathematics 30, 225–239 (1999)
58. Markov, S.M., Popova, E., Ullrich, C.: On the Solution of Linear Algebraic Equations Involving Interval Coefficients. In: Margenov, S., Vassilevski, P. (eds.) Iterative Methods in Linear Algebra II. IMACS Series in Computational and Applied Mathematics, vol. 3, pp. 216–225 (1996)
59. Moore, R.E.: Interval analysis. Prentice-Hall, Englewood Cliffs (1966)

60. Muzzioli, S., Reynaerts, H.: Fuzzy linear systems of the form $A_1x+b_1=A_2x+b_2$. Fuzzy Sets and Systems 157, 939–951 (2006)
61. Neumaier, A.: Interval Methods for Systems of Equations. Cambridge University Press, Cambridge (1990)
62. Nieto, J.J., Rodríguez-López, R.: Existence of extremal solutions for quadratic fuzzy equations. Fixed Point Theory Appl. 3, 321–342 (2005)
63. Ning, S., Kearfott, R.B.: A comparison of some methods for solving linear interval equations. SIAM J. Numer. Anal. 34, 1289–1305 (1997)
64. Perrone, G.: Fuzzy multiple criteria decision model for the evaluation of AMS. Comput. Integrated Manufacturing Systems 7, 228–239 (1994)
65. Rzeźuchowski, T., Wasowski, J.: Solutions of fuzzy equations based on Kaucher arithmetic and AE-solution sets Source. Fuzzy Sets and Systems 159, 2116–2129 (2008)
66. Sendov, B.: Segment arithmetic and segment limit. C.R. Acad. Bulgare Sci. 30, 955–958 (1977)
67. Sevastjanov, P., Dymova, L.: Fuzzy solution of interval linear equations. In: Wyrzykowski, R., Dongarra, J., Karczewski, K., Wasniewski, J. (eds.) PPAM 2007. LNCS, vol. 4967, pp. 1392–1399. Springer, Heidelberg (2008)
68. Sevastjanov, P., Dymova, L.: A new method for solving interval and fuzzy equations: linear case. Information Sciences 17, 925–937 (2009)
69. Shary, S.P.: Solving the Tolerance Problem for Interval Linear Systems. Interval Computations 2, 6–26 (1994)
70. Shary, S.P.: Controllable Solution Sets to Interval Static Systems. Applied Mathematics and Computation 86, 185–196 (1997)
71. Shary, S.P.: A New Technique in Systems Analysis under Interval Uncertainty and Ambiguity. Reliable Computing 8, 321–418 (2002)
72. Shieh, B.-S.: Infinite fuzzy relation equations with continuous t-norms. Information Sciences 178, 1961–1967 (2008)
73. Ward, T.L.: Discounted fuzzy cash flow analysis. In: Proceedings of Fall Industrial Engineering Conference, pp. 476–481 (1985)
74. Wu, C.C., Chang, N.B.: Grey input-output analysis and its application for environmental cost allocation. European Journal of Operational Research 145, 175–201 (2003)
75. Yager, R.A.: On the measure of fuzziness and negation. Part 1. Membership in the Unit Interval. Int. J. Gen. Syst. 5, 221–229 (1979)
76. Zadeh, L.A.: Fuzzy sets. Inf. Control 8, 338–353 (1965)
77. Zheng, B., Wang, K.: General fuzzy linear systems. Applied Mathematics and Computation 181, 1276–1286 (2006)

Index

P

Q

R

S

T

U

W

Y

Z